Introductory Statistical Mechanics

Introductory Statistical Mechanics

Second Edition

ROGER BOWLEY

Department of Physics, University of Nottingham

and

MARIANA SÁNCHEZ

CLARENDON PRESS · OXFORD

To Alexander Edward and Peter Anthony Xavier.

We do not inherit the earth from our parents. We borrow it from our children.
American Indian proverb.

Preface to the first edition

Our aim in writing this book is to provide a clear and accessible introduction to statistical mechanics, the physics of matter made up of a large number of particles. Statistical mechanics is the theory of condensed matter. The content of our book is based on a course of lectures given to physics undergraduates at Nottingham University by one of us (R.M.B.). The lectures have been substantially revised by both of us partly to remove repetition of material—repetition that may be effective in a lecture is redundant in print—and partly to improve the presentation. Some of the history of the subject has been included during the revision, so that the reader may be able to appreciate the central importance of statistical mechanics to the development of physics. Some of the concepts of statistical mechanics are old, but they are as useful and relevant now as when they were first proposed. It is our hope that students will not regard the subject as old-fashioned, and that our book will be of use to all students, not only to those at Nottingham University.

The reader is assumed to have attended introductory courses on quantum mechanics and kinetic theory of gases, as well as having a knowledge of calculus. In Nottingham, students would have also taken a course in thermodynamics before studying statistical mechanics. The first few lectures of the course review the first and second laws of thermodynamics—they have been expanded to form Chapters 1 and 2. Not all students would have had a course in probability and statistics so that also forms part of the lecture course, and is the basis of Chapter 3.

In Chapters 4 and 5 the main ideas and techniques of statistical mechanics are introduced. The essential idea, Boltzmann's hypothesis, is that the entropy of a system is a function of the probability that it is in a particular microscopic state. Using this idea Planck, Einstein, and others were able to connect the entropy of an isolated system to the number of states that were accessible. First we have to specify more closely the idea of the microscopic state of the system, and for that we need the techniques and language of quantum mechanics. The remaining problem is to count the number of such states, or rather to do everything possible to avoid counting them. In Chapter 4 we show how the entropy can be determined if we can count the number of accessible states. Moreover, we can give a theoretical basis to the laws of thermodynamics. In Chapter 5 we show that by placing the system in contact with a heat bath we can get the thermodynamics of the system in a more direct way without actually having to count states. Using this method we derive the Boltzmann distribution for the probability that the system is in a particular quantum state, and connect the

Helmholtz free energy to the quantum states of the system.

The rest of the book is concerned with successively more complicated examples of these ideas as they are applied to quantum systems: Chapter 6 deals with the problem of identifying independent quantum states for identical particles and how this affects the thermodynamic properties of the system; Chapter 7 is concerned with the quantum description of particles as waves; Chapter 8 is concerned with the thermodynamics of waves themselves, in particular with black body radiation and sound waves in solids; Chapter 9 deals with systems with varying number of particles; Chapter 10 deals with the proper treatment of identical particles at low temperatures.

Statistical mechanics can appear to be too mathematical, but there is a reason: there must be a sufficient mathematical backbone in the subject to allow our ideas about the behaviour of a collection of a large number of particles to stand up to verification by experiment. Ideas that cannot be tested are not part of science. Nevertheless, in an introductory treatment the ideas must not be swamped by too much mathematics: we have tried to simplify the treatment sufficiently so that the reader is not overburdened by equations.

At the end of each chapter there are problems, graded so that the easier appear first, the more difficult later. These problems form an important part of the book: only by trying to solve them will you get the full benefit. Do not despair if they appear too hard—brief answers are given in Appendix E which should help you over any difficulties. The aim is to develop your understanding, skill, and confidence in tackling problems from all branches of physics.

Acknowledgements

Many people have helped us in preparing the book. Our thanks go: to Julie Kenney for typing an early draft; to Terry Davies for his great skill in producing the diagrams; to Martin Carter for his guidance in the use of LaTeX; to several colleagues who were kind enough to read and criticize parts of the manuscript, in particular Keith Benedict, Laurie Challis, Stefano Giorgini, Mike Heath, and Peter Main; to students who have told us about mistakes in earlier versions of the problems; to Donald Degenhardt and one of his readers at Oxford University Press, both of whom were sources of constructive criticism and advice. We are, of course, entirely responsible for any errors of fact or emphasis, or for any typographical mistakes which have crept in.

Figure 8.7 is taken from *Astrophysics Journal* **354**, L37–40, and is reprinted by permission of the COBE Science Working Group. Figure 7.6, which is taken from Miller and Kusch *Physical Review* **99**, 1314–21, and Figure 10.7, which is taken from Landau, Tough, Brubaker, and Edwards *Physical Review* **A2**, 2472–82, are reprinted by permission of the Americal Physical Society.

Nottingham
September 1995

R. M. B.
M. S.

Preface to the second edition

According to our dictionary the word 'introductory' implies a preliminary discourse or an elementary treatise. When preparing the second edition, one of our aims was to introduce slightly more advanced material on statistical mechanics, material which students should meet in an undergraduate course. As a result the new edition contains three new chapters on phase transitions at an appropriate level for an undergraduate student. We hope the reader will still regard our book as introductory.

Our other aim was to increase the number of problems at the end of each chapter: we give brief solutions to the odd-numbered problems in Appendix G; the even-numbered questions have only numerical answers where appropriate. We hope that this modification will make the book more useful to teachers of Statistical Mechanics: they can now set problems from the book without helpful hints being available to students.

Finally, we have included two extra appendices, one on the manipulation of partial derivatives, the other on a derivation of the van der Waals equation from the ideas of classical statistical mechanics.

Acknowledgements

When preparing the extra chapters we were strongly influenced by a set of lecture notes on first-order phase transitions by Philippe Nozières given at College de France. Brian Cowan, a man who savours books on thermal physics as others savour wine, gave us trenchant criticisms of an early draft. Andrew Armour kindly road-tested the material in the new chapters whilst writing his thesis. None of the above are responsible for errors or inaccuracies that may have crept into this final version.

There were some errors in the first edition despite our best efforts. We are thankful to one of R.M.B.'s students, Raphael Hirschi, who, with Swiss precision, spotted most of them.

Terry Davies showed great patience, care, and skill when creating the diagrams. We are very grateful to him for his time and his excellent sense of humour.

Nottingham
November 1998

R. M. B.
M. S.

Contents

1
The first law of thermodynamics

A theory is the more impressive the greater the simplicity of its premises, the more different kinds of things it relates, and the more extended its area of applicability. Therefore the deep impression that classical thermodynamics made upon me. It is the only physical theory of universal content which I am convinced will never be overthrown, within the framework of applicability of its basic concepts.   *Albert Einstein*

There are two approaches to the study of the physics of large objects. The older, or classical, approach is that of classical thermodynamics and is based on a few empirical principles which are the result of experimental observation. For example, one principle is that heat and work are both forms of energy and that energy is always conserved, ideas which form the basis of the first law of thermodynamics; another principle is that heat flows from a hotter to a colder body, an observation which is the basis of the second law of thermodynamics. The principles of thermodynamics are justified by their success in explaining experimental observations. They use only macroscopic concepts such as temperature and pressure; they do not involve a microscopic description of matter. The classical approach is associated with the names of Kelvin, Joule, Carnot, Clausius, and Helmholtz.

There are two objectives in the statistical approach: the first is *to derive the laws of thermodynamics*, and the second, *to derive formulae for properties of macroscopic systems*. These are the twin aims of this book: to explain the foundations of thermodynamics and to show how a knowledge of the principles of statistical mechanics enables us to calculate properties of simple systems. The former requires that you have some knowledge of thermodynamics. We start with a brief summary of the essential ideas of thermodynamics.

1.1   Fundamental definitions

The theory of thermodynamics is concerned with systems with a large number of particles which are contained in a vessel of some kind. The system could be a flask of liquid, a canister of gas, or a block of ice enclosed between walls. We prevent energy entering the system by lagging the walls of the container with an insulating material. Of course in practice no insulator is perfect, so in reality we cannot achieve the theoretical ideal in which no energy enters or leaves the system. An ideal insulating wall, which is a purely theoretical notion, is called an *adiabatic wall*. When a system is surrounded by adiabatic walls it is said to be *thermally isolated*. No energy enters such a thermally isolated system. In contrast, walls that allow energy to pass through them are called *diathermal*.

When we try to apply the theory of thermodynamics to real experiments, the walls are not perfect insulators; they are diathermal. However, they may be sufficiently good that the system tends to a state whose properties—viscosity, thermal conductivity, diffusion coefficient, optical absorption spectrum, speed of sound, or whatever—do not appear to change with time no matter how long we wait. When the system has reached such a state it is said to be in *equilibrium*.

It is a matter of experience that a given amount of a pure gas in equilibrium is completely specified by its volume, its pressure, and the number of particles in the gas, provided there are no electric or magnetic fields present. Suppose the gas is contained in a cylinder with a movable piston. The volume of the gas can then be adjusted by moving the piston up or down. Let us specify the volume at some value. By making thermal contact between the cylinder and some other hotter (or colder) system, we can alter the pressure in the gas to any desired value. The gas is left for a while until it reaches equilibrium.

Experience shows that it does not matter too much how we get to the final state of the system: whether we alter the volume first and then the pressure; or alter the pressure first and then the volume; or any other combination of processes. The past history of the gas has no effect; it leaves no imprint on the equilibrium state of the gas.

The same observation can be made not just for a pure gas, but for any system made up of a large number of particles: it applies to mixtures of gases, to liquids, to solids, to reacting gases, and to reacting elementary particles in stars. It even applies to electromagnetic radiation in a cavity which is in equilibrium with the walls, even though it is not immediately clear what constitute the particles of electromagnetic radiation.

Consider now two thermally isolated systems; each system is enclosed in adiabatic walls; each system has its own pressure and volume. Suppose we place the two systems so that they have a wall in common, and we open an area of thermal contact between two previously thermally isolated systems, a contact which allows energy to flow from one to the other. In other words we change the area of contact from an adiabatic to a diathermal wall. Of course, if the thermal contact is poor, energy will flow very slowly and it will be hard to detect any change of measurable properties with time. The thermal contact has to be sufficiently large that the energy flows freely, so that the new state of equilibrium is reached in a reasonable time. The flow of energy allows the composite system to evolve and to come to a new state of equilibrium with different values for its properties (such as the viscosity and thermal conductivity) in each of the two systems.

The question can then be posed: are there any circumstances when the new state of thermal equilibrium for the composite system is the same as the initial state of the two systems? Suppose we measure the properties of the first system, call it system A, and of the second system, B, both before and after the contact is made. Under some circumstances the properties of system A do not change with time either before or after thermal contact is made. It is an experimental observation that when this happens, the properties of system B do not change with time. Nothing changes. We say that the two systems are in equilibrium with each other.

We can build upon this notion by considering another experimental observation which is now called the *zeroth law of thermodynamics*. It says the following: if systems A and B are separately in equilibrium with system C, then systems A and B are in equilibrium with each other. It is this observation that lies behind the use of a thermometer. A *thermometer* (system C) is an instrument that enables us to tell easily if one of its properties is evolving with time because energy is being passed into it. Suppose when we place system C (the thermometer) in thermal contact with system A and nothing changes in the thermometer; suppose also that, when we place the thermometer in contact with system B, we detect no change in the thermometer. The zeroth law indicates that, when systems A and B are placed in contact with each other, no property of either system A or B changes with time. This observation indicates that no energy flows between them.

Suppose now we consider a reference system, call it A, and a test system, call it X. We keep system A at a constant volume and pressure. The state of system A is kept constant for all time. Any system which is in equilibrium with system A is said to be *isothermal* with it. We can tell if system X is isothermal by placing it in contact with A and seeing if any of the measurable properties of X change with time. If they do not, then A and X are said to be isothermal.

Now let us adjust—without changing the number of particles—the volume, V, and the pressure, P, of the test system X in such a way that it always is in thermal equilibrium with reference system A. We may need to use trial and error, but such a procedure is possible experimentally. In this way we can generate a curve of pressure and volume. Such a curve is called an *isotherm*. Let us now repeat the procedure for other reference states of system A and thus generate a series of isotherms.

We can repeat this procedure with another reference system instead of A, a reference system which is also held at constant pressure and volume. Again we plot out the isotherms for the new reference system and the results we get for the isotherms are found to be independent of our choice of reference system. They are an intrinsic property of the test system, X.

By careful experiment we can map out a series of isotherms. An illustration of a set of isotherms is shown in Fig. 1.1. It is an experimental observation that the neighbouring isothermal curves run nearly parallel to each other. For gases the isotherms do not intersect, but for other systems this may not be the case (for example water near 4° C).

We can label each curve by a single number; let us call it θ. The labelling of the curves can be done arbitrarily, but it pays to do it systematically. Each isotherm is associated with a different number. We want the numbers to vary smoothly and continuously so that neighbouring isotherms are associated with neighbouring numbers. The numbers are not scattered around randomly. The

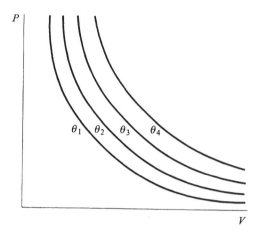

Fig. 1.1 A set of isotherms for a dilute gas. Each curve is labelled by a number representing the temperature of the gas.

number is called the temperature; its value depends on how we assigned numbers to the isotherms. There is no unique assignment. Although we have tried to be systematic, the whole process appears to be very arbitrary; it is. But provided we are systematic, there will exist a relationship of the form

$$\theta = F(P, V) \tag{1.1.1}$$

where P is the pressure of the test system and V is its volume. This is a very powerful mathematical statement: it says that the temperature has a unique value for any choice of pressure and volume; it does not depend on any other quantity.

Suppose now we repeat the whole programme for another test system, call it Y, and get another set of isotherms. It makes sense to adopt the same procedure in labelling the isotherms for system Y as for system X. If, according to the procedure chosen, both systems X and Y correspond to the same number, then the zeroth law tells us that the two systems will be in equilibrium when they are placed in contact with each other.

What is needed is an instrument which will generate a set of numbers automatically so that we can read the numbers off directly. The instrument is called a *thermometer*. Any measurable property of the system can be used. We can build instruments based on thermal expansion, on viscosity, velocity of sound, magnetization, and so on.

1.2   Thermometers

The first thermometer was invented by Galileo, who, when studying the effects of temperature, struck upon one of its fundamental properties: when things are given energy (that is heated) they expand. Galileo's thermometer was an instrument which had liquid at the bottom of a cylinder and air trapped at the top. He noticed that when air was heated it expanded and pushed the liquid down; when it was cooled it contracted and sucked the liquid back up.

The thermometer remained an imprecise instrument until Daniel Fahrenheit, a glass-blower, managed to blow a very fine, uniform capillary tube. He used the thermal expansion of mercury in a glass capillary to create a thermometer. His thermometer consisted of a bulb full of mercury connected to the fine capillary. The mercury is drawn into the capillary where it is clearly visible. As the bulb of mercury absorbs energy, the mercury expands and moves along the fine capillary. We can easily observe the effect of the expansion.

Such thermometers demonstrate qualitatively the effect of adding energy to a system, but to be useful they need to be made quantitative so that they give us numbers. To do that we need a *scale of temperature* based on how far the liquid has moved. We need to associate numbers with the concept of temperature to bring it to life. How should we design a temperature scale? To form a scale we need two reference points. For example, to make a scale of length we draw lines on a ruler a metre apart. What do we take as the reference points for our temperature scale? In the seventeenth century there was one temperature that scientists had established was constant: the point at which water turns to ice, which became the zero on the scale. Next they needed a fixed high temperature point. Again the answer came from water: at atmospheric pressure it always boils at the same temperature. We can draw a scale on the glass capillary with marks showing the position of the mercury in the capillary at 0 degrees, the temperature at which pure ice melts into water, and at 100 degrees, the temperature at which pure water boils into steam. Between these two marks are drawn lines which are equally spaced, say, at every five degrees, forming a scale against which the temperature is measured. This scale of temperature is called the centigrade temperature scale because there are exactly 100 degrees between the two fixed points.

The mercury thermometer is quite accurate and practical for ordinary purposes. To take the temperature we place the thermometer in contact with the system and leave it for a few minutes until the mercury stops expanding. The amount of energy transfer needed to cause the temperature to rise by one temperature unit is called the *heat capacity*. Ideally the heat capacity of the system should be very large in comparison with the heat capacity of the thermometer; when this is the case there is hardly any flow of energy into the thermometer before equilibrium is reached. The thermometer then gives an accurate reading of the initial temperature of the system.

Here are the qualities that we want for a good thermometer:
1. a property which varies markedly with temperature;
2. a small heat capacity;
3. a rapid approach to thermal equilibrium;
4. reproducibility so it always gives the same reading for the same temperature;
5. a wide range of temperatures.

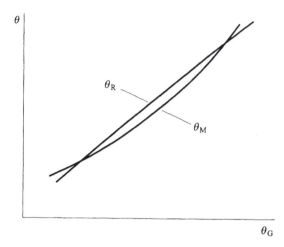

Fig. 1.2 The curves above result from plotting the temperature of a thermometer according to its own scale against the equivalent temperature on the perfect gas scale of temperature. All temperatures agree at the fixed points but need not agree elsewhere.

Mercury thermometers do not approach equilibrium very rapidly, nor do they work over a wide range of temperatures. However, they are very practical thermometers provided we want to measure the temperature of relatively large objects whose temperature changes very slowly with time. But we are not restricted just to mercury thermometers. We can build thermometers using a wide range of different properties such as: the resistance of a metal such as platinum; the thermocouple effect, such as that between copper and nickel; the resistance of a semiconductor such as silicon; the magnetic properties of a crystal such as cerium magnesium nitrate; the radiation from a hot body, such as from a brick in a kiln. One modern thermometer uses liquid crystals which change colour with temperature.

Each thermometer will agree on the value of the temperature at the fixed points by definition; away from the fixed points each thermometer will give a different value for the temperature. We denote the temperatures as θ_i where i is a label for a particular thermometer. For example we can write the temperature for a mercury thermometer as θ_M, for a constant volume gas thermometer we can write θ_G, for a platinum wire resistance thermometer we write θ_R.

If we plot say θ_M against θ_G, as shown in Fig. 1.2, we get a continuous line which passes through the two fixed points; but the line need not be straight. If we were to plot say θ_R against θ_G we would get a different line. The concept of temperature is well-defined for each thermometer based on its temperature scale, but there is no reason to prefer one particular thermometer over the rest.

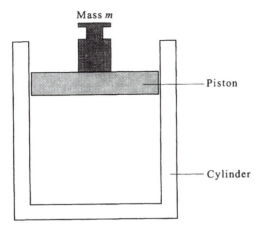

Fig. 1.3 A gas (or fluid) in a cylinder being compressed by the mass which is placed on top of the piston. When the piston stops moving, the force on the piston due to the pressure of the gas is equal to the weight exerted by the mass.

1.3 Different aspects of equilibrium

So far we have only considered flow of energy between systems. When the energy flow stops and all measurable properties are independent of time, the combined system is said to be in equilibrium. This idea can be extended to other sorts of changes in systems.

There are three different aspects of equilibrium between systems. The first is *thermal equilibrium*, where there is a flow of energy between systems which eventually balances and the flow stops. The second is called *mechanical equilibrium*, where the volume of each system changes until a balance is reached and then the volume changes stop. The third is *chemical equilibrium*, where there is a flow of particles between the systems until a new chemical balance is reached and the flow stops. We say that the system is in *thermodynamic equilibrium* when thermal, mechanical, and chemical equilibrium have been reached. Let us consider each in turn.

1.3.1 Mechanical equilibrium

Consider a cylinder with a piston on it as shown in Fig. 1.3. Place a mass on top of the piston. The piston goes down, and squashes the gas in the cylinder until the pressure in the gas exactly balances the downward force of the mass on the piston. When this happens the piston stops moving, the sign that mechanical equilibrium has been reached.

1.3.2 Thermal equilibrium

The analogous thermal problem occurs when two systems, A and B, are placed in thermal contact and thermally isolated from everything else. System A starts at one temperature, system B at another; as heat energy flows from one to the other the temperatures change until they both reach the same final temperature. When the temperature of each does not change with time, that is the sign which indicates that they are in thermal equilibrium.

1.3.3   Chemical equilibrium

Chemical equilibrium concerns systems where the number of particles can change. For example the particles can react together. Consider the reaction

$$C + D \rightleftharpoons CD. \tag{1.3.1}$$

This notation means that one particle of C reacts with one of D to form the bound state CD. The reaction could be a chemical reaction involving atoms and molecules, it could be a reaction of elementary particles, or it could be a reaction of an electron and a proton to form a hydrogen atom. Any process whereby C reacts with D to form a compound CD is described here as a *chemical reaction*.

If there is too much C and D present, then the reaction proceeds to form more of the compound CD. If there is too much CD then the reaction proceeds to form C and D. In chemical equilibrium there is a balance between the two rates of reaction, so that the number of particles of C, of D, and of CD remains constant. That is the sign that chemical equilibrium has been reached.

Chemical equilibrium concerns systems where the number of particles is free to change. It is not solely concerned with reactions. Sometimes the particles can coexist in different phases. Condensed matter can exist as a liquid, as a gas, or as a solid; these are different *phases of matter*. Solids can exist in different structural phases, or in different magnetic phases such as ferromagnetic or paramagnetic.

As an example of a phase change, consider the situation where water particles exist as ice in thermal equilibrium with water at $0°$ C. As the ice melts, so the number of ice molecules decreases and the number of water molecules increases. The number of particles in each phase changes. When chemical equilibrium is reached the number of particles in each phase remains constant and there is no change in the proportion of ice and water with time.

When all three equilibria have been reached the system is said to be in *thermodynamic equilibrium*. The system then has a well-defined pressure, temperature, and chemical potential. When the system is not in thermodynamic equilibrium these quantities are not well defined.

It may help to have a mental picture of a system where all three processes occur. Imagine a mixture of gases in a cylinder with adiabatic walls as in Fig. 1.3. The gases can react together to form molecules; as they react some energy is released which makes the gases expand, causing the piston to move in the cylinder. Also the temperature changes as the energy is released. We put sensors in to measure the volume of the gas, the temperature of the gas, and a spectroscopic probe to measure the amount of each type of particle. After a sufficient length of time none of the measured quantities changes. All the energy has come out of the reaction, the gas has fully expanded, and all the gas in the cylinder has come to the same temperature. There are no convection currents of the gas in the cylinder. The system is then said to be in thermodynamic equilibrium.

1.4 Functions of state

When the system is in thermodynamic equilibrium, properties of the system only depend on thermodynamic coordinates, such as the pressure and volume; they are independent of the way the system was prepared. All memory of what has happened to produce the final equilibrium state is lost. The history of the preparation of the gases is irrelevant. No matter what we do to the gases or the order we do it, the system always comes to the same final state with a particular temperature, volume, pressure, and number of particles of each sort. This is a profound statement which has far-reaching implications.

For simplicity, let us now consider a pure gas with no possibility of chemical reactions between the particles or with the containing vessel. We restrict the analysis for the present to systems with a constant number of particles. The temperature of the gas, as measured by a particular thermometer, depends only on the volume and pressure of the system through the relationship

$$\theta = F(P, V). \tag{1.1.1}$$

The temperature does not depend upon the previous history of the system. Whenever a quantity only depends on the present values of macroscopic variables such as the pressure and volume through a formula like eqn (1.1.1) we say that the quantity is a *function of state*. Therefore, when a system is in equilibrium its temperature is a function of state. The equation which describes this state of affairs, eqn (1.1.1), is called an *equation of state*

There are other equations of state. Instead of writing $\theta = F(P, V)$ we could write the pressure in the form

$$P = G(\theta, V). \tag{1.4.1}$$

In this case the pressure is a function of state which is a function of volume and temperature. Alternatively we could write the volume as

$$V = H(\theta, P). \tag{1.4.2}$$

Then the volume is a function of state which depends on temperature and pressure. No one way of writing the equation of state is preferable to any other. In an experiment we may want to control the temperature and the volume; with this choice, the equation of state fixes the pressure.

Generally, equations of state are very complicated and do not give rise to a simple mathematical formula. The ideal gas is an exception. The isotherms for an ideal gas are given by the equation

$$PV = \text{const.} \tag{1.4.3}$$

We can choose the temperature of the ideal gas, θ_G, to be proportional to the product PV; with this choice neighbouring isotherms have neighbouring values of temperature. All we need to define a temperature scale is to choose a value for the constant of proportionality for a mole of the gas. In this way we can define the *ideal gas scale of temperature*, denoted as θ_G, as

$$\theta_G = \frac{PV}{R} \tag{1.4.4}$$

where R is the constant of proportionality, called the *gas constant*. For n moles of gas the equation of state is

$$PV = nR\theta_G.$$

It must be stressed that for substances other than the perfect gas, such as a liquid or an imperfect gas, the equation of state for the temperature is a complicated function which is not readily expressible by a mathematical formula. For example, we might guess that the equation of state for the pressure of a gas is given by *van der Waals equation*, which can be written for 1 mole of gas as

$$\theta R = \left(P + \frac{a}{V^2} \right) (V - b) \tag{1.4.5}$$

where a and b are constants for a particular gas, and θ is the temperature of the gas. This equation can be rewritten quite easily as an equation of state for P.

Both the ideal gas equation and the van der Waals equation are only approximations, albeit often very good ones, to the function of state of a real gas. The real function of state can be measured and tabulated. The function of state exists, even if our best efforts at calculating or measuring it are imprecise.

All functions of state can be written in the form

$$G = g(x, y). \tag{1.4.6}$$

For example, the temperature of a system with a fixed number of particles is given by the equation

$$\theta = F(P, V).$$

To change it into eqn (1.4.6) we substitute G for θ, x for P, y for V, and g for F. The thermodynamic coordinates P and V are then turned into 'position' coordinates x and y.

Any small change in the function of state G is given by the equation

$$dG = \left(\frac{\partial g(x, y)}{\partial x} \right)_y dx + \left(\frac{\partial g(x, y)}{\partial y} \right)_x dy \tag{1.4.7}$$

where the notation $(\partial g(x, y)/\partial y)_x$ means differentiating $g(x, y)$ with respect to y keeping x constant. Such derivatives are called *partial derivatives*.

Equation (1.4.7) can be rewritten as:

$$dG = A(x,y)dx + B(x,y)dy, \qquad (1.4.8)$$

where

$$A(x,y) = \left(\frac{\partial g(x,y)}{\partial x} \right)_y$$

and

$$B(x,y) = \left(\frac{\partial g(x,y)}{\partial y} \right)_x.$$

Usually functions of state are smooth and continuous with well-defined partial derivatives at each point (x,y). The partial derivatives can be first order, as above, or they can be higher order. For example the second-order derivatives come in two types: a simple second-order derivative is of the form

$$\frac{\partial^2 g(x,y)}{\partial x^2} = \frac{\partial}{\partial x} \left(\frac{\partial g(x,y)}{\partial x} \right)_y ;$$

a mixed derivative is of the form

$$\frac{\partial^2 g(x,y)}{\partial x\, \partial y} = \left(\frac{\partial}{\partial x} \left(\frac{\partial g(x,y)}{\partial y} \right)_x \right)_y.$$

This notation means that we first differentiate $g(x,y)$ with respect to y keeping x constant, and then we differentiate with respect to x keeping y constant.

For functions of state which are continuous everywhere, the order of the mixed derivatives is not important, so that

$$\left(\frac{\partial}{\partial x} \left(\frac{\partial g(x,y)}{\partial y} \right)_x \right)_y = \left(\frac{\partial}{\partial y} \left(\frac{\partial g(x,y)}{\partial x} \right)_y \right)_x. \qquad (1.4.9)$$

Such functions of state are called *analytic*. It follows that

$$\left(\frac{\partial B(x,y)}{\partial x} \right)_y = \left(\frac{\partial A(x,y)}{\partial y} \right)_x. \qquad (1.4.10)$$

Consequently, changes in an analytic function of state can be written in the form of eqn (1.4.8) with the quantities $A(x,y)$ and $B(x,y)$ satisfying eqn (1.4.10).

Now let us look at the situation the other way around. Suppose we derive an equation of the form

$$dG = A(x,y)\, dx + B(x,y)\, dy. \qquad (1.4.8)$$

Can we integrate this equation to obtain a function of state? The answer is that we can derive a function of state provided eqn (1.4.10) is satisfied. We call this our first assertion. When this is the case the quantity dG is said to be an *exact differential*, that is it only depends on the difference in the function of state between two closely spaced points *and not on the path between them*. This is our second assertion.

To prove these assertions let us construct a two-dimensional vector:

$$\mathbf{G} = \hat{\mathbf{i}}\, A(x,y) + \hat{\mathbf{j}}\, B(x,y)$$

and a small 'real space' vector

$$\mathbf{dr} = \hat{\mathbf{i}}\, \mathrm{d}x + \hat{\mathbf{j}}\, \mathrm{d}y.$$

Here $\hat{\mathbf{i}}$ and $\hat{\mathbf{j}}$ are unit vectors in the x and y directions respectively. The line integral of $\mathbf{G} \cdot \mathbf{dr}$ around a closed contour in the x–y plane is

$$\oint \mathbf{G} \cdot \mathbf{dr} = \iint \mathrm{d}\mathbf{S} \cdot \operatorname{curl} \mathbf{G}$$

according to Stokes' theorem. But the z component of $\operatorname{curl} \mathbf{G}$ is

$$(\operatorname{curl} \mathbf{G})_z = \left(\frac{\partial B(x,y)}{\partial x}\right)_y - \left(\frac{\partial A(x,y)}{\partial y}\right)_x = 0$$

if eqn (1.4.10) is satisfied. It follows that the integral of $\mathbf{G} \cdot \mathbf{dr}$ around *any* closed contour in the x–y plane is zero.

The integral from $\mathbf{r}_1$ to $\mathbf{r}_2$ along two paths must satisfy the relation

$$\left(\int_{\mathbf{r}_1}^{\mathbf{r}_2} \mathbf{G} \cdot \mathbf{dr}\right)_{\text{path 1}} + \left(\int_{\mathbf{r}_2}^{\mathbf{r}_1} \mathbf{G} \cdot \mathbf{dr}\right)_{\text{path 2}} = \oint \mathbf{G} \cdot \mathbf{dr} = 0.$$

It follows that

$$\left(\int_{\mathbf{r}_1}^{\mathbf{r}_2} \mathbf{G} \cdot \mathbf{dr}\right)_{\text{path 1}} = \left(\int_{\mathbf{r}_1}^{\mathbf{r}_2} \mathbf{G} \cdot \mathbf{dr}\right)_{\text{path 2}}$$

We can choose any path between $\mathbf{r}_1$ and $\mathbf{r}_2$ and the answer is the same. The integral between $\mathbf{r}_1$ and $\mathbf{r}_2$ is independent of path.

Now let us derive the function of state. We choose one point, say $\mathbf{r}_1$, to be a reference point. Then we can define the function of state by the equation

$$g(x,y) = \int_{\mathbf{r}_1}^{\mathbf{r}} \mathbf{G} \cdot \mathbf{dr}$$

where $\mathbf{r} = \hat{\mathbf{i}}x + \hat{\mathbf{j}}y$. Clearly the integral is a function of x and y alone, so it is a function of state. We have proved our first assertion.

To prove the second assertion consider two points $\mathbf{r}$ and $\mathbf{r} + \mathbf{dr}$ which are very close together. The change in G is

$$\mathrm{d}G \;=\; g(x + \mathrm{d}x, y + \mathrm{d}y) - g(x,y)$$

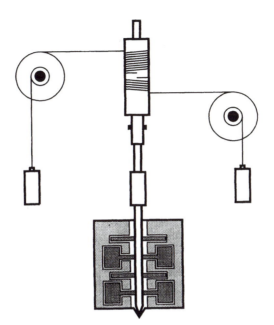

Fig. 1.4 Joule's paddle-wheel experiment. The weights fall from a known height and release their potential energy to the system. The falling weights turn paddles which do work on the liquid by forcing it to flow through baffles. Joule showed that the amount of energy absorbed depends on the total amount of work that was done, not on how it was done.

$$= \int_{\mathbf{r}_1}^{\mathbf{r}+d\mathbf{r}} \mathbf{G} \cdot d\mathbf{r} - \int_{\mathbf{r}_1}^{\mathbf{r}} \mathbf{G} \cdot d\mathbf{r} = \int_{\mathbf{r}}^{\mathbf{r}+d\mathbf{r}} \mathbf{G} \cdot d\mathbf{r}.$$

This integral depends only on the points $\mathbf{r}$ and $\mathbf{r}+d\mathbf{r}$ and *is independent of the path between them*. Hence dG is an exact differential, our second assertion.

We can only derive the function of state if $A(x,y)$ and $B(x,y)$ satisfy eqn (1.4.10) so that dG is an exact differential. Sometimes $A(x,y)$ and $B(x,y)$ do *not* satisfy eqn (1.4.10). When this is the case we say that đG is an *inexact differential*. To remind us that it is inexact we put a line through the đ. We will see that small additions of work and heat are represented by inexact differentials.

1.5 Internal energy

Consider an experiment which is designed to measure the mechanical equivalent of heat, such as *Joule's paddle-wheel experiment*. The apparatus used in the experiment is illustrated in Fig. 1.4. A mass of water is enclosed in a calorimeter which is thermally insulated as well as possible. Weights, of total mass M, fall through a distance h, thereby causing the paddle-wheels to rotate in the calorimeter, doing work against the viscosity of the water. The work done, W, is equal to Mgh, where g is the acceleration due to gravity. As a result of the work, the temperature of the water (and calorimeter) is found to change. From a careful series of such experiments it can be deduced how much work needs to be done to effect a given change in temperature.

An alternative experiment involves a resistance, r, placed inside the water in the calorimeter. A measured current, I, is passed through the resistance for a known length of time, t. The work done pushing the current through the resistance is given by $I^2 rt$. Experiment shows that the amount of work needed to effect a given change in temperature is the same for both experiments, provided the system starts at the same temperature. The same result is found when other forms of work are done on the system.

In short: *when the state of an isolated system is changed by the performance of work, the amount of work needed is independent of the means by which the work is performed.*

We can go further than this. Suppose we do the work partly by dropping weights and partly by electrical work. Either we do work against gravity first and then electrical work, or the other way around. The total work done does not depend on any intermediate state of the system. No matter in which order we choose to add the work, the final equilibrium state of the system depends *only* on the total amount of work done on the system and not on the order in which it was done.

The crucial point is that we can associate numbers with the work done—it is just the distance fallen by the mass times its weight—and hence we can associate numbers, numerical values, with the change in the internal energy. We can add and subtract numbers in any order we like and the answer is the same. Add 2 joules of electrical work first and then 3 joules of mechanical work through falling weights, and we get a total amount of work of 5 joules. If the work is done in the reverse order, first 3 joules of mechanical work and then 2 joules of electrical work, we add 5 joules of work to the internal energy of the system and the final state of the system is the same. The order of these processes does not matter.

This statement is non-trivial: in daily life we can tell whether we put our socks on first and then our shoes, or the other way round. The order of the addition matters, for it leaves a record. No record is left of the order in which work is done.

Where does the work go? In classical mechanics we can show that the total energy of the system, that is the sum of the kinetic energy and potential energy of all the particles, is a constant which does not vary with time. But in Joule's experiment the potential energy associated with the masses decreases as they fall. It appears that the energy of the system has decreased. If we can specify where the energy has gone, we may be able to preserve the notion that *energy is always conserved.*

We have just observed that to effect a given change in the equilibrium state of the system, the amount of work needed does not depend on the intermediate stages through which the system passes. The system can be taken through any attainable intermediate state, by any series of process we choose, and the final equilibrium state is the same. It is as a consequence of this observation that we can define a quantity called the *internal energy* of the system, denoted by the symbol U. We choose U so as to ensure that the total energy of the system is constant. If a thermally isolated system is brought from one state to another by the performance of work, W, then the internal energy is said to be increased by an amount ΔU which is equal to W. Any loss in potential energy, say by a weight falling through a distance h, is matched by the gain in internal energy.

The total work done is independent of the path chosen from the initial to the final equilibrium states; it follows that the quantity U is a function of the initial and final states only. The quantity U is therefore a function of state. In the initial state the pressure and volume are well-defined quantities, so the internal energy can be written as $U(P, V)$. Alternatively it could be written as $U(P, \theta)$, or as $U(V, \theta)$, since if we know P and V we know the temperature θ, from eqn (1.1.1).

To specify U completely we must choose a reference value of internal energy, U_0, and then we can obtain the internal energy for any other equilibrium state as $U = U_0 + \Delta U$. The choice of a reference energy is arbitrary. This arbitrariness is not surprising; after all there is the same difficulty in defining an absolute value of the potential energy.

The analysis so far has been restricted to systems which are thermally isolated. Can we extend it to systems which are not thermally isolated? The internal energy of a system which is not thermally isolated can be altered because energy can enter or leave the system. We can change the state of a beaker of water by placing a lit Bunsen burner under it. The Bunsen burner does no work, but the state of the system changes (its temperature rises) which indicates that the internal energy has changed. In order to preserve the law of conservation of energy we choose to write the change in internal energy from its initial value U_i to its final value U_f as

$$\Delta U = U_f - U_i = W + Q \qquad (1.5.1)$$

where Q is defined as the *heat* which is added to the system, and W is the work done on the system. The change in the internal energy is the sum of the work done on the system and the heat added. The extra term, Q, is included so that we can preserve the law of conservation of energy for systems which are not thermally isolated. *The first law of thermodynamics then says that energy is conserved if heat is taken into account.*

As an example, consider a motor which produces 1.5 kJ of energy per second as work and loses 0.4 kJ of energy per second as heat to the surroundings. What is the change in the internal energy of the motor and its power supply in one second? The energy lost as work W is negative and equal to -1.5 kJ; the heat is lost from the motor so $Q = -0.4$ kJ; hence the change in the internal energy of the motor is -1.9 kJ in one second.

In real experiments we may know how energy was transferred to a system, but the system itself cannot tell whether the energy was transferred to it as work or as heat. Heat energy and work energy, when added to a system, are stored as internal energy and not as separate quantities.

1.6   Reversible changes

Consider the system shown in Fig. 1.3 in which a piston moves without any friction and the gas in the cylinder is perfectly insulated so no heat can enter. In the initial state the gas in the cylinder is in equilibrium. The mass m on the piston compresses the contents of the cylinder until mechanical equilibrium has been reached. Now we slowly place a tiny mass, δm, on the piston and alter the pressure on the piston infinitesimally. The gas in the cylinder is compressed slightly as the piston moves down, the volume of the gas decreases, and the temperature rises. If we were to remove the tiny mass, the gas would expand back to its original volume and the temperature would fall back to its original value. The process is reversible.

Now imagine that the cylinder walls allow heat in easily. The surrounding temperature is changed very slowly by a tiny amount, $\delta\theta$. Energy flows in through the walls and the gas in the cylinder expands slightly as its temperature rises. Now we slowly cool the surroundings back to the original temperature, and we find that the gas contracts back to the original volume. In both cases we get back to a final state which is the same as the initial state; nothing has altered in the surroundings, so the process is reversible. A *reversible process* is defined as one which may be exactly reversed to bring the system back to its initial state with no other change in the surroundings.

A reversible process does not have to involve small changes in pressure and temperature. We can make up a large change in pressure by slowly adding lots of tiny weights, δm, to the piston; as each addition is made the gas compresses and its temperature rises slightly; we wait for mechanical equilibrium to be reached, at the same time changing the temperature of the environment so that no energy flows through the imperfectly insulated walls; then the next tiny weight is added. We imagine a change in the system which is made in such a way that *the system remains in equilibrium at all times*. Such a change is called *quasistatic*. A reversible process involving a large change proceeds through a series of quasistatic states; each change is made so that the system remains in equilibrium at all stages. Undoing the process step by step would return the system to the same original state. Similarly we can make up large, reversible changes in temperature from lots of tiny changes, $\delta\theta$.

A system which contains a large number of particles could be described by a large number of microscopic coordinates and momenta, all of which vary rapidly, on a microscopic time-scale, τ_C, typically the collision time between particles. There is another set of coordinates such as the pressure and temperature of the system which varies slowly, over a time-scale $\tau_r \gg \tau_C$. To a first approximation the system is at each moment in a thermodynamic state which is in instantaneous equilibrium—this is what we mean by a quasistatic state. The instantaneous equilibrium state is achieved with an error of order τ_C/τ_r. The difference between the equilibrium state and a quasistatic state is small, but, conceptually, it is essential since it is this difference which leads to internal energy dissipation in the system and irreversibility. If we can ensure that τ_C/τ_r is as small as possible we approach the ideal of a reversible process.

For tiny reversible changes in volume there is a very simple expression for the work done on the system. To be specific we will take the system to be a gas in the cylinder. Suppose the system starts with a mass m on the piston, and the system is in mechanical and thermal equilibrium so that everything is stationary. We add a tiny mass, δm; as a result the piston drops down by a small distance, dz. We add such a tiny mass at each stage, say a grain of sand, that the system is only slightly perturbed from equilibrium. If we continue slowly adding tiny grains of sand, keeping the system in equilibrium at each stage, then we make a quasistatic change of the system. The area of the piston is A, so the change in the volume of gas is

$$dV = -A\,dz. \tag{1.6.1}$$

The force acting downwards on the gas is $-(m + \delta m)g$. The work done by the total mass when it moves down by a distance dz is

$$\begin{aligned} đW &= (m + \delta m)g\,dz \\ &= -\frac{(m + \delta m)g}{A}\,(-A\,dz) = -\frac{(m + \delta m)g\,dV}{A}. \end{aligned} \tag{1.6.2}$$

(Notice the line through the d in $đW$ denoting an inexact differential.) Since the original mass, m, is in mechanical equilibrium, the quantity mg/A is the pressure of the gas, P. Provided δm is negligible compared to m, the work done on the system is

$$đW = -P\,dV. \tag{1.6.3}$$

For reversible processes involving a small change in volume, *the first law of thermodynamics* can be written as

$$dU = đQ_{\text{rev}} - P\,dV \tag{1.6.4}$$

where $đQ_{\text{rev}}$ is the small amount of heat which is added reversibly.

The total work done is the integral of $-P\,dV$ between the initial and final states; thus

$$\Delta W = -\int_{V_1}^{V_2} P\,dV. \tag{1.6.5}$$

We can imagine the system evolving in a P–V diagram through a series of quasistatic states from the initial state with volume V_1 to the final state with volume V_2. The evolution can occur along different paths. The area under the curve between V_1 and V_2, which is equal to $-W$, depends on the path taken. If the system goes along path 1 in Fig. 1.5, clearly the work done is different from that done when going along path 2.

Work is not a function of state: it depends not only on the initial and final equilibrium states but also on the path taken. If we were to take the system

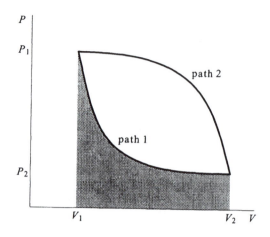

Fig. 1.5 Two different paths for the system to evolve from P_1, V_1 to P_2, V_2. Both paths require that the system remains in equilibrium at all stages so that the pressure is well defined. The area under each curve is different, which tells us that the work done on the system depends on the path taken as well as the initial and final equilibrium states. Work is not a function of state.

around a cycle back to the starting point there would be no change in the internal energy, but we can get work out of a system during the cycle. That is what an engine does. But notice: there is a sign convention depending upon which way the system is taken around the cycle. When the system goes around the cycle clockwise in the P–V diagram, the area enclosed is positive and so the work done on the system is negative; the system is doing work on its surroundings. If the system is taken anticlockwise around the cycle in the P–V diagram, the area enclosed is negative and so the work is positive; the system has work done on it.

Furthermore *heat is not a function of state*. The change in the internal energy due to reversible processes is the sum of the heat added and the work done. In a cycle the change in the internal energy is zero since internal energy is a function of state. The change in the sum of the heat and work over the cycle must be zero. If net work is done by a system during the cycle, net heat must be added during the cycle. Since the total heat added to the system in a cycle need not be zero, heat is not a function of state.

Whenever we consider a tiny amount of work or heat we write it as đW or đQ, where the line through the d indicates that the quantity is an inexact differential, a differential which depends on the path in the space of thermodynamic coordinates as well as the initial and final states.

For reversible processes the work done on the system is a well-defined quantity. For an ideal gas in the cylinder in Fig. 1.3, the work done is đ$W = -P\,\mathrm{d}V$ with $P = nR\theta_G/V$ for n moles of an ideal gas. Therefore the total work done in an isothermal change (constant temperature) is

$$\Delta W_{1,2} = -nR\theta_G \int_{V_1}^{V_2} \frac{dV}{V} = -nR\theta_G \ln\left(\frac{V_2}{V_1}\right). \qquad (1.6.6)$$

We use the symbol '$\ln(x)$' to denote the natural logarithm.

For *irreversible processes* the work done on the system for a tiny change in volume is not given by $-P\,dV$. Suppose that the initial mass on the piston, m, is in equilibrium with the gas; the initial pressure of the gas is $P = mg/A$. Imagine that we now place a mass, Δm, on the piston, a mass which is not tiny in comparison with m. The piston drops down and comes to a new equilibrium position. The work done by the two masses obeys the inequality

$$đ W = -\frac{(m + \Delta m)g\,dV}{A} \geq -P\,dV. \qquad (1.6.7)$$

Remember that $-P\,dV$ is positive since the volume shrinks when we add the weight. In general the work is larger than $-P\,dV$; it is only equal to $-P\,dV$ for reversible changes.

Work can be done in ways other than the expansion of the volume of a system. As an example, we can do work by magnetizing a system. A magnetic field, $\mathbf{H}$, is generated by passing a current through a coil placed around the system. If we change the magnetic induction field, $\mathbf{B}$, in a reversible way by an amount $d\mathbf{B}$, then we do work $\mathbf{H}.d\mathbf{B}$ per unit volume, an assertion which is proved in Appendix A. Another form of work occurs when an electric field, $\mathbf{E}$, is applied to a dielectric medium. The electric field can be generated by charging a capacitor and placing the system between the plates of the capacitor. If we change the electric displacement, $\mathbf{D}$, in a reversible way by an amount $d\mathbf{D}$, then we do work $\mathbf{E}.d\mathbf{D}$ per unit volume.

The system can be two-dimensional, for example a soap film. When the area of the film changes reversibly by an amount dA, the work done on the system is $\gamma_S\,dA$, where γ_S is the surface tension of the film. Alternatively the system can be one-dimensional such as a thin wire under tension F. When the length of the wire changes reversibly by an amount dl the work done is $F\,dl$.

For the present we ignore these other forms of work and concentrate on the work done on the system caused by changes in volume. It is straightforward to repeat the analysis for other forms of work. For the work done on a stretched wire it turns out that all that one has to do is to replace dV by dl (say) and $-P$ by F.

1.7 Enthalpy

Many experiments are carried out at constant pressure. Often they are done at atmospheric pressure because that is the most convenient pressure to use in the laboratory. In this circumstance any increase in the volume of a system leads to the system doing work on the environment. If the system changes its volume reversibly by an amount dV then the change in internal energy is

$$dU = đQ_{rev} - P \, dV. \tag{1.6.4}$$

The heat added to the system at constant pressure can then be expressed as

$$đQ_{rev} = dU + P \, dV = d(U + PV) \tag{1.7.1}$$

which follows because $dP = 0$ since the pressure is constant.

The quantity $U + PV$ is called *the enthalpy*; it is denoted by the symbol H. Since $U, P,$ and V are functions of state, it is clear that enthalpy is a function of state as well. A reversible, small change in the enthalpy is given by

$$dH = d(U + PV) = đQ_{rev} + V \, dP. \tag{1.7.2}$$

When the system is at constant pressure, so that $dP = 0$, any change in H is equal to the heat added to the system provided that there is no other form of work.

Changes in H at constant pressure can be worked out from a knowledge of the change in internal energy, ΔU, and the change in the volume of the system, ΔV. Thus

$$\Delta H = \Delta U + P \, \Delta V. \tag{1.7.3}$$

As an example, consider 1 mole of $CaCO_3$ as calcite at a pressure of 1 bar $(= 10^5 \, N \, m^{-2})$. If it were to change to the aragonite form of $CaCO_3$ the density would change from $2710 \, kg \, m^{-3}$ to $2930 \, kg \, m^{-3}$, and the internal energy would increase by 210 J. The mass of 1 mole of $CaCO_3$ is 0.1 kg. Hence the volume of 1 mole changes from $37 \times 10^{-6} \, m^3$ to $34 \times 10^{-6} \, m^3$. The change in enthalpy is

$$\Delta H = 210 + 10^5 \times (34 - 37) \times 10^{-6} = 209.7 \, J.$$

Note that ΔH is very nearly the same as $\Delta U = 210 \, J$. The PV term makes a small contribution at atmospheric pressure for solids. Usually the difference between internal energy and enthalpy can be neglected as being small; it is not small when the system is at very high pressures or when gases are involved.

1.8   Heat capacities

When a tiny amount of heat energy is added to a system it causes the temperature to rise. The *heat capacity* is the ratio of the heat added to the temperature rise. Thus

$$C = \frac{đQ_{rev}}{d\theta}. \tag{1.8.1}$$

However, this ratio is not well defined because it depends on the conditions under which the heat is added. For example, the heat capacity can be measured with the system held either at constant volume or at constant pressure. The values measured turn out to be different, particularly for gases. (See Table 1.1 for the molar heat capacities of some gases.)

Table 1.1 The molar heat capacity of some gases at 25 K

Gas	C_P	C_V	$C_P - C_V$
He	20.9	12.6	8.3
Ar	20.9	12.5	8.4
Hg	20.9	12.5	8.4
O_2	29.3	20.9	8.4
CO	29.3	21.0	8.3
Cl_2	34.1	25.1	9.0
SO_2	40.6	31.4	9.2
C_2H_6	51.9	43.1	8.8

First let us consider a system which is held at constant volume so that $P\,dV = 0$, and suppose that the system cannot do any other kind of work. The heat that is required to bring about a change in temperature $d\theta$ is

$$\text{d}Q_V = C_V \text{d}\theta \qquad (1.8.2)$$

where C_V is the heat capacity at constant volume. When there is no work done on the system the change in the internal energy is caused by the heat added, so that $dU = \text{d}Q_V$. The heat capacity at constant volume is

$$C_V = \left(\frac{\partial U}{\partial \theta} \right)_V \qquad (1.8.3)$$

where the partial derivative means that we take the ratio of a tiny change in internal energy to a tiny change in temperature with the volume of the system held constant.

As an example, consider an *ideal monatomic gas*. In a monatomic gas the molecules only have kinetic energy associated with translational motion for they cannot vibrate or rotate. According to the kinetic theory of gases the internal energy of n moles of such a gas is given by $U = 3nR\theta_G/2$. The heat capacity at constant volume is $C_V = 3nR/2$.

Now consider the heat capacity at constant pressure. If the system is held at constant pressure, it usually expands when heated, doing work on its surroundings, $P\,dV$, where dV is the increase in volume on expansion. If no other work is done, the heat required to raise the temperature by $d\theta$ is

$$\text{d}Q_P = C_P \text{d}\theta \qquad (1.8.4)$$

where C_P is the heat capacity at constant pressure. Some of the energy which is fed in as heat is returned to the surroundings as work; this part of the heat is *not* used to raise the temperature. More heat is needed to raise the temperature per degree than if the volume were held constant. Provided the system expands on heating, C_P is larger than C_V.

At constant pressure the heat added is the change in enthalpy, $đQ_P = dH$. Thus

$$C_P = \left(\frac{\partial H}{\partial \theta}\right)_P = \left(\frac{\partial U}{\partial \theta}\right)_P + \left(\frac{\partial (PV)}{\partial \theta}\right)_P. \tag{1.8.5}$$

For a monatomic ideal gas $U = 3nR\theta_G/2$, and so $(\partial U/\partial \theta_G)_P = 3nR/2$ which is the same as the heat capacity at constant volume, C_V. A single mole of a perfect gas obeys the equation $PV = R\theta_G$, so that eqn (1.8.5) becomes

$$C_P = C_V + R. \tag{1.8.6}$$

The difference between C_P and C_V for 1 mole of any gas is always about $R = 8.3\,\mathrm{J\,K^{-1}}$, the gas constant. This statement can easily be proved for diatomic and triatomic gases as well as for monatomic gases. The evidence in its favour is shown by the values given in Table 1.1.

The reason for the large difference between C_P and C_V for gases is that they expand significantly on heating; in contrast, solids and liquids hardly change their volumes on heating so the difference between C_P and C_V is much less pronounced.

It is an experimental observation that the heat capacity of a gas, or indeed any system, is proportional to the number of particles present. This means that if we double the number of particles and keep the density and the temperature of the system the same, then the heat capacity is doubled.

1.9   Reversible adiabatic changes in an ideal gas

For a small change in 1 mole of a perfect gas

$$P\,dV + V\,dP = R\,d\theta_G. \tag{1.9.1}$$

The temperature change can be related to the internal energy by

$$dU = C_V\,d\theta_G. \tag{1.9.2}$$

An *adiabatic change* is defined as a change in which no heat enters the system. For such a reversible change $dU = -P\,dV$. The change in the internal energy is the work done on the system. Therefore

$$P\,dV + V\,dP = \frac{R}{C_V}dU = -\frac{R}{C_V}P\,dV, \tag{1.9.3}$$

and it follows that

$$\frac{dP}{P} + \gamma\frac{dV}{V} = 0 \tag{1.9.4}$$

where γ is the ratio of heat capacities $\gamma = C_P/C_V$. The quantity γ is a constant. Integration of eqn (1.9.4) gives the equation

$$PV^\gamma = \text{constant} \qquad\qquad (1.9.5)$$

for reversible adiabatic changes. If we write the pressure as $P = R\theta_G/V$ then we get the relationship between temperature and volume,

$$\theta_G V^{(\gamma-1)} = \text{constant} \qquad\qquad (1.9.6)$$

for reversible adiabatic changes.

1.10 Problems

1. Rewrite the van der Waals equation in the form $P = G(\theta, V)$.
2. By writing the internal energy as a function of state $U(T, V)$ show that

$$\dslash Q = \left(\frac{\partial U}{\partial T}\right)_V dT + \left[\left(\frac{\partial U}{\partial V}\right)_T + P\right] dV.$$

3. Which of the following is an exact differential:
   (a) $dx = (10y + 6z)\,dy + 6y\,dz$,
   (b) $dx = (3y^2 + 4yz)\,dy + (2yz + y^2)\,dz$,
   (c) $dx = y^4 z^{-1}\,dy + z\,dz$?

4. The equations listed below are not exact differentials. Find for each equation an integrating factor, $g(y, z) = y^\alpha z^\beta$, where α and β can be any number, that will turn it into an exact differential.
   (a) $dx = 12z^2\,dy + 18yz\,dz$,
   (b) $dx = 2e^{-z}\,dy - ye^{-z}\,dz$.

5. Differentiate

$$x = z^2 e^{y^2 z}$$

   to get an expression for $dx = A\,dy + B\,dz$. Now divide by $ze^{y^2 z}$. Is the resulting equation an exact differential?

6. For an ideal gas $PV = nRT$ where n is the number of moles. Show that the heat transferred in an infinitesimal quasistatic process of an ideal gas can be written as

$$\dslash Q = \frac{C_V}{nR}V\,dP + \frac{C_P}{nR}P\,dV.$$

7. Suppose the pressure of a gas equals $\alpha V^{-\gamma}$ for an adiabatic change. Obtain an expression for the work done in a reversible adiabatic change between volumes V_1 and V_2.

8. An explosive liquid at temperature 300 K contains a spherical bubble of radius 5 mm, full of its vapour. When a mechanical shock to the liquid causes adiabatic compression of the bubble, what radius of the bubble is required for combustion of the vapour, given that the vapour ignites spontaneously at 1100°C? The ratio C_V/nR is 3.0 for the vapour.

9. A system consists of a gas inside a cylinder as shown in Fig. 1.3. Inside the cylinder there is a $10\,\Omega$ resistor connected to an electrical circuit through which a current can pass. A mass of $0.1\,kg$ is placed on the piston which drops by $0.2\,m$ and comes to rest. A current of $0.1\,A$ is passed through the $10\,\Omega$ resistor for $2\,s$. What is the change in the internal energy of the system when it has reached equilibrium?

10. Consider n moles of an ideal monatomic gas at initial pressure P_1 and volume V_1 which expands reversibly to a new volume $V_2 = V_1(1 + \varepsilon)$ with ε small. Use a $P - V$ diagram (refer to Fig. 1.5) to establish in which of the following conditions the system does most and least work; check your answer by calculating the work done by the system in each of these circumstances. [Hint: work to second order in ε.]

    (a) isobaric (the system expands at constant pressure);
    (b) isothermal (the system expands at constant temperature);
    (c) adiabatic (no heat enters when the system expands).

11. A hot water tank contains 175 kg of water at 15°C.

    (a) How much heat is needed to raise the temperature to 60°C? [The specific heat capacity of water is $4200\ J\ kg^{-1}K^{-1}$.]
    (b) How long does it take for a 5 kW electric immersion heater to achieve this task?

12. An energy-saver light-bulb of 11 W gives the same illumination as a conventional 60 W light-bulb.

    (a) If a typical light-bulb is used on average for five hours a day, how much energy is saved per year by one light-bulb?
    (b) If each household used eight such light-bulbs, and there are 20 million households in a country, how much energy would that country save annually?

2

Entropy and the second law of thermodynamics

2.1   A first look at the entropy

For reversible processes involving the expansion of the system we can write the first law of thermodynamics in the form

$$đQ_{rev} = dU + P\,dV. \qquad (1.6.4)$$

Remember that $đQ_{rev}$ is an *inexact differential*: the heat added to a system depends upon both the initial and final equilibrium states *and* the path between them. This equation cannot be integrated to yield a function of state. Sometimes it is possible to find an *integrating factor*, a function which multiplies the inexact differential and turns it into an exact differential. As an example consider the equation

$$đx = z^3\,dy + (2z + yz^2)\,dz. \qquad (2.1.1)$$

Clearly this is an inexact differential, as can be checked by writing $A(y, z) = z^3$ and $B(y, z) = 2z + yz^2$ and substituting them into eqn (1.4.9). However, if we were to multiply $đx$ by e^{yz} we would get the equation

$$
\begin{aligned}
e^{yz}\,đx &= z^3\,e^{yz}\,dy + (2z + yz^2)\,e^{yz}dz \\
&= d(z^2\,e^{yz}) \qquad\qquad\qquad\qquad (2.1.2)
\end{aligned}
$$

which is an exact differential. The function e^{yz} is called the integrating factor for this equation.

The question can be posed: can we find an integrating factor which multiplies $đQ_{rev}$ and produces an exact differential? If we can do this then we can construct a new function of state and call it the entropy. To give you some confidence that this is a realistic proposal, we show how it can be achieved for an ideal gas.

Suppose the system consists of n moles of an ideal monatomic gas whose internal energy, according to the kinetic theory of gases, is $3nR\theta_G/2$. The internal energy can be changed by altering the temperature. A small change in internal energy is given by $dU = 3nR\,d\theta_G/2$. An ideal gas in a volume V obeys the equation $P = nR\theta_G/V$, so that $P\,dV = nR\theta_G dV/V$. Therefore, for reversible changes

$$
\begin{aligned}
đQ_{rev} &= dU + P\,dV \\
&= \frac{3nR\,d\theta_G}{2} + \frac{nR\theta_G\,dV}{V}.
\end{aligned}
\tag{2.1.3}
$$

If we divide eqn (2.1.3) by θ_G we get

$$
\frac{đQ_{rev}}{\theta_G} = \frac{3nR\,d\theta_G}{2\theta_G} + \frac{nR\,dV}{V}
\tag{2.1.4}
$$

which can be integrated directly between initial equilibrium state, i, and final equilibrium state, f, to give

$$
\begin{aligned}
\int_i^f \frac{đQ_{rev}}{\theta_G} &= \int_{\theta_G^i}^{\theta_G^f} \frac{3nR\,d\theta_G}{2\theta_G} + \int_{V_i}^{V_f} \frac{nR\,dV}{V} \\
&= \frac{3nR}{2}\ln\left(\frac{\theta_G^f}{\theta_G^i}\right) + nR\ln\left(\frac{V_f}{V_i}\right).
\end{aligned}
\tag{2.1.5}
$$

For a reversible process, the quantity $đQ_{rev}/\theta_G$ is an exact differential which defines a small change in a new function of state. Let us denote the function of state as S and call it the *entropy*. Then the small change in S is $dS = đQ_{rev}/\theta_G$. If we integrate this function between initial and final equilibrium states we get

$$
S_f - S_i = \int_i^f \frac{đQ_{rev}}{\theta_G}.
\tag{2.1.6}
$$

Let us take S_0 to be the entropy of the n moles of an ideal gas in a volume V_0 and at a temperature θ_G^0. In a volume V and at a temperature θ_G the entropy is

$$
S = \frac{3nR}{2}\ln\left(\frac{\theta_G}{\theta_G^0}\right) + nR\ln\left(\frac{V}{V_0}\right) + S_0.
\tag{2.1.7}
$$

We have created a new function of state for an ideal gas, a function of state called the entropy.

The calculation shows that it is possible to obtain the entropy function, but it does not tell us how to generate it for other systems. For this calculation we have used as an integrating factor the quantity θ_G^{-1}, the inverse of the temperature according to the ideal gas scale. Should we use as an integrating factor the inverse of the temperature for all systems, and if so which temperature? Remember we have not yet settled on a precise way of defining the temperature.

The aim of generating in a systematic way a new function of state seems to be a very ambitious task at present. Yet the task is possible, as we shall show, and it leads to a precise definition of the absolute temperature.

We shall show this in a rather involved way: first we state the second law of thermodynamics and use it to prove that the most efficient engine operating between two heat baths is a Carnot engine; then we show that efficiency of all Carnot engines running between two heat baths is the same; next we argue that the temperature scale, T, can be defined from the ratio of heat flows in one cycle of the Carnot engine; any large reversible cycle can be broken down into a collection of tiny reversible Carnot cycles, so that the integral of $đQ/T$ over the cycle is zero; finally we show that a function of state called the entropy can be defined as the integral

$$S = \left(\int_{i}^{f} \frac{đQ_{rev}}{T} \right)$$

where i is some initial reference state.

2.2 The second law of thermodynamics

When two systems are placed in thermal contact with each other they tend, if left to themselves, to come to equilibrium with each other, with each system ending in a new equilibrium state. The reverse process in which the two systems revert back to their initial state never occurs in practice, no matter how long we wait. If it did the very notion of equilibrium would lose its meaning. But here let us enter a distinction: processes which are reversible could revert back to their initial state; irreversible processes cannot. In reality, reversible processes are the exception, the ideal to which we might aspire; they do not occur in nature. The natural behaviour of systems is more or less irreversible and therefore shows a preferred direction.

It is an experimental observation that energy prefers to flow from a body with a higher temperature to one with a lower temperature. We associate the higher temperature with the hotter body and the lower temperature with the colder body. Temperature is associated with the degree of hotness of a body. It is our everyday observation that heat energy flows from a hot body to a cold one. This is the essence of *the second law of thermodynamics*.

Thermodynamics as a science starts from such mundane experimental observations, so commonly observed that they appear to hold no mystery. All disciplines of science start by collecting the facts. But that is not all that is involved; to quote Henri Poincaré: 'Science is built up of facts as a house is built up of stones; but an accumulation of facts is no more science than a heap of stones is a house.'

We start with a definition. A *thermal reservoir* (also known as a *heat bath*) is defined as a body which has such a huge heat capacity that its temperature does not change when energy is added to it. For example, a large block of copper can act as a thermal reservoir for small additions of heat energy.

Let us build on this a definition of the second law of thermodynamics. The simplest expressions of the second law relate to a cycle of processes in which heat flows between hot and cold reservoirs producing work. The statement of the second law due to Kelvin and Planck is: '*It is impossible for an engine, operating in a cycle, to take heat from a single reservoir, produce an equal amount of work, and to have no other effect.'* This statement denies the possibility of constructing an engine which takes heat from the sea, for example, and converts all the heat into work, leaving no heat left over. Another statement of the second law is due to Clausius: *'It is impossible to construct a machine which, operating in a cycle, produces no other effect than to transfer heat from a colder to a hotter body.'* These two statements of the second law can be shown to be equivalent to each other.

The phrase operating in a cycle is important. For cyclical processes the system always ends up in the same initial state. We can devise machines which transfer heat from a colder to a hotter body. For example, let us place a gas in a cylinder in contact with a cold reservoir and expand the gas; the gas takes heat from the reservoir. Now we remove the cylinder from the cold reservoir and we place it in contact with a hotter reservoir; we then compress the gas so that the energy flows into the hotter reservoir. In this way heat has been transferred from a colder to a hotter reservoir; but the process is not cyclical so we have not violated Clausius' statement.

2.3   The Carnot cycle

The way we obtain the entropy for a general system is by analysing energy flows in an ideal reversible engine. We describe the working of an engine as a cycle of processes, the net result of which is to take heat energy out of a hotter thermal reservoir, convert some of the energy into work, with the remainder entering the colder thermal reservoir. When the cycle is complete the system has returned to its initial condition ready to start the cycle again. A real engine produces work more or less continuously. In contrast, we will mentally imagine the cycle in discrete stages.

There are two sorts of changes which are particularly important. An *isothermal change* is one in which the pressure and volume are altered in such a way that the temperature remains constant; an *adiabatic change* is one in which no heat enters the system when the volume and pressure change. During an adiabatic process the temperature of the system changes even though no heat enters the system.

The *Carnot cycle* is made up of four reversible processes. First there is an *isothermal expansion* of the system with heat Q_1 being absorbed by the system from a thermal reservoir at constant temperature θ_1 (the process is shown by the line from A to B in Fig. 2.1); then there is an *adiabatic expansion* of the system with the temperature falling from temperature θ_1 to θ_2 (this is indicated by the line from B to C); next, an *isothermal compression* of the system at temperature θ_2 with heat Q_2 flowing into the second thermal reservoir (C to D); and finally, an *adiabatic compression* of the system with the temperature rising from θ_2 to θ_1 (D to A).

(Note: for reasons of simplicity of presentation, in this section and section 2.4 we use Q_2 instead of $-Q_2$ for the heat leaving the system and going into the colder reservoir.)

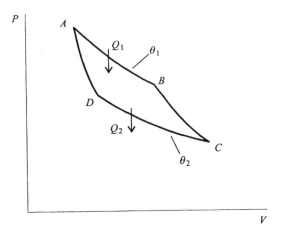

Fig. 2.1 Two isothermal processes (A to B and C to D) and two adiabatic processes (B to C and A to D) making up a Carnot cycle. During the isothermal processes heat flows into or out of the system; there is no heat flow in the adiabatic processes.

The net work done during the cycle, W, is the area enclosed by the lines joining A, B, C, and D in Fig. 2.1.

At the end of the cycle the system returns to its initial state so there is no change in the internal energy. Since energy is conserved in the cycle, $W = Q_1 - Q_2$. The two heat flows and the work done by the system are shown symbolically in Fig. 2.2.

The *efficiency* of an engine is defined as the work output per cycle divided by the heat input per cycle from the hotter reservoir. Thus for any engine, the efficiency, η, is given by

$$\eta = \frac{W}{Q_1} = 1 - \frac{Q_2}{Q_1}. \qquad (2.3.1)$$

If the engine were perfectly efficient, all the heat would be converted into work and the efficiency would be one. Nature is not so kind; even for the most efficient engine η is always less than one.

The Carnot engine is completely reversible which means that it can run forwards or backwards with equal facility. When the engine runs backwards we put work, W, *into the cycle*, place heat Q_1 into the hotter reservoir, and take heat Q_2 from the colder one.

Suppose a Carnot engine, running backwards, is joined to another engine which is irreversible and running forwards so as to make a new composite engine. The irreversible engine produces work, W, which is used to drive the Carnot engine backwards. All the work which is generated by one engine is consumed by the other, which means that no work is done on the rest of the universe. (Here

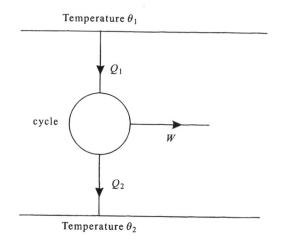

Temperature θ_1

Q_1

cycle

W

Q_2

Temperature θ_2

Fig. 2.2 Symbolic representation of an ideal reversible engine operating between two thermal reservoirs which are at constant temperatures, showing the two heat flows and the work done by the system during one cycle.

the word universe with a lower case 'u' means a finite portion of the world plus the system; it does not have any cosmic significance.)

The arrangement of the composite engine is illustrated in Fig. 2.3. We take the heat bath at temperature θ_1 to be hotter than the bath at temperature θ_2. The efficiency of the irreversible engine is $\eta_I = W/Q'_1$, where Q'_1 is the heat out of the hotter reservoir into the irreversible engine. The efficiency of the Carnot engine is $\eta_R = W/Q_1$, where Q_1 is the heat into the hotter reservoir.

We will now prove that the reversible Carnot engine is the most efficient engine. To prove this statement we demonstrate that no irreversible engine has a greater efficiency.

Suppose that the irreversible machine is more efficient than the Carnot machine so that η_I is greater than η_R. If this were true it would follow that Q'_1 would be smaller than Q_1. But if this were the case it would mean that, for both machines taken together as a unit, heat would have to flow from the colder (θ_2) to the hotter (θ_1) body, with no external work done on the rest of the universe. *This is contrary to everyday experience.* Heat always flows from a hotter to a colder body, an idea which is the basis of the second law of thermodynamics.

Q_1 must be smaller than Q'_1 for otherwise heat flows from a colder body to a hotter body with no other change to the rest of the universe. The premise that an irreversible machine is more efficient than a Carnot engine violates Clausius' statement of the second law of thermodynamics. Hence any irreversible engine is less efficient than a Carnot engine. The most efficient machine is a Carnot engine. Carnot's theorem can be stated thus: *no engine operating between two given reservoirs can be more efficient than a Carnot engine operating between*

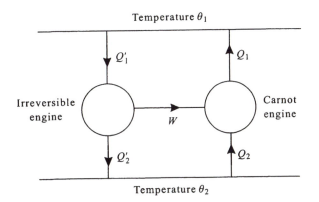

Fig. 2.3 A Carnot engine connected to an irreversible engine in such a way that all the work done by the irreversible engine is used to drive the Carnot engine. The latter runs backwards. No work is done on the rest of the universe. The two engines taken together cannot cause heat to flow from the colder to the hotter reservoir.

the same two reservoirs.

Suppose now two Carnot engines are joined together so that when one gives out work, W, the other absorbs this work completely. Both engines are completely reversible. Let one engine take heat Q_1' from the hotter reservoir and the other engine put heat Q_1 into the hotter reservoir. If these energies are not equal then by running the engine either forwards or backwards we can extract heat from the colder reservoir and transfer it to the hotter one leaving the rest of the universe unchanged. Hence the two energies Q_1' and Q_1 must be equal. The efficiency of one engine is W/Q_1', the efficiency of the other is W/Q_1. The efficiency of both engines is the same. Any two reversible engines operating between the same two temperature have the same efficiency.

It does not matter what material we use to construct the Carnot engine, or how the work is done. The efficiency of the Carnot engine operating between two reservoirs is a universal function of the two operating temperatures. The efficiency of a reversible engine, $\eta = 1 - Q_2/Q_1$, only depends on the two temperatures θ_1 and θ_2. Hence we can write

$$\frac{Q_1}{Q_2} = f(\theta_1, \theta_2). \tag{2.3.2}$$

Suppose that there are three heat baths with temperatures θ_1, θ_2, and θ_3. We join one Carnot cycle to the baths at temperatures θ_1 and θ_2 and another Carnot cycle to the baths at temperatures θ_2 and θ_3. The arrangement is shown in Fig. 2.4. The heat that flows into the thermal reservoir at temperature θ_2 passes directly into the second Carnot cycle. Thus we have $Q_2/Q_3 = f(\theta_2, \theta_3)$. The two Carnot engines together can be considered as a single reversible Carnot

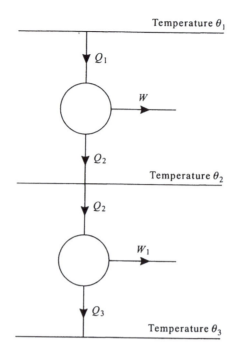

Fig. 2.4 Two Carnot engines and three heat reservoirs operate in such a way as to allow the heat channelled into the cooler reservoir of the first cycle to be fed directly into the second Carnot engine. No heat is lost to the rest of the universe.

engine. Since

$$\frac{Q_1}{Q_3} = \frac{Q_1}{Q_2}\frac{Q_2}{Q_3} \qquad (2.3.3)$$

it follows that

$$f(\theta_1, \theta_3) = f(\theta_1, \theta_2)f(\theta_2, \theta_3). \qquad (2.3.4)$$

The functions $f(\theta_1, \theta_2)$ and $f(\theta_2, \theta_3)$ must take the form

$$f(\theta_1, \theta_2) = \frac{\phi(\theta_1)}{\phi(\theta_2)}; \quad f(\theta_2, \theta_3) = \frac{\phi(\theta_2)}{\phi(\theta_3)} \qquad (2.3.5)$$

in order to ensure that θ_2 disappears from eqn (2.3.4). We can therefore write

$$\frac{Q_1}{Q_2} = \frac{\phi(\theta_1)}{\phi(\theta_2)}. \qquad (2.3.6)$$

Kelvin suggested that *the absolute temperature, T, is defined as*

$$T = \phi(\theta). \qquad (2.3.7)$$

As a consequence of this definition, the ratio of heat flows in a Carnot cycle is

$$\frac{Q_1}{Q_2} = \frac{T_1}{T_2} \qquad (2.3.8)$$

where T_1 and T_2 are the absolute temperatures of the two reservoirs.

Let us set the temperature of the lower reservoir at some fixed point (say the triple point of water where water, ice, and water vapour coexist) and vary the temperature of the upper reservoir. We can determine the ratio of the two absolute temperatures of the thermal reservoirs by measuring the ratio of the heat flows in the Carnot cycle. Knowing the reference temperature (the triple point of water), we could calculate the other temperature. The Carnot cycle could be used as a thermometer if we could measure the heat flows in a cycle. However, it appears to be an impractical thermometer since we cannot build a truly reversible cycle in reality. How can the definition of the absolute temperature be turned into a practical thermometer?

The constant volume gas thermometer is thought to be one of the most accurate ways of measuring temperature. We now show that it is possible to make the absolute temperature scale agree precisely with the ideal gas scale of temperature. All gases which are sufficiently dilute behave as if they were ideal gases. Consequently we can use a constant volume gas thermometer containing a dilute gas to measure the absolute temperature.

2.4 The equivalence of the absolute and the perfect gas scale of temperature

Consider a Carnot engine which contains a perfect gas, and which goes around the cycle shown in Fig. 2.1. The processes A to B and C to D are isothermal, and those from D to A and from B to C are adiabatic. Remember that in adiabatic processes no heat enters or leaves the system, in isothermal processes the temperature is constant. Heat flows only occur in isothermal processes.

For a perfect gas the internal energy only depends on the temperature of the system and the number of moles of gas present. For an isothermal process the internal energy is constant so we have $đQ_{rev} = P\,dV$. The total reversible heat added in the isothermal stage of a Carnot cycle involving a perfect gas is found by integrating this expression along the path AB:

$$\begin{aligned}
Q_1 &= \int_A^B P\,dV = nR\theta_1 \int_{V_A}^{V_B} \frac{dV}{V} \\
&= nR\theta_1 \ln\left(\frac{V_B}{V_A}\right).
\end{aligned} \qquad (2.4.1)$$

Q_1 is the heat entering the engine at the temperature θ_1. (In this section we omit the suffix G for the gas temperature to simplify the notation.)

The heat leaving the system at temperature θ_2 can be calculated from the corresponding integral along the path CD. In fact it is better to *reverse* the path in the cycle and to consider the path DC; this changes the sign of the integral, a manipulation which is needed as Q_2 is the heat given *out* by the engine. Thus

$$Q_2 = nR\theta_2 \int_{V_D}^{V_C} \frac{dV}{V} = nR\theta_2 \ln\left(\frac{V_C}{V_D}\right). \tag{2.4.2}$$

The ratio of the heat flows is in the ratio of the absolute temperatures. Thus

$$\frac{Q_1}{Q_2} = \frac{T_1}{T_2} = \frac{\theta_1 \ln(V_B/V_A)}{\theta_2 \ln(V_C/V_D)}. \tag{2.4.3}$$

In order to simplify this expression we need to know the relation between volume and temperature for adiabatic processes. This problem was solved in section 1.9 where we showed that for reversible adiabatic changes the quantity $\theta V^{(\gamma-1)}$ is a constant (eqn (1.9.6)), where γ is the ratio of heat capacities of the gas, C_P/C_V. The process B to C is adiabatic so that $\theta V^{(\gamma-1)}$ is a constant along the adiabatic path from B to C. The end-points, B and C, must satisfy the relation

$$\theta_1 V_B^{(\gamma-1)} = \theta_2 V_C^{(\gamma-1)}. \tag{2.4.4}$$

The process D to A is also adiabatic so that $\theta V^{(\gamma-1)}$ is a constant for any point along the path from D to A. Thus we get for the end-points D and A

$$\theta_1 V_A^{(\gamma-1)} = \theta_2 V_D^{(\gamma-1)}. \tag{2.4.5}$$

By eliminating the temperatures between these two equations and by taking logarithms we get

$$\ln\left(\frac{V_B}{V_A}\right) = \ln\left(\frac{V_C}{V_D}\right). \tag{2.4.6}$$

As a consequence, when we return to eqn (2.4.3) we find

$$\frac{T_1}{T_2} = \frac{\theta_1}{\theta_2}, \tag{2.4.7}$$

an equation which shows that the absolute temperature, T, is proportional to the gas temperature, θ. The two temperatures are exactly equal if the constant of proportionality is taken to be one. From now on this choice is made, and there is no need to differentiate between the perfect gas temperature scale and the absolute temperature scale. Both are denoted by the symbol T. The fixed points are the absolute zero of temperature and the triple point of water. Provided the water is pure, and the three phases are present in thermodynamic equilibrium, the triple point is completely defined. This scale is called the *Kelvin temperature scale*.

2.5   Definition of entropy

From now on we revert to the more natural definition of Q_2 as the heat given by the thermal reservoir at temperature T_2 to the system. All we have to do to

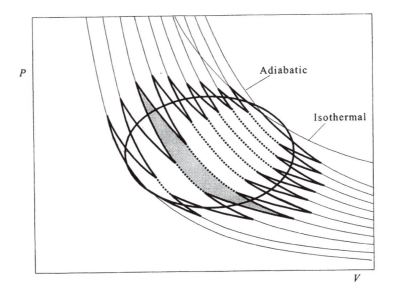

Fig. 2.5 A reversible cycle superimposed on a set of adiabatic lines. Neighbouring adiabatic lines are connected by lines representing isothermal processes in such a way that they form a collection of tiny Carnot cycles which approximates the original reversible cycle. The shaded area represents a single Carnot cycle. As the adiabatic lines are drawn closer together so the set of Carnot cycles approximates the reversible cycle more closely.

effect this change is to reverse the sign of Q_2. With this convention we have for reversible processes

$$\frac{-Q_2}{Q_1} = \frac{T_2}{T_1}. \tag{2.5.1}$$

Another way of writing this equation for the heat flows in the Carnot cycle is

$$\frac{Q_1}{T_1} + \frac{Q_2}{T_2} = 0. \tag{2.5.2}$$

We now assert that any arbitrary reversible cycle can be thought of as a collection of tiny Carnot cycles. Figure 2.5 shows an arbitrary reversible cycle superimposed on a collection of isotherms. The actual cycle can be approximated by connecting the isotherms by suitably chosen adiabatic lines, thereby breaking the cycle up into a collection of tiny Carnot cycles. In the limit that the isotherms become very narrowly spaced, the jagged contour shown in Fig. 2.5 more closely approximates the original contour. By making the temperature interval between adjacent contours sufficiently small the actual cycle is reproduced as closely as we could wish.

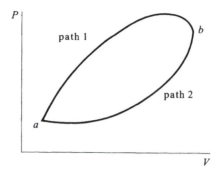

Fig. 2.6 Two possible paths from a to b representing reversible processes.

Notice that adjacent Carnot cycles have a common adiabatic line along which they travel in opposite directions. There is substantial cancellation between these adiabatic lines concerning the work done. The heat is added in the isothermal process, so there is a finite heat transfer đQ for each tiny Carnot cycle.

For each tiny cycle we can write

$$\frac{đQ_1}{T_1} + \frac{đQ_2}{T_2} = 0. \tag{2.5.3}$$

Here đQ_1 is the tiny amount of heat out of the heat bath at temperature T_1 and $-đQ_2$ is the tiny amount of heat lost from the engine to the heat bath at temperature T_2. When we add together lots of tiny Carnot cycles the result is

$$\sum_i \frac{đQ_i}{T_i} = 0. \tag{2.5.4}$$

In the limit that each of these heat flows becomes tiny the sum becomes an integral

$$\oint \frac{đQ_i}{T} = 0. \tag{2.5.5}$$

The integral of đQ_i/T over a cycle is zero if all the processes are reversible.

Suppose the system is taken around a cycle reversibly so that it comes back to the original starting place. Equation (2.5.5) tells us that the total amount of đQ_{rev}/T for the cycle is conserved; the conserved quantity is called the entropy, the function of state we want.

Suppose the above integral is evaluated between two points a and b as shown in Fig. 2.6 along two paths. The sum of the integrals

$$\left(\int_a^b \frac{đQ_{\text{rev}}}{T} \right)_{\text{path 1}} + \left(\int_b^a \frac{đQ_{\text{rev}}}{T} \right)_{\text{path 2}} = 0. \tag{2.5.6}$$

The two paths 1 and 2 make up a closed contour, that is a path that ends where it began. Equation (2.5.6) can be written as

$$\left(\int_a^b \frac{\text{d}Q_{\text{rev}}}{T} \right)_{\text{path 1}} = \left(\int_a^b \frac{\text{d}Q_{\text{rev}}}{T} \right)_{\text{path 2}}. \tag{2.5.7}$$

The choice of paths is completely arbitrary. Any pair of paths can be chosen. Consequently, the integral of $\text{d}Q_{\text{rev}}/T$ is independent of path and depends only on the end-points, a and b. This implies that a function of state $S(b,a)$ can be defined such that

$$S(b,a) = \left(\int_a^b \frac{\text{d}Q_{\text{rev}}}{T} \right). \tag{2.5.8}$$

$S(b,a)$ is called the *entropy*. It is a function of state because it does not matter how the system is taken from a to b as long as the process is reversible; the history of the journey leaves no imprint on S. We have therefore achieved the goal set in section 2.1: to define the function of state called the entropy.

Only entropy differences are defined. Point a is taken to be a reference point and the entropy is defined with respect to it. The reference point is usually chosen as the lowest temperature that can be achieved—ideally it should be the absolute zero of temperature, but no real system can ever be cooled to the absolute zero of temperature. The entropy of a system at absolute temperature T_{f} is defined as the integral of $\text{d}Q_{\text{rev}}/T$ from the reference temperature up to the temperature T_{f}. All the heat must be added reversibly. In practice this means tiny amounts of heat are added very slowly to the system starting from the lowest temperature available using current technology.

For a tiny reversible change (corresponding to a small addition of heat)

$$\text{d}S = \frac{\text{d}Q_{\text{rev}}}{T}. \tag{2.5.9}$$

For tiny reversible changes the heat added is $T\,\text{d}S$. It should be emphasized that only reversible processes give this result. The *differential form of the first law* for reversible processes can be expressed as

$$\text{d}U = T\,\text{d}S - P\,\text{d}V. \tag{2.5.10}$$

The definition of entropy allows us to derive a new formula for the heat capacity. Since $\text{d}Q_{\text{rev}} = T\,\text{d}S$ we have for reversible changes in temperature at constant volume

$$C_V = T\left(\frac{\partial S}{\partial T}\right)_V \tag{2.5.11}$$

and for temperature changes at constant pressure

$$C_P = T\left(\frac{\partial S}{\partial T}\right)_P. \tag{2.5.12}$$

2.6 Measuring the entropy

Suppose heat is added reversibly to a system. The entropy change is

$$\Delta S = S(T_f) - S(T_i) = \int_i^f \frac{\mathrm{d}Q_{rev}}{T}. \qquad (2.6.1)$$

In a reversible process at constant pressure $\mathrm{d}Q_{rev} = C_P\,\mathrm{d}T$; by integrating eqn (2.6.1) from T_i to T_f we get

$$S(T_f) - S(T_i) = \int_{T_i}^{T_f} C_P(T)\frac{\mathrm{d}T}{T} \qquad (2.6.2)$$

provided the only work involved is that due to the expansion (or contraction) of the system. This formula allows us to obtain the difference in entropy between any two temperatures provided we measure $C_P(T)$ over this temperature interval. As a simple example, let us take $C_P(T)$ to be a constant. Then

$$S(T_f) = S(T_i) + C_P \ln\left(\frac{T_f}{T_i}\right). \qquad (2.6.3)$$

The entropy need not be a continuous function of the temperature. It is an experimental observation that whenever there is a change in the phase of a substance there is generally a change in the entropy of the system, as happens when a liquid freezes or boils, or when a superconductor becomes a normal metal. The entropy change between different phases of a substance can be measured by the heat that is added to convert from one phase to the other.

Consider a system in which the liquid and solid phase of a substance are in equilibrium. This occurs at a particular temperature, T_p, the phase transition temperature. As heat is added to the system the solid melts but the temperature remains constant. When heat ΔQ_P is added to the system at a constant pressure there is a change in entropy of the system of

$$\Delta S = \frac{\Delta Q_P}{T_p} = \frac{\Delta H}{T_p} \qquad (2.6.4)$$

where ΔH is the change in the enthalpy at the phase transition, usually called the *latent heat*. If ΔH is positive the process is said to be endothermic; if it is negative the process is said to be exothermic. For liquids with simple bonding the entropy change on vaporization is about $30\,\mathrm{kJ\,mol^{-1}}$, an observation known as Trouton's rule. Some liquids, such as water, do not obey this rule because there is complicated hydrogen bonding between the molecules.

Can the *absolute value of the entropy* be measured? Let us call the lowest temperature we can attain, T_0. Starting from T_0 we measure $C_P(T)$ as a function of temperature for all temperatures up to T_f. The solid melts at T_{fus} and there is a change in enthalpy ΔH_{fus}; the liquid vaporizes at the temperature T_{boil} and there is a change in the enthalpy ΔH_{boil}. If there are no other phase transitions then

$$
\begin{aligned}
S(T_{\text{f}}) \quad = \quad & S(T_0) + \int_{T_0}^{T_{\text{fus}}} C_P(T)\frac{\mathrm{d}T}{T} \\
& + \frac{\Delta H_{\text{fus}}}{T_{\text{fus}}} + \int_{T_{\text{fus}}}^{T_{\text{boil}}} C_P(T)\frac{\mathrm{d}T}{T} \\
& + \frac{\Delta H_{\text{boil}}}{T_{\text{boil}}} + \int_{T_{\text{boil}}}^{T_{\text{f}}} C_P(T)\frac{\mathrm{d}T}{T} \quad\quad (2.6.5)
\end{aligned}
$$

where T is the temperature variable over which we integrate. Everything in this expression, except $S(T_0)$, can be measured using calorimetry. All the integrals can be evaluated numerically once $C_P(T)$ has been measured. After long, painstaking work, tables of values of $S(T)$ can be compiled as a function of temperature for each substance. We can then look up the values of entropy in standard reference books.

If the lowest temperature which could be reached were the absolute zero, then T_0 would be zero, and we could get a consistent set of values of $S(T)$. There is one problem with this picture: in practice nobody can reach the absolute zero of temperature. Suppose T_0 is the lowest temperature that can be reached. We need to make an extrapolation of the entropy from T_0 to the absolute zero so as to be able to pretend that entropies are measured from the absolute zero. A correction has to be made.

For most non-metallic solids for temperatures below $10\,\text{K}$ the heat capacity varies as aT^3; we can fit the experimental data to the equation $C_P = T(\partial S/\partial T)_P = aT^3$ at low temperatures and find the constant, a. At low temperatures the entropy varies as

$$
S(T) = S(0) + \frac{aT^3}{3}. \quad\quad (2.6.6)
$$

This expression allows us to extrapolate the entropy back to $T = 0$ and to estimate the correction that arises from starting the measurements at a finite temperature.

As an example, the molar heat capacity of gold at a temperature of $10\,\text{K}$ is $0.43\,\text{J}\,\text{K}^{-1}\,\text{mol}^{-1}$. Let us assume that the heat capacity of gold varies with temperature as $C_P = aT^3$ at $10\,\text{K}$ so that the constant $a = 0.43 \times 10^{-3}\,\text{J}\,\text{K}^{-4}\,\text{mol}^{-1}$. It then follows that

$$
S(T = 10) = S(0) + \frac{a \times 10^3}{3} = S(0) + 0.143\,\text{J}\,\text{K}^{-1}\,\text{mol}^{-1}.
$$

We can extrapolate back to obtain the entropy at the absolute zero in terms of the entropy at $T = 10\,\text{K}$.

Of course we could be completely wrong if there is another, unrevealed source of entropy awaiting to be discovered at lower temperatures. Nature is full of such surprises. For example, the nuclear spin degrees of freedom give a contribution to the entropy which becomes small only at extremely low temperatures where the nuclear spins become ordered. In metals, the heat capacity has a contribution which is proportional to T, which at very low temperatures is larger than the contribution varying as T^3. Great care has to be taken to extrapolate correctly to the absolute zero of temperature.

Even if we believe in the extrapolation procedure, there is still the problem of choosing $S(0)$, the value of the entropy at the absolute zero of temperature. Nernst suggested that the entropy change of a transformation between phases approaches zero as the temperature approaches zero.

Such an idea can be tested experimentally. A simple example is the entropy change from monoclinic sulphur with entropy $S(\beta, T)$ to orthorhombic sulphur with entropy $S(\alpha, T)$. The phase transition occurs at $369\,\mathrm{K}$ and the enthalpy change is $-402\,\mathrm{J\,mol^{-1}}$. Hence

$$\Delta S = S(\alpha, 369) - S(\beta, 369) = -\frac{402}{369}\,\mathrm{J\,K^{-1}\,mol^{-1}}.$$

The two entropies at the transition temperature can be measured and are found to be

$$S(\alpha, 369) = S(\alpha, 0) + 37\,\mathrm{J\,K^{-1}\,mol^{-1}}$$

and

$$S(\beta, 369) = S(\beta, 0) + 38\,\mathrm{J\,K^{-1}\,mol^{-1}}.$$

From these values we can infer that there is no significant difference in the entropies between the two phases within experimental error at the absolute zero.

It is experimental observations like this that lead to the third law of thermodynamics. It can be stated as follows: *If the entropy of every element in its stable state at $T = 0$ is taken to be zero, then every substance has a positive entropy which at $T = 0$ may become zero, and does become zero for all perfectly ordered states of condensed matter.* The third law is restricted to perfectly ordered substances; it excludes disordered ones like glasses or solid solutions. Choosing the entropy to be zero is really a matter of convenience; it is the simplest way of avoiding confusion.

From numerous experiments we know that the heat capacity and latent heats of a system are proportional to the number of particles in the system. It follows that the entropy is also proportional to the number of particles. This property of the entropy turns out to be very important when we develop the ideas of statistical mechanics.

Once we have specified $S(0)$, it is possible to obtain values of the entropy by experiment. We are then in a position to create other functions of state. For example the quantity

$$F = U - TS \qquad (2.6.7)$$

is a function of state since it is composed of other functions of state. F is called the *Helmholtz free energy*. The free energy turns out to be a quantity that can be calculated directly using the techniques of statistical mechanics.

A small change in F is

$$\mathrm{d}F = \mathrm{d}U - T\,\mathrm{d}S - S\,\mathrm{d}T. \qquad (2.6.8)$$

When the only form of work is due to changes in volume, any small change in the internal energy can be written as $\mathrm{d}U = T\,\mathrm{d}S - P\,\mathrm{d}V$, so that

$$\mathrm{d}F = -S\,\mathrm{d}T - P\,\mathrm{d}V. \qquad (2.6.9)$$

We can derive relations between partial derivatives starting with this equation. Let us write the Helmholtz free energy as a function of T and V as $F(T,V)$. Then

$$\mathrm{d}F = \left(\frac{\partial F}{\partial T}\right)_V \mathrm{d}T + \left(\frac{\partial F}{\partial V}\right)_T \mathrm{d}V$$

which implies that

$$S = -\left(\frac{\partial F}{\partial T}\right)_V ; \qquad P = -\left(\frac{\partial F}{\partial V}\right)_T .$$

If we assert that $F(T,V)$ is an analytic function we have

$$\frac{\partial}{\partial V}\left(\frac{\partial F}{\partial T}\right) = \frac{\partial}{\partial T}\left(\frac{\partial F}{\partial V}\right)$$

or

$$\left(\frac{\partial S}{\partial V}\right)_T = \left(\frac{\partial P}{\partial T}\right)_V .$$

This is one of the four Maxwell relations.

Another useful quantity is the *Gibbs free energy*

$$G = H - TS. \qquad (2.6.10)$$

A small change in G can be written as

$$\mathrm{d}G = -S\,\mathrm{d}T + V\,\mathrm{d}P. \qquad (2.6.11)$$

We can derive a Maxwell relation from this equation. Let us write the Gibbs free energy as a function of T and P as $G(T,P)$.

$$\mathrm{d}G = \left(\frac{\partial G}{\partial T}\right)_P \mathrm{d}T + \left(\frac{\partial G}{\partial P}\right)_T \mathrm{d}P.$$

The argument that $G(T,P)$ is an analytic function gives in this case

$$\left(\frac{\partial V}{\partial T}\right)_P = -\left(\frac{\partial S}{\partial P}\right)_T .$$

There are two other Maxwell relations, one of which can be found by expressing a small change in the internal energy as $\mathrm{d}U = T\,\mathrm{d}S - P\,\mathrm{d}V$ and then writing

$U(S, V)$; the last Maxwell relation can be found by expressing a small change in the enthalpy as $dH = T\,dS + V\,dP$ and then writing $H(S, P)$. In this way we get

$$\left(\frac{\partial T}{\partial V}\right)_S = -\left(\frac{\partial P}{\partial S}\right)_V,$$
$$\left(\frac{\partial V}{\partial S}\right)_P = \left(\frac{\partial T}{\partial P}\right)_S.$$

2.7   The law of increase of entropy

Consider the situation represented in Fig. 2.3 in which a Carnot cycle and an irreversible engine are connected to the same thermal reservoirs at temperature T_1 and T_2 (these temperatures replace θ_1 and θ_2 respectively in the figure). The efficiency of a Carnot engine is greater than that of the irreversible engine:

$$\eta_{\mathrm{R}} > \eta_{\mathrm{I}}. \tag{2.7.1}$$

From this inequality it follows that for the two engines operating between the two thermal reservoirs

$$\left(1 - \frac{(-Q_2)}{Q_1}\right) > \left(1 - \frac{(-Q_2')}{Q_1'}\right). \tag{2.7.2}$$

(Remember we have changed the sign of Q_2.) From the definition of the thermodynamic scale of temperature we know that the heat flows in the Carnot cycle obey the equation

$$\frac{-Q_2}{Q_1} = \frac{T_2}{T_1}. \tag{2.5.1}$$

Therefore

$$\frac{-Q_2'}{Q_1'} > \frac{T_2}{T_1}. \tag{2.7.3}$$

Now consider small heat flows $đQ_1$ and $đQ_2$ as part of an irreversible cycle subdivided into lots of tiny cycles, the same subdivision that was done in Fig. 2.5 for a reversible cycle. For one tiny cycle inequality (2.7.3) can be written as

$$\frac{đQ_1'}{T_1} + \frac{đQ_2'}{T_2} < 0. \tag{2.7.4}$$

Adding together all the tiny cycles to make up the total cycle gives the inequality

$$\oint \frac{đQ_{\mathrm{irrev}}}{T} < 0 \tag{2.7.5}$$

where T is the temperature of the bath that supplies the heat.

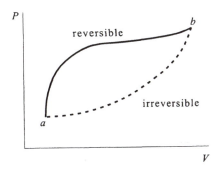

Fig. 2.7 Two possible paths from a to b representing a reversible and an irreversible process.

Suppose the cycle can be divided up into two parts: a path from a to b which is irreversible, and a path from b to a which is reversible, as illustrated in Fig. 2.7. Then

$$\int_a^b \frac{\text{đ}Q_{\text{irrev}}}{T} < \int_a^b \frac{\text{đ}Q_{\text{rev}}}{T}. \tag{2.7.6}$$

The integral along the reversible path can be rewritten in terms of the entropy difference between the end-points using eqn (2.5.8):

$$\int_a^b \frac{\text{đ}Q_{\text{irrev}}}{T} < S(b,a). \tag{2.7.7}$$

If a and b are very close together (their temperatures are almost equal) we write $S(b,a) = \delta S$ with δS small. Inequality (2.7.7) becomes

$$\frac{\delta Q_{\text{irrev}}}{T} < \delta S \tag{2.7.8}$$

where $\delta Q_{\text{irrev}} = \int_a^b \text{đ}Q_{\text{irrev}}$ is the total heat added in going from a to b. Inequality (2.7.8) above is known as *Clausius' inequality*: for an irreversible process the increase in the entropy is greater than $\delta Q_{\text{irrev}}/T$. Clausius' inequality is a direct consequence of the second law of thermodynamics.

For a thermally isolated system $\delta Q_{\text{irrev}} = 0$ in which case the inequality becomes simply $\delta S > 0$; this is known as *the law of increase entropy*. All changes in a thermally isolated system must lead to an increase in entropy, or keep the entropy unchanged if the process is reversible. When a thermally isolated system is approaching equilibrium the entropy must increase; the final equilibrium configuration of such a system is the one with the maximum entropy. At the maximum in the entropy there are no changes, for the entropy cannot decrease as a function of time. All of this follows from the ordinary observation that heat can only flow from a hotter to a colder body and not the other way around.

The statement that the entropy always increases or stays the same implies that there is a direction for time: positive time is the direction in which the entropy increases in a thermally isolated system. Newton's laws do not give a direction for time because they are invariant under the transformation $t \rightarrow -t$. Most physical phenomena are microscopically reversible. Thermodynamics offers no clue as to why there should be *an arrow of time*, but indicates that the entropy is an increasing function of time. Statistical mechanics gives a straightforward explanation of why this occurs.

2.8   Calculations of the increase in the entropy in irreversible processes

Entropy truly is a function of state because it only depends on thermodynamic coordinates such as the energy and volume of the system and not on how the system was prepared. We have defined entropy through reversible processes, but entropy differences are well-defined even when the changes that produced them involve irreversible processes.

For reversible processes we have derived eqn (2.5.10), which can be written as

$$T\,dS = dU + P\,dV. \tag{2.8.1}$$

Equation (2.8.1) states a relationship between functions of state only. Hence we can generalize it by declaring that *it can be applied to any differential change, both reversible and irreversible.*

There could be a tiny irreversible process from a to b, in which case the heat and work are not given by $T\,dS$ and $-P\,dV$. The internal energy change is well-defined, for internal energy is a function of state. We can, in principle, find a reversible path from a to b, and for this path measure the heat added, $đQ_{\text{rev}} = T\,dS$, along the path as well as the temperature. Hence we can determine dS along the path and integrate it to get the change in the entropy. In this sense the entropy change can be measured.

Alternatively the entropy change can be calculated. In this section we will calculate entropy changes for several irreversible processes.

2.8.1   Two systems at different temperatures thermalize

Imagine that there are two identical blocks of material held at constant volume with a constant heat capacity, C_V. One is at a temperature of 320 K, the other at a temperature of 280 K. The two blocks are thermally isolated and placed in contact with each other. The hotter block loses heat to the colder. The total internal energy at temperature T is $C_V T$. The internal energy is conserved so the final temperature of the two systems must be 300 K.

The entropy change of the hotter body is

$$\Delta S_1 = \int_{320}^{300} \frac{C_V\,dT}{T} = C_V \ln\left(\frac{300}{320}\right)$$

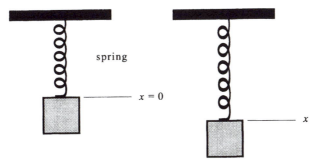

Fig. 2.8 The extension of a spring when a mass is added. In the diagram on the left the spring is not extended so $x = 0$; in the diagram on the right the mass has been released, causing the spring to extend by an amount x.

It decreases its entropy as it loses heat. The entropy change of the cooler body is

$$\Delta S_2 = \int_{280}^{300} \frac{C_V \mathrm{d}T}{T} = C_V \ln\left(\frac{300}{280}\right).$$

It increases its entropy. The entropy change of the combined system is

$$\Delta S = \Delta S_1 + \Delta S_2 = C_V \ln\left(\frac{300^2}{320 \times 280}\right) = 0.045 C_V.$$

We have calculated the entropy change by imagining a reversible flow of heat into each body. The net result is an increase in entropy. All irreversible processes lead to an increase in entropy.

2.8.2 Extending a spring

Imagine putting a mass M onto an unextended spring which is thermally isolated in air at a temperature T. The situation is illustrated in Fig. 2.8. The mass is released and oscillates with ever decreasing amplitude, and eventually comes to rest. The process is irreversible, so what is the entropy change of the universe?

Let us measure all displacements from the position of the unextended spring. Suppose the spring has a spring constant K. The potential energy of the spring when it is extended by x from its unextended position is $Kx^2/2$. When the spring is extended, the mass falls a distance x against gravity, and the change in gravitational potential energy is $-Mgx$. Therefore the total potential energy of the mass on the spring is

$$V(x) = \frac{Kx^2}{2} - Mgx.$$

The condition for mechanical equilibrium is that the potential energy is a minimum. The position where mechanical equilibrium occurs is $x_e = Mg/K$, and the potential energy there is $-Kx_e^2/2$. As the mass falls and eventually comes to rest at the equilibrium position x_e, energy $Kx_e^2/2$ is liberated as heat to the surroundings at temperature T. Provided the temperature of the surroundings does not change, the increase of entropy of the universe is $Kx_e^2/2T$.

2.8.3   Expanding a gas into a vacuum

Suppose we have a gas in a volume V. Imagine that there is an equal volume which is empty placed next to volume with the gas. A hole is made in the wall between the volumes so the gas expands into the empty volume so that it occupies a volume $2V$. This process is clearly irreversible, for the change is not quasistatic: at no intermediate stage in the expansion of the gas is the system in equilibrium. What is the entropy change associated with this process?

In section 2.1 we calculated the entropy of an ideal gas as

$$S = \frac{3nR}{2} \ln\left(\frac{\theta_G}{\theta_G^0}\right) + nR\ln\left(\frac{V}{V_0}\right) + S_0. \tag{2.1.7}$$

For an ideal gas θ_G is the same as the absolute temperature. Joule observed by careful experiments that an ideal gas does not change its temperature on expansion into a vacuum, provided no work is done on expansion and no heat is added. Since the temperature does not change, the increase in the entropy is

$$\Delta S = nR\ln\left(\frac{2V}{V}\right) = nR\ln(2).$$

In each of the three cases studied the change in the entropy for an irreversible process can be calculated.

2.9   The approach to equilibrium

Consider two systems, A and B, which are not in thermodynamic equilibrium with each other. Weak thermal contact is made between them so that energy can pass from one to the other. The two systems are thermally isolated from the rest of the universe so that the total energy of the two systems, $U_T = U_A + U_B$, is constant. Because the contact between the systems is weak, the entropy of the two systems is the sum of the entropies of each system. Thus

$$S(U_A) = S_A(U_A) + S_B(U_T - U_A). \tag{2.9.1}$$

The total entropy varies with the energy of system A. When systems A and B are in thermal equilibrium the entropy of the joint system is a maximum. If they are not in thermal equilibrium, the systems evolve so as to maximize the entropy.

The *rate of change of entropy with time*, which must be positive according to *Clausius' principle*, depends on the rate at which energy passes from one system to the other. Simple differentiation of $S = S_A + S_B$ gives

$$\frac{\mathrm{d}S}{\mathrm{d}t} = \frac{\mathrm{d}U_A}{\mathrm{d}t}\left(\frac{\partial S_A}{\partial U_A}\right) + \frac{\mathrm{d}U_B}{\mathrm{d}t}\left(\frac{\partial S_B}{\partial U_B}\right);$$

but $U_T = U_A + U_B$ and $\mathrm{d}U_T/\mathrm{d}t = 0$ because the total energy is constant. Therefore $\mathrm{d}U_B/\mathrm{d}t = -\mathrm{d}U_A/\mathrm{d}t$ and we get

$$
\begin{aligned}
\frac{\mathrm{d}S}{\mathrm{d}t} &= \frac{\mathrm{d}U_A}{\mathrm{d}t} \left\{ \left(\frac{\partial S_A}{\partial U_A} \right) - \left(\frac{\partial S_B}{\partial U_B} \right) \right\} \\
&= \frac{\mathrm{d}U_A}{\mathrm{d}t} \left\{ \frac{1}{T_A} - \frac{1}{T_B} \right\} \geq 0.
\end{aligned}
\tag{2.9.2}
$$

If $T_A = T_B$ then S is a constant and the system is in thermal equilibrium. If T_A is larger than T_B, we must have $\mathrm{d}U_A/\mathrm{d}t < 0$ so that energy leaves system A. The hotter system loses energy to the colder one.

Suppose now that there is a movable piston which separates two systems in thermal equilibrium. As the piston moves, the volume of system A increases, the volume of system B decreases, but the total volume of the two systems remains constant. The argument is the same as before, but we consider changes in volume instead of energy.

The total entropy can be written as

$$
S(V_A) = S_A(V_A) + S_B(V_T - V_A).
\tag{2.9.3}
$$

The rate of change of entropy with time is

$$
\begin{aligned}
\frac{\mathrm{d}S}{\mathrm{d}t} &= \frac{\mathrm{d}V_A}{\mathrm{d}t} \left\{ \left(\frac{\partial S_A}{\partial V_A} \right) - \left(\frac{\partial S_B}{\partial V_B} \right) \right\} \\
&= \frac{\mathrm{d}V_A}{\mathrm{d}t} \left\{ \frac{P_A}{T_A} - \frac{P_B}{T_B} \right\} \geq 0.
\end{aligned}
\tag{2.9.4}
$$

The two systems are in thermal equilibrium so that $T_A = T_B$. If the two pressures are equal then S is a constant, and the system is in mechanical equilibrium. If P_A is larger than P_B then $\mathrm{d}V_A/\mathrm{d}t > 0$ so that the side with the higher pressure expands, which makes perfect sense.

Finally let us consider a situation where the number of particles can change between systems A and B. Perhaps there is a membrane separating the two systems which allows particles through. The total number of particles in the two systems is constant.

The total entropy can be written as

$$
S(N_A) = S_A(N_A) + S_B(N_T - N_A).
\tag{2.9.5}
$$

The rate of change of entropy with time is

$$
\frac{\mathrm{d}S}{\mathrm{d}t} = \frac{\mathrm{d}N_A}{\mathrm{d}t} \left\{ \left(\frac{\partial S_A}{\partial N_A} \right) - \left(\frac{\partial S_B}{\partial N_B} \right) \right\}.
\tag{2.9.6}
$$

We define *the chemical potential*, μ, as

$$\mu = -T \left(\frac{\partial S}{\partial N} \right)_{U,V}.$$ (2.9.7)

It follows that

$$\frac{dS}{dt} = -\frac{dN_A}{dt} \left\{ \frac{\mu_A}{T_A} - \frac{\mu_B}{T_B} \right\} \geq 0.$$ (2.9.8)

Suppose the two systems are in thermal equilibrium so that $T_A = T_B$. If the two chemical potentials are equal then S is a constant, and the system is in *chemical equilibrium*. If μ_A is larger than μ_B then $dN_A/dt < 0$ so that the side with the higher chemical potential loses particles.

In general, the entropy is a function of energy, volume, and the number of particles. We write it as $S(U, V, N)$. A small change in entropy is given by an equation which can be written as

$$\begin{aligned} dS &= \left(\frac{\partial S}{\partial U} \right)_{V,N} dU + \left(\frac{\partial S}{\partial V} \right)_{U,N} dV + \left(\frac{\partial S}{\partial N} \right)_{U,V} dN \\ &= \frac{1}{T} dU + \frac{P}{T} dV - \frac{\mu}{T} dN. \end{aligned}$$ (2.9.9)

The term $\mu \, dN$ can be thought of as *chemical work*. We will deal with systems with variable numbers of particles in more depth in Chapter 9.

2.10 Questions left unanswered

This brief examination of the principles of thermodynamic systems has left several questions unanswered:

What is the microscopic basis of the entropy?
What is the physical basis of the second law which ensures that entropy increases as a function of time?
How do we calculate the equation of state for a system?

These questions are answered by statistical mechanics. But before we get to answer these question we need to review simple ideas about probability. The development of the ideas of statistical mechanics then follows more naturally.

2.11 Problems

1. At $10 \, \mathrm{K}$ the molar heat capacity of gold is $0.43 \, \mathrm{J \, K^{-1} \, mol^{-1}}$. Assume that C_P of gold varies as $C_P = aT$ over the range 0 to $10 \, \mathrm{K}$. Determine the entropy at $10 \, \mathrm{K}$ assuming that $S(T = 0) = 0$.
2. An engine operating between $300°\mathrm{C}$ and $60°\mathrm{C}$ is 15% efficient. What would its efficiency be if it were a Carnot engine?
3. What is the maximum possible efficiency of an engine that obtains heat at $250°\mathrm{C}$ and loses the waste heat at $50°\mathrm{C}$?
4. A steam turbine works at $400°\mathrm{C}$ and the exhaust gases are emitted at $150°\mathrm{C}$. What is the maximum efficiency of the turbine?

5. A Carnot engine takes 1 kJ of heat from a reservoir at a temperature of 600 K, does work, and loses the rest of the heat into a reservoir at a temperature of 400 K. How much work is done?

6. An ideal gas with $\gamma = 1.5$ is used as the working substance in the cylinder of an engine undergoing a Carnot cycle. The temperature of the hot reservoir is 240°C, that of the cold, 50°C. The gas is expanded isothermally at 240°C from a pressure of 10 atm (1 atm $= 1.01 \times 10^5$ N m^{-2}) and a volume of 1 litre to a pressure of 4 atm and a volume of 2.5 litres. Between what limits of pressure and volume does the gas operate when it is in thermal equilibrium with the cold reservoir?

7. A kilogram of water has a constant heat capacity of 4.2 kJ K^{-1} over the temperatures range 0°C to 100°C. The water starts at 0°C and is brought into contact with a heat bath at 100°C. When the water has just thermalized at 100°C, what is the change in the entropy of the water, and that of the universe?

8. A certain amount of water of heat capacity C is at a temperature of 0°C. It is placed in contact with a heat bath at 100°C and the two come into thermal equilibrium.

   (a) What is the entropy change of the universe?
   (b) The process is now divided into two stages: first the water is placed in contact with a heat bath at 50°C and comes into thermal equilibrium; then it is placed in contact with the heat bath at 100°C. What is the entropy change of the universe?
   (c) The process is divided into four stages with heat baths at 25, 50, 75, and 100°C. What is the entropy change of the universe in this case?
   (d) If we were to continue this subdivision into an infinite number of heat baths what would be the entropy change of the universe?

9. Two identical blocks of copper, one at 100°C, the other at 0°C, are placed in thermal contact and thermally isolated from everything else. Given that the heat capacity at constant volume of each block is C, independent of temperature, obtain an expression for the increase in the entropy of the universe when the two blocks of copper are in thermal equilibrium.

10. Assume that the heat capacity at constant volume of a metal varies as $aT + bT^3$ for low temperatures. Calculate the variation of the entropy with temperature.

11. Two identical bodies with heat capacities at constant volume which vary linearly with temperature, $C = bT$, are thermally isolated. One is at a temperature of 100 K, the other is at 200 K. They are placed in thermal contact and come to thermal equilibrium. What is the final temperature of the two bodies, and what is the entropy change of the system?

12. Three designs are proposed for a heat engine which will operate between thermal reservoirs at temperatures of 450 and 300 K. To produce 1 kJ of work output, design 1 claims to require a heat input per cycle of 0.2 kcal;

design 2, 0.6 kcal; and design 3, 0.8 kcal. Which design would you choose and why? [1 joule $\equiv 2.39 \times 10^{-4}$ kcal.]

13. An ideal refrigerator—a Carnot engine running backwards—freezes ice-cubes at a rate of $5\,\mathrm{g\,s^{-1}}$ starting with water at its freezing point. Energy is given off to the room which is at 30°C. The fusion energy of ice is $320\,\mathrm{J\,g^{-1}}$. At what rate must electrical energy be supplied as work, and at what rate is energy given off to the room?

14. An unextended spring is in equilibrium. A mass M is placed at the end of it; the force of gravity causes the spring to extend. After a time a new equilibrium position is reached. If x is the extension of the spring, the force of tension is Kx. Obtain a formula for the extension of the spring when it is in equilibrium. What is the change in the potential energy of the mass? What is the increase in the internal energy of the spring? How much heat is released? What is the change in the entropy when a mass of 100 g causes the spring to extend by 2.5 cm at room temperature $(= 25°C)$?

15. A spring which is extended by an amount x has a tension Kx. A mass M is placed on the spring; gravity causes the spring to extend. The system starts with the spring extended by an amount $Mg/2K$. After a long time the mass and spring reach a new equilibrium position. Given that the spring and its surroundings are at a temperature T, obtain an expression for the increase in the entropy of the universe.

16. A current of 0.5 A flows through a $2\,\Omega$ resistor for 100 s. The system whose initial temperature is 300 K is thermally isolated. The heat capacity of the resistor is $0.24\,\mathrm{J\,K^{-1}}$ for a wide range of temperatures. The temperature of the resistor changes appreciably. What is the entropy change of the system?

17. A capacitor of capacitance $1\,\mu\mathrm{F}$ is charged from a 9 V battery. When the capacitor is charged, what is the internal energy associated with the charge on the capacitor? What is the work done by the battery? If the experiment is done at a temperature of 300 K, what is the entropy change of the universe?

18. Show with the help of Maxwell's relations that

$$T\,\mathrm{d}S = C_V\,\mathrm{d}T + T\left(\frac{\partial P}{\partial T}\right)_V \mathrm{d}V$$

and

$$T\,\mathrm{d}S = C_P\,\mathrm{d}T - T\left(\frac{\partial V}{\partial T}\right)_P \mathrm{d}P.$$

Prove that

$$\left(\frac{\partial U}{\partial V}\right)_T = T\left(\frac{\partial P}{\partial T}\right)_V - P.$$

19. Two identical finite bodies of constant volume and of constant heat capacity at constant volume, C_V, are used to drive a heat engine. Their initial temperatures are T_1 and T_2. Find the maximum amount of work which can

be obtained from the system. [Hint: The maximum amount of work can be drawn when there is no entropy change.]

20. The differential of the internal energy of a surface of a liquid with surface tension γ and area A may be written as

$$dU = T\,dS + \gamma\,dA.$$

Write down the corresponding form of the Helmholtz free energy, $F = U - TS$. Using the fact that these equations involve exact differentials derive the Maxwell relation

$$\left(\frac{\partial S}{\partial A}\right)_T = -\left(\frac{\partial \gamma}{\partial T}\right)_A.$$

The internal energy and the entropy are proportional to the area A. Show that the internal energy per unit area is

$$u(T) = \frac{U}{A} = \gamma - T\left(\frac{\partial \gamma}{\partial T}\right)_A.$$

3

Probability and statistics

The race is not always to the swift, nor the battle to the strong, but that's the way to bet.
Damon Runyon

3.1 Ideas about probability

The notion of probability in everyday life is the measure of belief in a possibility that an individual horse will win a race or that a defendant is guilty of a crime. On the basis of the known form of the horse (or the criminal) and of our knowledge and reasoning ability we reach a judgement that the horse will win the race or the criminal will go to jail. This is not the interpretation of probability which is used in science.

There are two notions of probability which are used in science. The first is the *classical notion of probability* in which we assign, a priori, equal probabilities to all possible outcomes of an event. The second notion is *statistical probability* in which we measure the relative frequency of an event and call this the probability.

The subject of probability arose from the analysis of games of chance. When a pair of dice is thrown the outcome is not known for certain. Each throw of the dice is called a *trial*. The outcomes of trials are called *events*. For example, when we throw two dice, one possible outcome is a pair of sixes. If we measure the radioactive decay of an element, an experiment (or trial) involves measuring the count rate of a Geiger counter over, say, a ten second interval. Different integer counts can be measured by the Geiger counter. Each of these possible counts corresponds to a different event. If we get five counts then that is one event, if it is six it is another. These are *simple events*.

From these simple events we can build up *compound events* which are aggregates of simple events. We could group all the counts in the range five to nine in the Geiger counter experiment as one compound event. A compound event can be broken down into simpler parts; a simple event cannot.

Compound events need not be *mutually exclusive*. By mutually exclusive we mean that if one event occurs another cannot. For example the compound events (counts in range five to nine) and (counts in range seven to eleven) are clearly not exclusive since a count of eight satisfies both conditions. In contrast, simple events are mutually exclusive, for if one occurs the other cannot.

Simple events can be pictured as a discrete set of points in an abstract space as in Fig. 3.1. The points are called *sample points* and denoted by the symbol

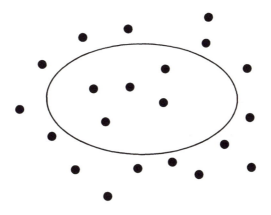

Fig. 3.1 Points in the sample space. A compound event is represented by the set of points inside the ellipse.

e_i. Compound events are made up of a collection of sample points. An example is the set of sample points inside the ellipse in Fig. 3.1.

The whole collection of all the sample points constitutes the *sample space*. Each point in the space has a non-negative number, $p_i = p(e_i)$, associated with it. These numbers are called the *probability of the event*. Any of the p_i can be zero. The set of numbers is chosen such that

$$\sum_i p_i = 1. \tag{3.1.1}$$

There remains the problem of estimating the magnitude of the probabilities, p_i. There are two simple ways to do this.

3.2 Classical probability

The rule here is very simple: list all the points in the sample space and count them; assign them equal probabilities; if there are W points in the sample space then the probability for each is $1/W$. Thus

$$p_i = \frac{1}{W}. \tag{3.1.2}$$

For example, if we toss a coin there are two possible outcomes, heads or tails. According to the classical rule the probability of getting heads in one throw is 0.5 for there are two possible outcomes. Since $W = 2$ we have $p = 0.5$. If we were to toss the coin N times, there would be 2^N possible outcomes. According to the classical probability all the 2^N possible outcomes would be equally likely and the probability that any one of them occurs is $p = 1/2^N$.

The classical approach is to enumerate all possible simple events, and assign equal probabilities to them. This is a useful working hypothesis in the absence of any information. If we try to make a definition of the classical probability it would be this: the probability of a simple event, e_i, equals $1/W$ where W is the number of outcomes of the experiment. All that is needed is a careful enumeration of the different outcomes to work out the classical probability.

In statistical mechanics we use the idea of classical probability by assuming that all the accessible quantum states of the system are equally likely. If there are W such quantum states then $p = 1/W$.

There can be difficulties with this approach. A simple illustration of the difficulty concerns the sex of an unborn baby. There are two possible outcomes (boy or girl) so $W = 2$. A priori it seems that a boy or a girl is equally likely. We can assume in the absence of any information that $p_{\text{boys}} = 0.5$. But when we examine the local records we find that p_{boys} is about 0.512 in England. More boys are born than girls in most countries.

Here is a more extreme case. The question is posed: how many days a year does it rain in Valparaíso, Chile? The possibilities are that it could rain on a given day or it could not. There are two possibilities so $W = 2$. In the absence of any knowledge of the geography of Chile we might propose that both possibilities are equally likely. Then in a leap year of 366 days we would expect to have 183 day of rain. Clearly this is wrong for it rains very much less than this. (One of us knows this from experience for she grew up there.)

The source of this difficulty arises because we are treating compound events as if they were simple events. If the event is compound, made up of say three simple events, then the compound event should have a larger probability than if it were truly a simple event. We need to test experimentally whether we have made the correct a priori assignments of probabilities, for they could be wrong.

3.3   Statistical probability

Statistical probability is concerned with the experimental method of assigning probabilities to events by measuring the relative frequency of occurrence. Suppose we measure the radioactive decay of a sample by measuring the counts recorded by a Geiger counter in a ten second interval. Let there be N trials measuring the decay rate over ten seconds, and let n_i be the number of times event i occurs. If the sequence of relative frequencies (n_i/N) tends to a constant as N goes to infinity, then this limit is defined as the statistical probability, p_i, given by

$$p_i = \lim_{N \to \infty} \left(\frac{n_i}{N} \right). \tag{3.3.1}$$

Suppose compound event 1 corresponds to counts in the range zero to four in ten seconds, event 2 is the counts in the range five to nine, event 3 is the counts in the range ten to fourteen, and so on. We do the same experiment ten times and get $n_2 = 3$ where n_2 is the number of times event 2 occurs. The ratio n_2/N is 0.3. We do 50 experiments and get $n_2 = 16$; then $n_2/N = 0.32$. The ratio fluctuates, but as N increases, the ratio n_2/N tends to a well-defined limit, the statistical probability that event 2 occurs in a single trial. For random events,

Table 3.1 The number of counts in the range 5 to 9 for different numbers of trials

Number of trials	Number in range 5 to 9	n/N
10	3	0.3
25	7	0.28
50	16	0.32
100	35	0.35
250	81	0.324
500	159	0.318
1000	317	0.317

the fluctuations in the statistical probability diminish as $N^{-1/2}$. The larger N becomes the more accurately we know p_i.

The values obtained in an imaginary experiment are listed in Table 3.1. The ratio n_2/N fluctuates as the number N increases, but it seems to converge on a value lying between 0.31 and 0.32 for large N. From the few readings taken we might guess that as N increases still further, the ratio will converge to a well-defined limit, a value which does not drift around too much with increase in N. Of course to check this out we have to do more and more experiments so that N gets larger and larger. We have to check that fluctuations in the ratio n_2/N vary as $N^{-1/2}$. If the ratio tends to a limit and the fluctuations behave as expected then there is every reason to treat the system as if there were a well-defined probability for the event.

It is a widely accepted principle that the frequency of well-defined events tends to a limit so that a definite probability can be assigned to them. Of course this is an act of faith because life is too short to make an infinite number of measurements.

Statistical probabilities only are well-defined when the same event occurs identically lots of times, and when the ratio n_i/N tends to a limit. There are examples where the limit does not exist. You can ask: what is the statistical probability that the stock market will close with a share index in the range 2850 to 2860 tonight? The stock market moves up and down. You can take readings of the identical stock market night after night for a long period of time, but the share index will not tend to a well-defined limit. It can drift upwards steadily as the economy grows. There is no concept of statistical probability in this case.

Another example concerns horse-races. We cannot run identical horse-races again and again for the weather is different each time, maybe one horse is injured, one jockey is off colour, and so on. Although people talk about the odds of a horse winning a particular race, the idea of a statistical probability of a horse winning a given race has no meaning. Statistical probability only has a meaning when the same situation can be reproduced again and again, and the ratio n_i/N tends to a well-defined limit.

3.4   The axioms of probability theory

Suppose that we have generated a set of probabilities, p_i. Perhaps they have been generated using the method of classical probability where, for simple events, the probability is $1/W$ where W is the total number of outcomes. Alternatively they could have been generated as the statistical probability from N trials using $p_i = n_i/N$ in the limit that N tends to infinity.

For both definitions, *the probabilities are positive numbers or zero*. This is the *first axiom* of probability theory.

The probabilities are less than or equal to one. This is the *second axiom* of probability theory. This statement is easy to prove for classical or statistical probabilities.

The *third axiom* of probability theory is concerned with a compound event. The notation $(i + j)$ means that either event i occurs or event j occurs or *both* occur. If both events cannot occur in one trial then the events are said to be *mutually exclusive*. The axiom is that for mutually exclusive events the probability of event $(i + j)$ is

$$p_{(i+j)} = p_i + p_j. \tag{3.4.1}$$

As an example, suppose a single card is chosen from a pack. The card cannot be both the three of clubs *and* the two of hearts at the same time. They are exclusive events: if one card is chosen and it is the two of hearts, then this excludes it being the three of clubs. We can talk about the probability of getting either the three of clubs or the two of hearts. This is just the sum

$$p_{(2H+3C)} = p_{2H} + p_{3C}.$$

For mutually exclusive events we add probabilities to get the probability that either one or other event occurs. Two or more events are mutually exclusive if not more than one of them can occur in a single trial.

A set $(1, 2, \ldots, r)$ of mutually exclusive events has a set of individual probabilities $(p_1, p_2, \ldots, p_r)$. The probability that one of this set occurs is given by the sum

$$p = p_1 + p_2 + \cdots + p_r. \tag{3.4.2}$$

As an example, the probability of getting either an ace, or a king, or the seven of clubs when drawing a single card from a normal pack is

$$p = \frac{4}{52} + \frac{4}{52} + \frac{1}{52} = \frac{9}{52}.$$

If the events are not mutually exclusive then we should not add probabilities together. As an example consider the roll of an ideal die with each face having a probability $\frac{1}{6}$. The chance of getting an even number is $\frac{1}{2}$; the chance of a number less than 4 is $\frac{1}{2}$; the chance that either or both occur is $\frac{5}{6}$, which is less than the sum of the other two probabilities.

3.5   Independent events

The acid test for independence of two events is if the probability that both events occur is equal to the product of the probabilities of each event. If the probabilities for events i and j are p_i and p_j respectively, the probability that both occur must be

$$p_{i,j} = p_i p_j \qquad (3.5.1)$$

if the *events are independent*.

As an example, suppose you drew a card from a pack and got the ace of spades, and then threw a die and got a two. The probabilities are $\frac{1}{52}$ of getting the ace of spades and $\frac{1}{6}$ of getting a two on the die; the probability that both events occur is

$$p = \frac{1}{52} \times \frac{1}{6}$$

if the events are independent of each other. When the events are completely independent (statisticians say the events are uncorrelated) then we multiply the probabilities together.

Consider a conjuring trick where two people draw a card from two packs of 52 cards. The classical probability that the first person draws an ace is $\frac{4}{52}$ since there are four aces in the pack of cards. Similarly the probability that the second person draws an ace from the second pack is $\frac{4}{52}$. If the events are independent the probability that two aces are drawn is $\frac{1}{169}$.

Now suppose we cheat: we get the people to agree to choose the sixteenth card in their pack of cards; we stack the packs so that both have an identical order of cards. If the first person gets an ace then the second person will get an ace also. The probability of one getting an ace is $\frac{4}{52}$; this is the probability that they both get an ace.

Only if the events are independent is it correct to multiply the probabilities together. Such events are said to be *uncorrelated*. When the probability of two events occurring is not given by the product of the individual probabilities, the two events are said to be *correlated*. If the chain of events leading to a disaster, such as the meltdown of a nuclear reactor, relies on one person making two errors, then the events may be correlated, for when a person has made a catastrophic error, fear and panic cause other mistakes to be made.

There can be more than two events. Suppose $p_1, p_2, \ldots, p_r$ are the separate probabilities of occurrence of r independent events. The probability p that they all occur in a single trial is

$$p = p_1 \times p_2 \times \cdots \times p_r. \qquad (3.5.2)$$

As an example, the trial could be three people each taking a card from a different 52-card pack. They get an ace from the first, a king from the second, and the seven of clubs from the third pack. The probability of this compound event is

$$p = \frac{4}{52} \times \frac{4}{52} \times \frac{1}{52}.$$

We can use this idea to calculate the probability of getting at least one six when throwing three dice. At first sight the answer is a half since each of the three dice has a one-in-six chance of producing a six. Such a simple approach ignores the possibility that two dice (or all three) come up with a six. The correct approach is to work out the probability that no six is thrown. The chance that the first (or the second, or the third) die does not produce a six is 5/6. The probability that the three dice do not produce a six is

$$p\,(\text{no }6) = \frac{5}{6} \times \frac{5}{6} \times \frac{5}{6} = \frac{125}{216}.$$

Therefore the probability that at least one six is thrown is

$$p\,(\text{at least one }6) = 1 - \frac{125}{216} = \frac{91}{216}.$$

3.6   Counting the number of events

When we use classical probability it is necessary to work out the number of different events that can occur. Counting the number of ways in which various events can happen leads to the concepts of arrangements, permutations, and combinations.

3.6.1   Arrangements

When calculating the total possible events in a sample space and the number of sample points for an event, we make use of the following results.

The number of ways of arranging n *dissimilar objects* in a line is $n!$. For example, if we have ten sticks of chalk, each one different from the rest, then we have ten ways of choosing the first, nine ways of choosing the second, eight ways of choosing the third, and so on. The total number of arrangements of the ten sticks of chalk is

$$10 \times 9 \times 8 \times 7 \times 6 \times 5 \times 4 \times 3 \times 2 \times 1 = 10!.$$

The number of ways of arranging in a line n objects of which p are identical is $n!/p!$. Whenever there are *identical objects* we have to take extra care to enumerate the arrangements. As an example, if we were to arrange the letters A, A, B, and D in a line there are only twelve possible arrangements. These are:

AABD	AADB	ABAD	ABDA
ADAB	ADBA	DAAB	DABA
DBAA	BAAD	BADA	BDAA

This result can be generalized to the case where there are n objects in a line; of the n objects p are of one kind, q of another, r of another, et cetera. In this case the number of arrangements is $n!/p!q!r!\ldots$.

3.6.2   Permutations of r objects from n

Consider the number of ways of choosing three sticks of chalk out of ten. The first choice can be any of ten different sticks of chalk; the second choice can be any of the nine which remain; the third choice can be any of the eight left over. Assume that all three choices are independent of each other. The number of permutations is

$$
\begin{aligned}
10 \times 9 \times 8 \;&=\; \frac{10 \times 9 \times 8 \times 7 \times 6 \times 5 \times 4 \times 3 \times 2 \times 1}{7 \times 6 \times 5 \times 4 \times 3 \times 2 \times 1}\\[4pt]
&=\; \frac{10!}{7!}.
\end{aligned}
$$

This quantity is called the *permutation* of three objects out of ten. It is the number of ways of making the choice if the order of the choice is important. We write it as

$$
{}^{10}P_3 = \frac{10!}{7!}.
$$

For the general choice of r objects out of n the number of permutations is

$$
{}^{n}P_r = \frac{n!}{(n-r)!}. \tag{3.6.1}
$$

3.6.3   Combinations of r objects from n

In some circumstances we do not care about the order in which the choice is made. For example, if all that matters is which pieces of chalk are chosen, and not the order of the choice, then the number of ways is reduced. For the example above, let the choice be chalk A, chalk B, and chalk C. There are 3! orders of the selected letters. We could have ABC, ACB, BAC, BCA, CAB, and CBA. We define the combination so that $3!\,{}^{10}C_3 = {}^{10}P_3$, or

$$
{}^{10}C_3 = \frac{10!}{3! \times 7!}.
$$

In general, the number of *combinations* of r objects out of n is

$$
{}^{n}C_r = \frac{n!}{r!(n-r)!}. \tag{3.6.2}
$$

In the example above n was ten and r was three.

We can generalize this result. Let us choose any three pieces of chalk out of the ten and put them in box one; now choose another five and place them in box two; there are two left which go in the third box. How many ways can this be done? The first three are chosen in

$$
{}^{10}C_3 = \frac{10!}{3! \times 7!}
$$

number of ways; the second five from seven are chosen in

$$^{7}C_5 = \frac{7!}{5! \times 2!}$$

number of ways; the final two are chosen in

$$^{2}C_2 = \frac{2!}{2! \times 0!}$$

number of ways. Remember that $0! = 1$. These three events are independent so we multiply the numbers together. The total number of ways of making the choices is

$$W = \frac{10!}{3! \times 5! \times 2!}.$$

In general, the number of arrangements for N pieces of chalk with n_i in box i is

$$W = \frac{N!}{n_1! \times n_2! \times n_3! \times \cdots} \tag{3.6.3}$$

where the $\cdots$ implies that you carry on dividing by all the $n_i!$.

Equation (3.6.3) can be used to work out the total number of ways energy can be arranged between different quantum mechanical states levels in N identical systems. The n_i represent the number of times this quantum state is occupied.

3.7 Statistics and distributions

Suppose we place a Geiger counter next to a piece of radioactive material, do a large number of experiments, and construct a histogram of the results. We place those with no counts in column 0, those with one count in column 1, and so on. Let us call the number of counts x_i, so that $x_0 = 0$, $x_1 = 1$, $x_2 = 2$, et cetera. After doing the experiment say a thousand times there are n_0 in column 0, n_1 in column 1, and so on. A measure of the probabilities is obtained by counting the relative frequencies of each event. It is desirable to have a large number for each event in order to get a small relative error in the probabilities. The absolute error for a count of n_i is of order $n_i^{1/2}$ and the relative error is $n_i^{-1/2}$. When n_i is of order 10^4 the relative error is about 0.01. In order to get accurate values for the probabilities we need to take a large number of readings. If the number of readings is large then the fluctuation in the height of each column is quite small. As a result the histogram gives a relatively smooth *distribution*.

In Fig. 3.2 we plot the results of an experiment where 100 readings were taken of the background count over a ten second interval. The number in the first column corresponds to zero counts, the number in the second column corresponds to one count, and so on. The distribution of these counts is given by a Poisson distribution. The dashed line shows the theoretical Poisson distribution corresponding to an average number of counts of 4.4 in ten seconds. It is shown as a smooth continuous line only for purposes of illustration: the distribution

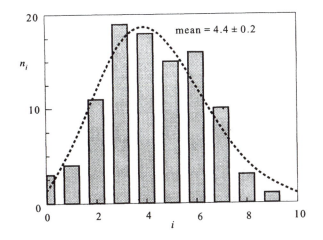

Fig. 3.2 The number of times out of a hundred trials that a background reading of 0, 1, 2,... was recorded in ten seconds by a Geiger counter. The dashed line shows the theoretical Poisson distribution that best fits the data.

only exists at integer values. Clearly the results of the experiment are roughly in agreement with the Poisson distribution, but many more experiments would be needed before we would be reasonably sure that the agreement is good.

What is the *average number* of counts? This is given by the formula

$$\bar{x} = \frac{\sum_i n_i x_i}{N} \qquad (3.7.1)$$

where $N = \sum_i n_i$, the sum of all the n_i. Equation (3.7.1) allows us to calculate the average value of x for any quantity described by a histogram; it is not restricted to the number of counts in the Geiger counter experiment. For example x_i could represent the energy of a system.

In the limit that N tends to infinity this average becomes

$$\bar{x} = \sum_i p_i x_i \qquad (3.7.2)$$

where p_i is the statistical probability that event i will occur.

Another quantity of interest is the size of the fluctuations around the mean value. This is expressed by the *standard deviation* in x (denoted by Δx), which is given by the expression

$$(\Delta x)^2 = \sum_i p_i (x_i - \bar{x})^2. \qquad (3.7.3)$$

Equation (3.7.3) can be applied to any quantity that can be represented by a

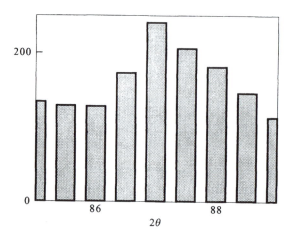

Fig. 3.3 The number of counts in ten seconds measured by a Geiger counter of X-rays scattered off a lithium fluoride crystal, as a function of the angle of deflection.

histogram; it is not restricted to the counts recorded by a Geiger counter. In thermal physics, an example is the fluctuations in the energy of a system which have a standard deviation given by

$$(\Delta E)^2 = \sum_i p_i (E_i - \bar{E})^2 \qquad (3.7.4)$$

where E_i is a particular energy of the system. Equation (3.7.4) can be used to evaluate the standard deviation of fluctuations in the energy of a system.

Suppose we take a huge number of readings so that the relative error in each column becomes small. At the same time we let the spacing between the columns become less and less. In the limit that the spacing between the values of x_i becomes tinier and tinier, the histogram tends to a smooth distribution.

To illustrate this procedure we show in Figs 3.3 and 3.4 some readings from an experiment using a Geiger counter which measures the count rate from X-rays scattered off a lithium fluoride crystal as a function of the angle of deflection, 2θ $(= x)$. The data in Fig. 3.3 were taken at every 0.5 degree over a period of ten seconds. In order to smooth out the distribution it is necessary to increase the number of counts, and to increase the resolution in the angle. For the data shown in Fig. 3.4 we have taken readings at every 1/6 of a degree and measured for a 100 seconds. The process can be repeated by decreasing the angle and increasing the time taken for the reading. In the limit that the spacing between the angles tends to zero and the number of counts tends to infinity, the histogram tends to a smooth distribution.

As the number in each column tends to infinity, the statistical probabilities become more accurate. We can replace p_i, the probability of being in column i,

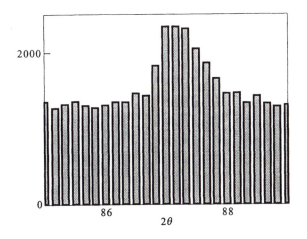

Fig. 3.4 This experiment is identical to the one which produced Fig. 3.3 except that we have increased the time interval to 100 seconds and the resolution in the angle from 1/2 degree to 1/6 degree.

by $p(x)\,\mathrm{d}x$ where $\mathrm{d}x$ is the width of the column and $p(x)$ is a smooth function of x. The continuous curve $p(x)$ is called a *probability distribution*. The sum over i becomes an integral over x. The condition

$$\sum_i p_i = 1 \tag{3.1.1}$$

becomes the integral

$$\int_{-\infty}^{\infty} p(x)\,\mathrm{d}x = 1. \tag{3.7.5}$$

There are several probability distributions which are common in probability theory, but the most important in statistical mechanics is the *normal* (or *Gaussian*) *probability distribution* which is defined by

$$p(x) = \frac{e^{-(x-\bar{x})^2/2\Delta x^2}}{(2\pi)^{1/2}\Delta x}. \tag{3.7.6}$$

There are just two parameters needed to define this distribution: the mean value, $\bar{x}$, and the standard deviation, Δx. A Gaussian probability distribution, $p(x)$, when plotted as a function of x, forms the bell-shaped curve shown in Fig. 3.5. The maximum is at $x = \bar{x}$. If Δx is small then the curve is sharply peaked; if Δx is large it is a broad, flat curve.

If we integrate the probability distribution from $(\bar{x} - \Delta x)$ to $(\bar{x} + \Delta x)$ we get the probability that the system will be found in this range of values. This probability has a value of 0.683; just over two-thirds of the distribution lies in

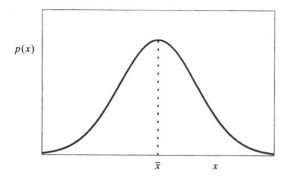

Fig. 3.5 The normal (or Gaussian) probability distribution. The distribution can be characterized by $\bar{x}$, the average value of x, and by the width of the distribution as given by the standard deviation, Δx.

the range $(\bar{x} - \Delta x)$ to $(\bar{x} + \Delta x)$. In practice this means that out of 100 readings, about 30 lie outside of one standard deviation from the mean. If we integrate from $(\bar{x} - 2\Delta x)$ to $(\bar{x} + 2\Delta x)$ we get 0.954; this means that out of 100 readings only about five lie outside two standard deviations from the mean. If we integrate from $(\bar{x} - 4\Delta x)$ to $(\bar{x} + 4\Delta x)$ we get 0.99994; this means that out of 10^5 readings only about six lie outside four standard deviations from the mean. The standard deviation sets the range over which the distribution varies.

Often it is simpler to break up a continuous variable x into discrete values which are equally spaced by an amount dx. The length dx is chosen to be sufficiently small that the probability distribution looks like a smooth curve, but sufficiently large so that the number in each column is huge. We will use this technique when considering the number of states a system has as a function of energy. The energy U is split up into equally spaced intervals of width dU so that the number of states in the range dU is very large and the statistical probability is well-defined, but the width dU is not so large that the distribution looks steplike and irregular.

3.8 Problems

1. What is the probability of getting either a seven or a six when throwing two dice?

2. If the probability of hitting a target is 0.1, and ten shots are fired independently, what is the probability that the target will be hit at least once?

3. What is the probability of first drawing the ace of spades and then drawing any of the three remaining aces from a 52-card pack?

4. If six different coloured inks are available, in how many ways can we select three colours for a printing job?

5. How many ways can one choose five objects out of twelve if either the order of the choice is important, or the order of the choice is not important, only the objects chosen?

6. In a population of fifty million people there are two hundred who have the same DNA profile.
   (a) What is the probability of two people having the same DNA profile?
   (b) What is the probability that the police, only knowing that a suspect from the population has a DNA profile that matches that found at the scene of a crime, has found the guilty party?

7. In how many ways can you choose from 12 objects, three subgroups containing 3, 4, and 5 objects respectively?

8. In the card game 'bridge' 52 distinct cards are dealt out to four players, each player receiving thirteen cards. Four of the cards are aces; one of them is called the ace of spades. When I pick up my hand:
   (a) what is the probability that I receive the ace of spades?
   (b) what is the probability that I receive both the ace of spades and the ace of clubs?
   (c) what is the probability that I receive all four aces?
   (d) what is the probability that I receive at least one ace?

9. What is the probability that out of five people, none have the same birthday? You may assume that there are 365 days in a year for simplicity. [Hint: start by working out the probability of two people not having the same birthday.]

10. I drop a piece of bread repeatedly. It lands either jam-side-up or jam-side-down. The probability of it landing jam-side-down is p.
    (a) What is the probability that the first n drops give jam-side-down?
    (b) What is the probability that it falls jam-side-up for the first time on the nth drop?
    (c) What is the probability that it falls jam-side-up for the second time on the nth drop?

11. If eight coins are thrown at random show that the probability of obtaining at least six heads is 37/256.

12. An apparatus contains six electronic tubes which all have to operate perfectly if it is to work. If the probability of each tube failing in a four hour period is 0.05, what is the probability that the apparatus will fail in this interval?

13. Three people, Alice, Bernard, and Charles, are playing a game for a prize. The first one to obtain heads when tossing a coin wins. They toss in succession starting with Alice, then Bernard, then Charles, et cetera. Show that Alice's chance of winning is 4/7. What are the chances of Bernard and of Charles winning?

14. A manufacturer knows that their resistors have values which are distributed as a Gaussian probability distribution with a mean resistance of 100 Ω and standard deviation of 5 Ω.

(a) What percentage of resistors have resistances between 95 and 105 Ω?

(b) What is the probability of selecting a resistor with a resistance less than 80 Ω?

15. The quantity x can take on values of $-20, -10$, and 30 with respective probabilities of $3/10$, $1/5$, and $1/2$. Determine the average value of x and the standard deviation, Δx.

16. In a factory a motor drives an electric generator; both need to be running for the factory to operate. During a 30-day period the possibility of the motor needing a repair is 15%, the possibility of the electric generator needing a repair is 20%. Both events are independent of each other. What is the probability that the combined unit will need a repair in a 30-day period due to the breakdown of either the motor, or the generator, or both?

17. A device uses five silicon chips. Suppose the five chips are chosen at random from a batch of a hundred chips out of which five are defective. What is the probability that the device contains no defective chip when it is made up from one batch?

18. Imagine tossing a coin a large number of times and recording a sequence of heads or tails. Let the probability of a head be P_H, that of a tail $P_T = 1 - P_H$. Now chose at random one of the recorded values, say the 257th. Let P_L be the probability that there is a cluster of L heads starting from the chosen one. Clearly $P_{L=0} = P_T$. What is the formula for the P_L for $L > 0$? Check that the sum of the probabilities for all lengths gives the correct answer.

19. Three dice are thrown. What is the likelihood that all faces are the same? What is the likelihood that the sum of the faces is 11?

20. In how many ways can the letters of the word 'statisticians' be arranged?

21. If the letters of the word 'minimum' are arranged in a line at random, what is the probability that the three 'm's are together at the beginning of the arrangement?

4

The ideas of statistical mechanics

But although as a matter of history, statistical mechanics owes its origin to the inves-
tigations in thermodynamics, it seems eminently worthy of an independent development, both on account of the elegance and simplicity of its principles, and because it
yields new results and places old truths in a new light in departments quite outside
thermodynamics.   *J. Willard Gibbs, Collected Works.*

4.1 Introduction

We now return to the questions listed at the end of Chapter 2. What is the
microscopic basis of the entropy? How do we calculate the equation of state for a
system? What is the physical basis of the second law which ensures that entropy
increases as a function of time? The structure of this chapter is based on these
three questions. We shall answer them in order. The answer to the first question
comes from *Boltzmann's hypothesis*: the entropy of a system is related to the
probability of its state. The basis of entropy is statistical. The answer to the
second question is given in section 4.4; the answer to the third is given in section
4.5.

To calculate thermal properties we use statistical ideas together with a knowledge of the laws of mechanics obeyed by the particles that make up a system.
Hence the name *statistical mechanics*. The beauty of statistical mechanics comes
from the very simplicity of the ideas it uses, and the power and generality of the
techniques involved.

There are two notions of probability that are commonly used: statistical and
classical probability. The more important is classical probability. We will use
this to predict what should be seen in experiments. The notion of statistical
probability is also useful. If we consider a series of identical measurements made
over a period of time on a single system, then we can obtain the statistical
probability of the outcome of a measurement from its relative frequency. Of
course this is only possible if a well-defined limit to the relative frequency exists,
and that will only happen if the system is in equilibrium. We can then test to
see if our predictions are realized experimentally.

To calculate the classical probability we must list all the possible outcomes of
the experiment and assign equal probabilities to each. There is a problem: how
do we know that these possible outcomes of an experiment correspond to simple
events and not compound ones? If the events are simple it is correct to assign
equal probabilities; if they are compound then the probabilities are unequal. We

need a way to define a simple event for the outcome of a single experiment. To be as thorough as possible, let us imagine a single experiment which defines the state of the system as completely as possible; we postulate that the outcome of such an experiment is a simple event. We assume that no further refinement of the experiment could reveal a deeper layer of knowledge which could be used to define the state of the system more completely.

The ideas on which statistical mechanics is based are these:

(a) *The state of the system is specified as completely as possible.* Each such state will be called a *quantum state*. We imagine doing a complete set of measurements that determines the position, momentum, energy, and spin of each particle, all at once, so as to specify the system as far as is possible at one instant.

For example, imagine a set of measurements which would allow us to instantaneously determine where every single particle is in a volume V. To achieve this we divide the space up into tiny cells of volume ΔV, and specify the position of each particle according to which cell it occupies.

(b) *In practice it is impossible to make a complete set of simultaneous measurements.* As a consequence, we cannot make a complete specification of the state of the system so as to specify the quantum state. It follows that we cannot measure the relative frequency of the occurrence of the quantum state in a series of trials, and so we cannot obtain the statistical probability. If we want to use statistical methods we are forced to use the classical probability.

(c) *We assign the classical probability to each quantum state.* It is convenient to imagine an *ensemble* of systems. An ensemble is a collection of similar systems all prepared identically with the same number of particles, energy, volume, shape, magnetic field, electric field, and so on. All these quantities are constrained to be the same. The quantum state of each system in the ensemble can be different, for there are a huge number of quantum states that are consistent with the way that they were prepared. We then ask: what is the probability that an experiment will give a particular answer corresponding to a particular quantum state of the system? The probability that a system is in a particular quantum state is given by the fraction of the ensemble in this state. What is this fraction?

(d) Suppose that there are W possible quantum states of the system which satisfy the set of constraints on energy, volume, and the rest. The essential assumption of statistical mechanics is that *all the quantum states are equally probable, a priori*. The probability that the system is in any of these quantum states is W^{-1}. By making this assumption we can calculate average values of any quantity we are interested in, such as the average kinetic energy of a particle. The average value of a quantity calculated this way is assumed to be exactly equal to the average value which would be obtained by a series of measurements over time (a time average) on a single system.

This set of ideas is not complete. There is one extra hypothesis which is needed to construct statistical mechanics. *Boltzmann's hypothesis* is that *the entropy of a system is related to the probability of its being in a quantum state*. The probability is $p = W^{-1}$ if there are W quantum states. We can express

Boltzmann's hypothesis in the following way:

$$S = \phi(W) \tag{4.1.1}$$

where $\phi(W)$ is some unknown function of W.

To determine $\phi(W)$ we use an approach due to Einstein (1905). Consider two systems A and B which are not interacting so they are independent of each other. Their entropies are

$$S_A = \phi(W_A) \tag{4.1.2a}$$
$$S_B = \phi(W_B). \tag{4.1.2b}$$

Instead of considering the two systems separately, we could consider the two systems as a single system with entropy S_{AB} and probability $p_{AB} = W_{AB}^{-1}$. Thus $S_{AB} = \phi(W_{AB})$. The total entropy is the sum of the entropies of the two systems

$$S_{AB} = S_A + S_B \tag{4.1.3}$$

since the entropy of independent systems is the sum of their individual entropies. Because the two systems are independent the total number of states is $W_{AB} = W_A W_B$. Hence

$$\phi(W_{AB}) = \phi(W_A W_B) = \phi(W_A) + \phi(W_B). \tag{4.1.4}$$

The only solution to eqn (4.1.4) is $\phi(W) = k_B \ln(W)$ where k_B is a universal constant. Hence Boltzmann's hypothesis leads to a mathematical expression for the entropy involving a new universal constant, k_B, now called *Boltzmann's constant*.

$$S = k_B \ln(W). \tag{4.1.5}$$

Let us illustrate the power of this equation by considering a simple model which could represent an ideal gas or a dilute solution. The argument is that of Einstein (1905). Consider N non-interacting molecules moving within a volume V. Imagine we specify the position of each molecule by subdividing the space into cells each of volume ΔV. The number of ways of placing one particular molecule in the volume is $W = V/\Delta V$. The number of ways of arranging N molecules is $(V/\Delta V)^N$ since each molecule goes in independently of the rest. Hence, according to eqn (4.1.5), the entropy is

$$S = N k_B \ln(V/\Delta V).$$

According to this expression the entropy depends on the volume of the cell, ΔV, something which is arbitrary. Remember that we only measure differences in entropy. If we were to alter the volume of the system but were to keep ΔV constant, the difference in entropy between the initial and final states would be

$$\Delta S = S_f - S_i = Nk_B \ln(V_f/V_i)$$

which is independent of ΔV.

The pressure of the system is given by $P = T (\partial S/\partial V)_U$. Our new expression for the entropy gives

$$P = \frac{Nk_B T}{V}.$$

This is the equation of state for an ideal gas. For one mole of gas we have the equation $PV = RT$. The equation above gives $PV = Nk_B T$. We now have a way of obtaining k_B from a knowledge of the gas constant, R, and *Avogadro's number*, N_A, as $k_B = R/N_A$.

In the derivation above the molecules in the volume V could be particles in a gas or they could be molecules dissolved in a solution. The expression $PV = Nk_B T$ describes the osmotic pressure, P, of molecules in solution. No assumption has to be made as to the law of motion of molecules in solution; all we have to assume is that the molecules go into solution independently of each other so that the number of arrangements is $(V/\Delta V)^N$. But this was known before Einstein's work: Van't Hoff had discovered experimentally this law for the osmotic pressure of solutions.

The success of such a simple calculation may convince you that the statistical method has merit. Simple counting of states gives the entropy. But when people tried to define the states more precisely they discovered the layer of reality which has to be described using quantum mechanics. One of the problems is that we can no longer specify the position of a particle and its momentum simultaneously with complete accuracy. Another problem is that the counting of states is more complicated if the particles are identical.

There is one source of confusion which we wish to clarify before proceeding. The phrase 'the state of the system' has two different meanings in thermal physics. In statistical mechanics a complete microscopic description of 'the state of the system' is called a *quantum state*. (Other authors call it a *microstate*, and Boltzmann called it a *complexion*.) In thermodynamics 'the state of the system' means a complete specification of the external constraints applied to the system: the system has a certain energy, number of particles, volume, magnetic field, and so on. We will call this state a *macrostate*. For example an 'equation of state' refers to the macrostate of the system. It is easy to get confused if we leave off the distinction, as we have done in the preceding paragraph. We should have written: 'Simple counting of [quantum] states gives the entropy. But when people tried to define the [quantum] states more precisely they discovered the layer of reality which has to be described using quantum mechanics.'

There are, in general, very many different quantum states that are consistent with a macrostate. Another way of saying this is that a macrostate contains very many quantum states. Before we can find out how many, we must define the concept of a quantum state more precisely.

4.2 Definition of the quantum state of the system

In the language of quantum mechanics the state of a system is described mathematically by a wavefunction. We use the notation $\phi(x)$ to describe a single-particle wavefunction. The quantity x represents the position of the particle. If we want to describe a single particle contained in a box, then the wavefunction must be zero everywhere outside the box. One way of ensuring that this condition is satisfied is to take $\phi(x) = 0$ on the boundaries of the box. We use the notation $\psi(x_1, x_2, x_3, \dots)$ for the wavefunction of a many-particle system where x_1 is the coordinate of the first particle and so on. The boundary conditions are then imposed on all these coordinates for a many-particle system.

In quantum mechanics, physical quantities such as energy and momentum are represented mathematically by *operators* which act on the wavefunction. The operator describing the energy of a system is called the *Hamiltonian*. An example is the operator which describes the kinetic energy of a single particle which is constrained to move in the x direction. It is represented by the operator

$$\hat{H} = -\frac{\hbar^2}{2m}\frac{\partial^2}{\partial x^2}.$$

Operators are denoted by symbols with a little inverted 'v' above them. The Hamiltonian operator $\hat{H}$ when acting on an arbitrary single-particle wavefunction $\phi(x)$ produces another wavefunction. For example if the wavefunction were $\phi(x) = e^{-ax^2}$ then

$$\begin{aligned}
\hat{H}\phi(x) &= -\frac{\hbar^2}{2m}\frac{\partial^2(e^{-ax^2})}{\partial x^2} \\
&= \frac{\hbar^2}{2m}(2a - 4a^2x^2)e^{-ax^2}.
\end{aligned}$$

The effect of the operator acting on a wavefunction is to produce a different wavefunction, $(2a - 4a^2x^2)e^{-ax^2}$. However, for particular wavefunctions, called *eigenfunctions*, the operator acting on the wavefunction produces the original wavefunction multiplied by a number; the number is called the *eigenvalue*.

$$\hat{A}\phi_i(x) = \alpha_i\phi_i(x). \tag{4.2.1}$$

Here $\phi_i(x)$ is an eigenfunction of the operator $\hat{A}$ with eigenvalue α_i.

As an example, let us determine eigenfunctions of the kinetic energy for a single particle in a box in one dimension with impenetrable walls at $x = 0$ and at $x = L$. A possible wavefunction is

$$\phi_n(x) = B\sin\left(\frac{n\pi x}{L}\right)$$

where n is chosen to be an integer because we want the wavefunction to vanish at $x = L$. We operate on this wavefunction with the kinetic energy operator and get

$$-\frac{\hbar^2}{2m}\frac{\partial^2}{\partial x^2}\left(B\sin\left(\frac{n\pi x}{L}\right)\right) = \frac{\hbar^2\pi^2 n^2}{2mL^2}B\sin\left(\frac{n\pi x}{L}\right).$$

The same wavefunction appears on the right-hand side multiplied by a number. The quantity $\hbar^2\pi^2 n^2/2mL^2$ is the eigenvalue of the kinetic energy operator and $B\sin(n\pi x/L)$ is the corresponding eigenfunction.

Another example concerns the spin of a particle. Suppose a measurement is made to determine the z component of the spin of an electron. The allowed values of spin for an electron are $\hbar/2$ and $-\hbar/2$. The spin operator has eigenstates ψ_1 and ψ_2 which obey the equations

$$\hat{s}_z\psi_1 = \frac{\hbar}{2}\psi_1; \qquad \hat{s}_z\psi_2 = -\frac{\hbar}{2}\psi_2.$$

In the theory of quantum mechanics, the physical quantities—energy, momentum, spin, and the like—are represented mathematically by operators. One of the postulates of quantum mechanics is that the values we measure experimentally are the eigenvalues of operators. If we were to measure the kinetic energy, for example, the values that would be obtained in an experiment are the eigenvalues of the operator $\hat{H}$. Other values are never obtained.

Another postulate of quantum mechanics says that when the state of a system is an eigenstate of the operator $\hat{A}$ with eigenvalue α_i, a measurement of the variable represented by $\hat{A}$ yields the value α_i with certainty. For this state we know the value of the physical quantity, for whenever we measure it there is only one possible outcome. There can be several physical quantities which can, in principle, be measured simultaneously so that we can know their values at the same time. For example, for a free particle we could know the kinetic energy and the momentum simultaneously. For a many-particle system we could know the values of momentum of each particle, their individual energies as well as the total energy of the system.

Imagine that we could list all the physical quantities which could be measured simultaneously for a many-particle system. If we knew the values of all the physical quantities then this would give us *the most complete description of the state of the system that is possible*. Each such state is assigned a wavefunction $\psi_i(x)$ and is called a *quantum state*. The label i is used to represent the values of all the physical quantities that can be measured simultaneously.

A quantum state is said to be *accessible* if it is compatible with the constraints specified for the system. The constraints are that the volume, the energy, and the number of particles are fixed for the isolated system, all fields applied to the system are constant, and the shape of the container is fixed. Only quantum systems of the correct energy need be considered; others are excluded and are said to be *inaccessible*. Similarly we only consider quantum states with the correct number of particles because the number of particles is fixed. (Sometimes certain quantum states are excluded from consideration because they cannot occur in the time-scale of the observation. These states are also said to be inaccessible. For example a metastable liquid can persist for a long time before a solid forms.

The quantum states associated with the solid phase are inaccessible except for very long times.)

Whenever there are many quantum states that have exactly the same energy, number of particles, volume, et cetera, we say that they are *degenerate*. The problem is to calculate their *degeneracy*, that is the number, W, of degenerate quantum states. The ensemble of these degenerate states is called the *microcanonical ensemble*.

4.3 A simple model of spins on lattice sites

How do we calculate W? The answer is to identify all the independent quantum states for the N-particle system and count them—it sounds a mammoth task. But it can be done. To illustrate what is involved for a relatively simple problem, we consider the different possible arrangements of spins on the lattice sites of a solid.

Suppose there are N particles placed on lattice sites with each particle having a spin of $1/2$ with an associated magnetic moment μ. In Fig. 4.1 we show a possible arrangement of spins. Let us place the spins in a magnetic field with magnetic induction B along the z direction. The spin can either point up, ↑, or down, ↓, relative to the z axis. The spin-up state has a spin eigenvalue $\hbar/2$, the spin-down state has a spin eigenvalue $-\hbar/2$. If the spin points up the particle has an energy eigenvalue $\varepsilon = -\mu B$, if the spin points down the particle has an energy eigenvalue $\varepsilon = \mu B$.

The quantum state is described by a wavefunction ψ_i, where i is a label which denotes all the quantum numbers which are known for that state. For example we can write the quantum state as

$$\psi_i = |\uparrow, \uparrow, \downarrow, \ldots\rangle \tag{4.3.1}$$

where the bracket contains a series of arrows. Reading from the left, this says the spin of the first particle is up, the spin of the second is up, the spin of the third is down, et cetera. This notation for the quantum state is a little different from that which is met in introductory courses in quantum mechanics. We choose to use it to emphasize those quantities that are known for the quantum state.

In a closed system containing many particles the energy and number of particles are fixed. Even so there are a huge number of quantum states which are consistent with these constraints. *The macrostate is degenerate.* Let n_1 particles be spin-up and have energy $-\varepsilon$ and let $n_2 = N - n_1$ particles be spin-down and have energy ε. The total energy is

$$U = -n_1\varepsilon + (N - n_1)\varepsilon = N\varepsilon - 2n_1\varepsilon. \tag{4.3.2}$$

If we specify U and N then both n_1 and n_2 are determined. The number of ways of choosing n_1 spin-up particles out of N is

$$W = \frac{N!}{n_1!(N - n_1)!}. \tag{4.3.3}$$

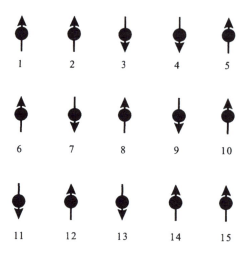

Fig. 4.1 One possible arrangement of spins on a lattice. There are only 15 particles shown—you have to imagine the rest of the lattice. Each particle has a spin of either up or down. We can distinguish each particle by its lattice site and number them according to some scheme. We have assumed that each particle is localized on its lattice site and cannot exchange places with its neighbours.

This is the degeneracy of the N-particle system with energy U. All these W quantum states satisfy the constraints on the system.

For the spin system the entropy is

$$S = k_{\rm B} \ln(W) = k_{\rm B} \ln \left(\frac{N!}{n_1!(N - n_1)!} \right). \tag{4.3.4}$$

If $N = 5$ and $n_1 = 3$ then the number W is easily calculated; it is $5!/3!2! = 10$. But we want to be able to calculate $N!$ for large N. This technical problem can be overcome by using *Stirling's approximation*.

Let us consider the general case for $M!$ with M an integer. Rather than work out $M!$ we take logarithms

$$
\begin{aligned}
\ln (M!) &= \ln (M) + \ln (M - 1) + \ln (M - 2) + \\
&\quad \cdots + \ln(2) + \ln(1) \\
&= \sum_{n=1}^{M} \ln (n). \tag{4.3.5}
\end{aligned}
$$

The sum can be done approximately by a useful trick: we replace the summation by an integral running from $\frac{1}{2}$ to $(M+\frac{1}{2})$. The procedure is discussed in Appendix B. Each term in the sum is represented by an integral over n. For example the pth term involves an integral from $(p - \frac{1}{2})$ to $(p + \frac{1}{2})$. By adding them together we turn the sum into an integral over the continuous variable, n.

$$\sum_{n=1}^{M} \ln(n) = \int_{\frac{1}{2}}^{M+\frac{1}{2}} \ln(n)\mathrm{d}n$$

$$\approx |n\ln(n) - n|_{\frac{1}{2}}^{M+\frac{1}{2}}. \tag{4.3.6}$$

The term we shall always use is

$$\ln(M!) \approx M\ln(M) - M. \tag{4.3.7}$$

All terms of lower order in M have been ignored. We can calculate W by using eqn (4.3.7) and so the technical problem is solved.

The number of quantum states for the spin system is given by the formula

$$\begin{aligned}\ln(W) &= \ln(N!) - \ln(n_1!) - \ln((N - n_1)!) \\ &\approx N\ln(N) - n_1\ln(n_1) - (N - n_1)\ln(N - n_1)\end{aligned}$$

which—by the artifice of adding and subtracting the term $n_1\ln(N)$, and then regrouping—can be written as

$$\ln(W) = -N\left\{\left(\frac{n_1}{N}\right)\ln\left(\frac{n_1}{N}\right) + \left(1 - \frac{n_1}{N}\right)\ln\left(1 - \frac{n_1}{N}\right)\right\}. \tag{4.3.8}$$

According to eqn (4.3.2) the energy per particle is $U/N = \varepsilon - 2\varepsilon n_1/N$. If we take the energy per particle to be independent of N, then n_1/N is independent of N and $\ln(W)$ is proportional to N. In other words when the system doubles the number of spins in such a way that the energy per spin is constant then $\ln(W)$ is proportional to N. Quantities which are proportional to N are said to be *extensive*. When the energy is extensive, so is $\ln(W)$.

We have achieved our immediate objective, to calculate the number of quantum states that satisfy the constraints on the system and so get the entropy. Now let us turn to our second main question: how do we get equations of state?

4.4 Equations of state

To show that we can derive *equations of state* we shall consider three examples: a spin system, vacancies in a crystal, and a crude model of a rubber band.

4.4.1 Spin system

For the spin system described in section 4.3 we proceed in the following way. From eqn (4.3.2) we can write the ratio $n_1/N = (1 - x)/2$ with $x = U/N\varepsilon$. The entropy of the spin system is

$$S = -Nk_B\left\{\frac{(1+x)}{2}\ln\left(\frac{1+x}{2}\right) + \frac{(1-x)}{2}\ln\left(\frac{1-x}{2}\right)\right\}. \tag{4.4.1}$$

We show this function in Fig. 4.2. However, from eqn (2.8.1) the temperature is

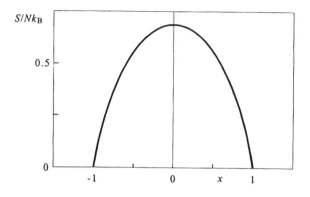

Fig. 4.2 The entropy per particle (in units of k_B) of a collection of non-interacting spin-half particles as a function of the variable $x = U/N\varepsilon$.

given by the equation

$$\frac{1}{T} = \left(\frac{\partial S}{\partial U}\right)_V = \left(\frac{\partial S}{\partial x}\right)_V \frac{\mathrm{d}x}{\mathrm{d}U} = \left(\frac{\partial S}{\partial x}\right)_V \frac{1}{N\varepsilon} \qquad (4.4.2)$$

so differentiation of eqn (4.4.1) gives

$$\frac{1}{T} = \frac{k_B}{2\varepsilon} \ln\left(\frac{1-x}{1+x}\right) \qquad (4.4.3)$$

which can be rewritten as

$$x = \frac{U}{N\varepsilon} = -\tanh\left(\frac{\varepsilon}{k_B T}\right). \qquad (4.4.4)$$

The total energy is $U = -n_1\varepsilon + n_2\varepsilon$. Therefore the difference in populations, $n_1 - n_2$, is equal to $-U/\varepsilon$, and so

$$n_1 - n_2 = -\frac{U}{\varepsilon} = N\tanh\left(\frac{\varepsilon}{k_B T}\right). \qquad (4.4.5)$$

The total magnetic moment of the system is the difference between μn_1, the net magnetic moment of the up-spins, and the net magnetic moment of the down-spins, μn_2. Suppose the spins of the particles are aligned by a magnetic induction field B giving rise to an energy of the up-spins of $-\mu B$ and an energy of the down-spins of μB. The magnetic moment per unit volume, the *magnetization*, is

$$M = \frac{\mu(n_1 - n_2)}{V} = \frac{\mu N}{V}\tanh\left(\frac{\mu B}{k_B T}\right). \qquad (4.4.6)$$

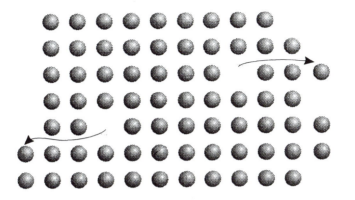

Fig. 4.3 Vacancies in a crystal formed by atoms leaving their lattice sites and migrating to the surface of the crystal.

This is the equation of state for the magnetization of the system made up of non-interacting spin 1/2 particles in a magnetic induction field, B.

4.4.2 Vacancies in a crystal

Suppose atoms are taken away from their lattice sites and forced to migrate to the surface of the crystal. Each time this happens a vacancy at the lattice site is created, something which is called a *Schottky defect*. This is illustrated schematically in Fig. 4.3. In what follows we assume that the volume of the solid is held constant when Schottky defects are created so there is no mechanical work done on the system.

Suppose there are n such defects in the solid, each with an energy ϵ, so the total energy is $n\epsilon$. A quantum state of the system can be defined by the positions of the vacancies in the crystal. Let us number the sites in some way. The vacancies could be, for example, at sites 7, 43, 219, 657, ... , in which case the quantum state could be written as

$$\psi_i = |7, 43, 219, 657, \ldots\rangle \,.$$

Each such arrangement of vacancies among the lattice sites corresponds to an independent quantum state.

There are N atoms in the crystal and n vacancies arranged at random on $N + n$ lattice sites. The number of configurations of vacancies is $(N+n)!/n!N!$. The entropy of the defects is therefore

$$S = k_B \left\{ (N+n)\ln(N+n) - n\ln(n) - N\ln(N) \right\} \,. \tag{4.4.7}$$

The internal energy is $U = n\varepsilon$ so the temperature is

Fig. 4.4 A collection of links representing a simplified model of a rubber band. The links lie in the $+z$ and $-z$ directions. We have drawn some of the links at an angle so that the arrangement may be seen more clearly.

$$\frac{1}{T} = \left(\frac{\partial S}{\partial U}\right)_V = \frac{1}{\varepsilon}\left(\frac{\partial S}{\partial n}\right)_V = \frac{k_B}{\varepsilon}\ln\left(\frac{N+n}{n}\right). \tag{4.4.8}$$

This equation can be rearranged to give

$$\frac{n}{N} = \frac{1}{e^{\varepsilon/k_B T} - 1} \tag{4.4.9}$$

a formula which makes sense only if n/N is much less than one, for otherwise the solid would be riddled with holes and would collapse. A typical value of ε is 1 eV which means that at room temperature $\varepsilon/k_B T$ is about 39 and n/N is about 10^{-17}. There are relatively few vacancies in the solid at room temperature.

4.4.3 A model for a rubber band

A simple picture of a rubber band is of a single long chain of linking groups of atoms, with the links oriented in any direction. When the rubber band is pulled so that the chain of atoms is completely linear there is only one possible arrangement and the entropy is zero; when the rubber band is all tangled up there is a huge number of arrangements of the links leading to a large entropy. Heat up a rubber band under tension and it wants to tangle up. Cool it down and it stretches.

Let us simplify matters by assuming that the links can lie in only two directions, either in the direction of increasing z or in the opposite direction. This is a caricature of the situation, but it contains the essence of the problem. A crude picture of such a chain of links is given in Fig. 4.4. The links lie in the direction of either $+z$ or $-z$. Start at one end of the chain (the end with the smaller value of z) and count how many lie along the $+z$ direction $(= n_+)$ and how many lie along the $-z$ direction $(= n_-)$.

There can be many different ways of ending up with n_+ links in the $+z$ direction. Each such arrangement can be represented as a different quantum state. Starting from the end we note a series of pluses and minuses to label each link. The successive arrangement of links along the chain defines the quantum state. One such state for 15 links is the state (see Fig. 4.4)

$$\psi_i = |+, -, +, +, +, +, -, +, +, +, -, +, -, +, +\rangle.$$

There are 11 positive links and four negative links. Each link has a length d. The total extension of the rubber band, the distance from one end of the chain to the other, is l. The quantum state shown in Fig. 4.4 has an extension of $7d$. There are many other quantum states ($15!/11!4!$ of them) with this length. Here are a few:

$$\psi_j = |+,+,-,+,+,+,-,+,+,+,-,+,-,+,+\rangle,$$
$$\psi_k = |+,+,+,-,+,+,-,+,+,+,-,+,-,+,+\rangle,$$
$$\psi_l = |+,+,+,+,-,+,-,+,+,+,-,+,-,+,+\rangle.$$

The complete set of quantum states of length $7d$ make up the macrostate. We are interested in much longer molecules with many more links than 15.

The work done on the rubber band when it is extended by an amount dl is $F\,dl$ where F is the tension in the band. The change in the internal energy is therefore given by

$$dU = T\,dS + F\,dl. \tag{4.4.10}$$

From this relation and the expression for the entropy, $S = k_B \ln(W)$, we get

$$\frac{F}{T} = -k_B \left(\frac{\partial \ln(W)}{\partial l} \right)_U. \tag{4.4.11}$$

In order to calculate $\partial \ln(W)/\partial l$ for constant U, we need to know $\ln(W)$ for our crude model and how it varies with l.

Let there be n_+ links going to the right (labelled by a right-pointing arrow) and n_- going to the left. For a chain of N links, W is given by

$$W(N, n_+) = \frac{N!}{n_+!(N - n_+)!}. \tag{4.4.12}$$

Here the relation $n_- = (N - n_+)$ has been used. $W(N, n_+)$ is the number of arrangements of the n_+ positive links, chosen out of the total number of links, N. The total extension is

$$l = (n_+ - n_-)d = (2n_+ - N)d. \tag{4.4.13}$$

Let us define a parameter x as

$$x = \frac{l}{Nd} = \left(\frac{2n_+}{N} - 1 \right). \tag{4.4.14}$$

Nd is the length that occurs when all the links point the same way, that is the maximum length of the chain molecule that makes up the rubber band. Thus x is a measure of how much the band has been stretched. Simple algebra gives

$$\frac{n_+}{N} = \frac{1 + x}{2} \tag{4.4.15a}$$

$$\frac{n_-}{N} = \frac{1-x}{2}. \tag{4.4.15b}$$

Now

$$\ln(W) = \ln(N!) - \ln(n_+!) - \ln\left((N-n_+)!\right). \tag{4.4.16}$$

The expression for $\ln(W)$ can be simplified by using Stirling's approximation.

$$\begin{aligned}
\ln(W) &= -N\left\{\left(\frac{n_+}{N}\right)\ln\left(\frac{n_+}{N}\right) + \left(1 - \frac{n_+}{N}\right)\ln\left(1 - \frac{n_+}{N}\right)\right\} \\
&= -N\left\{\left(\frac{1+x}{2}\right)\ln\left(\frac{1+x}{2}\right) + \left(\frac{1-x}{2}\right)\ln\left(\frac{1-x}{2}\right)\right\} \tag{4.4.17}
\end{aligned}$$

The tension, F, in the rubber band (eqn (4.4.11)) is given by

$$\frac{F}{T} = -k_{\mathrm{B}}\left(\frac{\partial \ln(W)}{\partial x}\right)_U \frac{\mathrm{d}x}{\mathrm{d}l} \tag{4.4.18}$$

with

$$\frac{\mathrm{d}x}{\mathrm{d}l} = \frac{1}{Nd}. \tag{4.4.19}$$

Hence the equation for the force of tension is

$$\frac{F}{T} = \frac{k_{\mathrm{B}}}{2d}\ln\left(\frac{1+x}{1-x}\right) = \frac{k_{\mathrm{B}}}{2d}\ln\left(\frac{Nd+l}{Nd-l}\right). \tag{4.4.20}$$

For small values of l/Nd we can make a power series expansion of eqn (4.4.20) to get in lowest order

$$F \approx \frac{k_{\mathrm{B}}Tl}{Nd^2}. \tag{4.4.21}$$

This equation indicates that the tension is proportional to the product of the extension and the temperature. For a rubber band under a constant tension, the extension will decrease on heating, a somewhat counterintuitive result. Of course our model is very crude and should not be trusted. It is a first approximation, a caricature of a rubber band which has the essence of the problem without being precisely correct. In reality rubber bands under tension do shrink when heated as can be demonstrated quite easily.

Figure 4.5 illustrates a simple experiment which you could do at home. You will need a rubber band, a long rod, a pivot, and a hair-drier. The rubber band is under tension and pulls on the rod which provides a torque about the pivot: the torque is the product of the force and the perpendicular distance of the force from the pivot. There is also a torque provided by the weight of the rod acting at the centre of mass of the rod. This torque is the product of the weight and the distance of the centre of mass from the pivot. When the rod is in equilibrium the two torques are equal and opposite.

When you raise the temperature of the rubber band it contracts with the result that the rod rotates about the pivot and moves up the scale on the right.

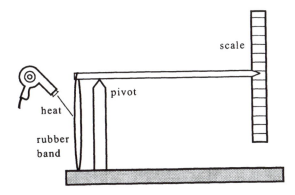

Fig. 4.5 The arrangement of a simple experiment to demonstrate that a rubber band under tension shrinks when heated. The hair-drier heats the rubber band. As the temperature of the rubber band rises it contracts causing the rod to rotate about the pivot and to move up the scale on the right.

The pivot acts to magnify the small contraction of the rubber band into a large movement of the rod. This a counterintuitive result: normally objects expand on heating. A rubber band under tension contracts. The physics is simple: when the band is under tension it is stretched and there are fewer arrangements possible; heat the band up and it wants to become more disordered amongst a greater number of states, something it can only do if the rubber band contracts.

The whole machinery of statistical mechanics can be put in operation for systems which can be described by sufficiently simple models. Given that we have a simple model, all we have to do is work out $\ln(W)$, and usually that is difficult. But once we have done that we can get to the thermodynamics.

4.5 The second law of thermodynamics

We can try to count W and use eqn (4.1.5) to calculate S; or we can take the value of S and calculate W. We will use the latter approach to explain the physical basis of the second law of thermodynamics and show why the entropy increases with time.

We invert eqn (4.1.5) to get

$$W = e^{S/k_B}. \tag{4.5.1}$$

The absolute value of S is not known since thermodynamic measurements give only differences in entropy. The consequence is that we can obtain W from eqn (4.5.1) apart from an overall multiplicative factor, something which is irrelevant to the argument.

In section 2.9 we showed that the entropy of two systems A and B, taken together but isolated from everything else, varies as a function of the energy of system A (or of the volume of system A, or of the number of particles in system A). We can write it as $S(U_A)$. The system evolves so that $S(U_A)$ is a maximum. The maximum value of S occurs when the energy of system A is U_A^*. Near the maximum, we can make a Taylor expansion of $S(U_A)$ as the series

$$
S(U_A) \;=\; S(U_A^*) + (U_A - U_A^*) \left(\frac{\partial S}{\partial U_A} \right)_{U_A = U_A^*}
$$
$$
+ \frac{(U_A - U_A^*)^2}{2} \left(\frac{\partial^2 S}{\partial U_A^2} \right)_{U_A = U_A^*} + \cdots .
$$

(4.5.2)

At the maximum of $S(U_A)$ the slope $(\partial S/\partial U_A)_{U_A^*} = 0$, and $S'' = (\partial^2 S/\partial U_A^2)_{U_A^*}$ is less than zero. Hence we get from eqn (4.5.1)

$$
W(U_A) = e^{S(U_A^*)/k_B} \, e^{(U_A - U_A^*)^2 S''/2k_B} .
$$

(4.5.3)

$W(U_A)$ is a Gaussian function of U_A with a maximum at U_A^* and a standard deviation ΔU_A given by (see eqn 3.7.6)

$$
\Delta U_A^2 = -\frac{k_B}{S''} .
$$

(4.5.4)

Statistical mechanics predicts that there are *fluctuations* in the energy of system A around the mean value U_A^*, with a typical fluctuation in energy of order ΔU_A. These fluctuations are inherent in the macrostate. The equilibrium state of the system does not just consist of the system staying at the maximum value of $S(U_A)$; it includes the fluctuations of the system around the mean energy.

The joint system can be in any of the W states of the system, but for systems with a large number of particles the majority of states are clustered around U_A^*, lying within a few standard deviations of energy.

Let $p(U_A)$ be the probability that the joint system is in a tiny energy range dU centred on U_A. Suppose that the number of quantum states in this range is $W(U_A)$. The classical probability is

$$
p(U_A) = \frac{W(U_A)}{\sum_i W(U_i)} .
$$

$W(U_A)$ is a maximum at U_A^*; the corresponding probability is

$$
p(U_A^*) = \frac{W(U_A^*)}{\sum_i W(U_i)} .
$$

Therefore the probability that the joint system is in a tiny energy range dU centred on U_A compared to the probability that it is in the energy range dU centred on U_A^* is

$$\frac{p(U_A)}{p(U_A^*)} = \frac{W(U_A)}{W(U_A^*)} = e^{-(U_A - U_A^*)^2/2\Delta U_A^2}. \tag{4.5.5}$$

For example, the ratio $p(U_A)/p(U_A^*)$ when system A has an energy of, say, $U_A^* + 20\Delta U_A$ is e^{-200}. The probability $p(U_A)dU$ is a very small number compared to $p(U_A^*)dU$. Very small is hardly an adequate expression, for it is unimaginably small. If the system were to sample a different state every 10^{-50} s then such a fluctuation could be expected every $10^{-50}e^{200} = 10^{38}$ s. This is way beyond the age of the Universe. Fluctuations of such magnitude happen so rarely they can be assumed never to occur.

Boltzmann expressed this idea as follows: 'One should not imagine that two gases in a 0.1 litre container, initially unmixed, will mix, then after a few days separate, then mix again, and so forth. On the contrary one finds ... that not until a time enormously long compared to years will there be any noticeable unmixing of the gases. One may recognize that this is practically equivalent to never ...'.

The scale of fluctuations in energy is ΔU_A, which depends on the second derivative of S with respect to the energy of the system. Let us estimate S''. Using $S = S_A + S_B$ we find that the first derivative at constant volume is

$$\left(\frac{\partial S}{\partial U_A}\right)_V = \frac{1}{T_A} - \frac{1}{T_B} \tag{4.5.6}$$

and the second derivative is

$$\begin{aligned} S'' &= \left(\frac{\partial^2 S}{\partial U_A^2}\right)_V = \left(\frac{\partial T_A^{-1}}{\partial U_A}\right) + \left(\frac{\partial T_B^{-1}}{\partial U_B}\right) \\ &= -\frac{1}{T_A^2}\left(\frac{\partial T_A}{\partial U_A}\right) - \frac{1}{T_B^2}\left(\frac{\partial T_B}{\partial U_B}\right) = -\frac{1}{T^2}\left(\frac{1}{C_A} + \frac{1}{C_B}\right). \end{aligned} \tag{4.5.7}$$

Here we have used eqn (1.8.3) for the heat capacity at constant volume. The second derivative of S_A with respect to U_A is negative and involves the sum of the inverse heat capacities of the two systems. If the heat capacity of system B is much larger than that of system A we get

$$\Delta U_A^2 = k_B T^2 C_A. \tag{4.5.8}$$

Experimentally, the heat capacity is found to be proportional to the number of particles in the system provided the number density is held fixed. Thermodynamic quantities which are proportional to the size of a system, and hence to the number of particles, are said to be *extensive*. The entropy and heat capacity of any system are extensive quantities. Hence ΔU_A is proportional to $N^{1/2}$. The internal energy at the maximum, U_A^*, is proportional to N, so it follows that the ratio $\Delta U_A/U_A^*$ varies as $N^{-1/2}$. As the number of particles in the system increases, the ratio of the fluctuations in energy to the average internal energy tends to zero. The distribution of states, when plotted in terms of the energy per particle, becomes sharper as N increases.

What then is the basis of the second law thermodynamics? Why does the entropy increase as a function of time? Oscar Wilde said of the Niagara Falls: 'It would have been more impressive if it flowed the other way.' What is it that makes natural processes flow one way and not the other?

Imagine a joint system, which is made up of subsystems A and B, thermally isolated from the rest of the universe so that the total energy, U_T, is constant. Let $W_A(U_A)$ be the number of states of energy in the range U_A to $U_A + dU$, and let $W_B(U_B)$ be the number of states of energy in the range U_B to $U_B + dU$. The energy U_A is split up into discrete, equally spaced intervals of width dU so that the number of states in the range dU is very large and the statistical probability is well-defined. (In other words dU is very much smaller than U_A but is large enough to contain sufficient quantum states so that the concept of a statistical probability is well-defined.)

If system A were isolated from system B, the number of states of the joint system would be

$$W^i = W_A(U_A^i)W_B(U_B^i) \qquad (4.5.9)$$

where U_A^i is the initial energy of system A and U_B^i is the initial energy of system B. The total energy is fixed at $U_T = U_A^i + U_B^i$. As a result of thermal contact energy flows from one system to the other, and the number of accessible states increases to

$$W^f = \sum_{U_A} W_A(U_A)W_B(U_T - U_A) \qquad (4.5.10)$$

where the sum is over all possible discrete energies U_A.

The number of accessible states increases as a result of thermal contact. The crucial effect is that the value of $W_A(U_A)W_B(U_T - U_A)$ where it is a maximum, $W_A(U_A^*)W_B(U_T - U_A^*)$, is enormously larger than the initial value, as shown in Fig. 4.6. Thus

$$W_A(U_A^*)W_B(U_T - U_A^*) \gg W_A(U_A^i)W_B(U_T - U_A^i). \qquad (4.5.11)$$

Since all accessible quantum states are equally likely, the system will most probably evolve from the initial configuration with energy U_A^i towards the configuration with energy U_A^* where the number of states is largest. Once there, the probability of returning is given roughly by the ratio

$$p = \frac{W_A(U_A^i)W_B(U_T - U_A^i)}{W_A(U_A^*)W_B(U_T - U_A^*)}. \qquad (4.5.12)$$

If this probability were about 0.01 say, there would be a reasonable chance of the system returning back to its initial energy due to random fluctuations. However, when the ratio is of the order of, say, e^{-400} the chances of such a fluctuation are so remote that in practice it can be ignored.

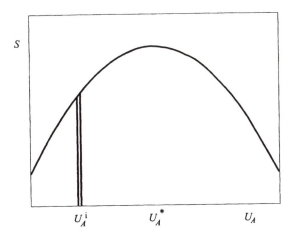

Fig. 4.6 The quantity $S = k_{\mathrm{B}} \ln(W)$ of the joint system, made up of subsystems A and B, has a maximum. (If we were to plot W instead of $k_{\mathrm{B}} \ln(W)$ the curve would have a very narrow peak, too sharp to draw.) Before thermal contact is made systems A and B both have a well-defined energy, for they are thermally isolated. Thermal contact between them allows energy to flow from one to the other with the total energy held constant. As a result of thermal contact more quantum states of the same total energy become accessible. The majority of these states lie within a few standard deviations in energy of the maximum in the entropy.

To visualize the magnitude of the numbers involved consider the following example. Suppose system A is at $300.2\,\mathrm{K}$ and system B is at $300\,\mathrm{K}$. The two systems are identical and have a heat capacity of $C = 1\,\mathrm{J\,K^{-1}}$. The two are placed in thermal contact and reach thermal equilibrium at $300.1\,\mathrm{K}$. The change in entropy of the two systems is

$$\Delta S = C \ln \left(\frac{300.1 \times 300.1}{300.2 \times 300} \right) = 1.1 \times 10^{-7}\,\mathrm{J\,K^{-1}},$$

a small amount. Or so it appears. But S can be expressed in terms of W by using Boltzmann's hypothesis. The change in entropy is

$$\Delta S = k_{\mathrm{B}} \ln(W_{\mathrm{f}}) - k_{\mathrm{B}} \ln(W_{\mathrm{i}}) = 1.1 \times 10^{-7}\,\mathrm{J\ K^{-1}}$$

where W_{f} is the final number of quantum states and W_{i} is the initial number. Nothing staggering, until you realize how small Boltzmann's constant is. We find

$$\ln \left(\frac{W_{\mathrm{f}}}{W_{\mathrm{i}}} \right) = \frac{1.1 \times 10^{-7}}{1.38 \times 10^{-23}} = 8 \times 10^{15}.$$

The increase in the number of accessible states is by a factor $\mathrm{e}^{8 \times 10^{15}}$. In other words we are not considering millions, we are considering the exponential of eight thousand million million.

The chances of the system fluctuating and going back to the initial temperature is $e^{-8\times10^{15}}$. It is not precisely zero, but effectively zero. Once the system has moved to the maximum in the entropy, the chance of it going back is negligible. This is the source of irreversibility and the basis of the second law of thermodynamics.

The calculation can be repeated for two systems having a difference in temperatures of, say, 2×10^{-6} K instead of 2×10^{-1} K. The increase in the number of states in this case is

$$\ln\left(\frac{W_f}{W_i}\right) = \frac{1.1 \times 10^{-17}}{1.38 \times 10^{-23}} = 8 \times 10^5.$$

The increase in the number of quantum states is the exponential of almost a million. If the difference in temperature is about 2×10^{-9} K then

$$\ln\left(\frac{W_f}{W_i}\right) = \frac{1.1 \times 10^{-23}}{1.38 \times 10^{-23}} = 0.8.$$

Such a change in the number of accessible quantum states is sufficiently small that a thermal fluctuation could produce it.

4.6   Logical positivism

Every statement in physics has to state relations between observable quantities.
Ernst Mach (1838–1916)

The modern view of scientific method was put forward by Karl Popper. He said: 'I shall certainly admit a system as empirical or scientific only if it is capable of being *tested* by experience. These considerations suggest that it is not the *verifiability* but the *falsifiability* of a system is to be taken as a criterion for demarcation ... *It must be possible for an empirical scientific system to be refuted by experience.*' According to this view, an understanding of the natural world is not obtained solely by accumulating experimental information and looking for patterns and relations amongst the data alone. Instead, the scientific method consists of experimentalists weeding out incorrect theories by careful experiments, and theorists proposing imaginative conjectures which might accord with reality. In this way we hope to get closer and closer to an approximation of the truth without ever conclusively getting there.

The older view of scientific method was proposed by Ernst Mach. Mach was a scientist and also a philosopher. His philosophy was that you should never introduce any quantity in science which could not be actually observed and measured. This is the *positivist philosophy*. The concern with what can actually be observed is the essence of positivism.

Lord Kelvin expressed the classical view when he wrote: 'I often say that when you can measure what you are speaking about, and express it in numbers, you know something about it; but when you cannot measure it, when you cannot express it in numbers, your knowledge is of a meagre and unsatisfactory kind'. Concepts have most meaning, according to the classical view, if they can be measured and counted. Clearly this is the view we are all happiest with. You can weigh yourself each day, and hence tell whether or not you are putting on weight. The rate of increase in weight can be measured. Measurement lies at the heart of science.

You can measure the entropy provided the measurements are made in such a way that the addition of heat energy is completely reversible. To avoid error you must add the heat very slowly, bit by bit, and measure the heat capacity as the temperature rises.

But can we observe W? We cannot observe it, but we can sometimes calculate it. Boltzmann's hypothesis goes completely against Mach's philosophy. But Boltzmann's hypothesis works brilliantly.

The positivists were against all notions that could not be tested experimentally. One hypothesis they disliked was the idea that matter is made of atoms. The concepts of atoms and molecules come from chemistry. John Dalton introduced the idea that the atoms of a chemical element are completely alike in all their properties. Chemical reactions involve the combination of different atoms to form bound states called molecules. It was not easy to quantify this notion until Avogadro introduced the idea that ideal gases at the same temperature and pressure contain equal numbers of molecules.

One of the earliest theories based on the notion of atoms is the kinetic theory of gases. Heat in gases is regarded as molecular motion. Simple kinetic theory shows that the pressure of an ideal gas is related to the average kinetic energy of the molecules. Maxwell worked out a theory of collisions and introduced the notion that there is a range of molecular speeds.

Boltzmann went one stage further. He introduced the idea that molecular disorder is related to the entropy of the system. Here were two ideas that Mach disliked: the idea of molecules which could not be seen and the idea of the disorder (through the large number of independent microscopic descriptions, W) which could not be measured. Boltzmann's ideas were not widely accepted in his lifetime.

Science can allow for the presence of concepts which cannot be measured, as long as there is no contradiction with experiment. Our preference is to introduce as few concepts as possible. 'It is vain to do with more what can be done with less.' This notion is called *Occam's razor*. Remove all concepts that can be shown to be unnecessary. For example, the concept of an aether is unnecessary— electromagnetic waves can travel without a medium. The positivists' idea was to use Occam's razor as a scythe cutting down unnecessary ideas, reducing speculation to a minimum. A valuable service unless the ideas that are cut down are correct.

Boltzmann's hypothesis was needed in order to give some theoretical basis for thermodynamics. We need a theory to explain the experimental results that are encapsulated in the laws of thermodynamics. Boltzmann's hypothesis does this brilliantly. But it was rejected by the positivists out of hand.

The current view is that Mach's philosophy is too extreme, although at the time it provided a useful service. When Boltzmann put his hypothesis forward Mach's view held sway. Mach's philosophy had a big influence on the development of quantum mechanics made by Heisenberg. The philosophy had an enormous influence on Einstein when he formulated the theory of relativity: Einstein only considered those things which people could observe in their frame of reference. Observables are one of the central ideas of quantum mechanics. We construct theories based on what people can measure.

Boltzmann's hypothesis contains a quantity we cannot measure, W, and so it is contrary to Mach's positivist philosophy. Perhaps this conflict with Mach (and also with Ostwald) contributed to the bout of depression that caused Boltzmann to commit suicide. Yet his hypothesis is now fully accepted. Here are two quotations from Steven Weinberg (1993): 'In rejecting [Statistical Mechanics] the positivists were making the worst sort of mistake a Scientist can make: not recognizing success when it happens.' and 'Despite its value to Einstein and Heisenberg, positivism has done as much harm as good'. Weinberg also quotes an extract from a letter which Einstein wrote to Heisenberg: 'Perhaps I did use such a philosophy earlier [positivism], and also wrote it, but it is nonsense all the same'.

4.7 Problems

1. Take 10^{-6} J of heat from a system at a temperature of 300 K and add it to a system at 299 K. What is the total entropy change of the two systems, and by what factor does the number of accessible states increase?

2. An isolated, macroscopic system at temperature 300 K absorbs a photon from the visible part of the spectrum (say $\lambda = 500$ nm). Obtain a value for the relative increase, $\Delta W/W$, in the number of accessible states of the system.

3. If 10^{-7} J of heat is added to a system which is so large that its temperature does not change, by what factor will the number of accessible states of the system increase if the temperature is 30 K?

4. How much heat must be added to a system at 298 K for the number of accessible states to increase by a factor of 10^6?

5. Suppose that W varies as $W = A\,e^{\gamma(VU)^{1/2}}$ with γ a constant. How does the temperature vary as a function of U? For what value of the energy is the temperature zero?

6. The entropy of an ideal paramagnet is give by $S = S_0 - CU^2$ where U is the energy, which can be positive or negative, and C is a positive constant. Determine the equation for U as a function of T and sketch your result.

7. Two systems A and B are arranged so that their total volume is constant, but their individual volumes can change. Calculate the standard deviation of fluctuations in the volume in system A for constant energy. (Hint: see section 4.5 for the analogous calculation of the fluctuation in energy.)

8. A paramagnet in one dimension can be modelled as a linear chain of $N+1$ spins. Each spin interacts with its neighbours in such a way that the energy is $U = n\epsilon$ where n is the number of domain walls separating regions of up spins from down as shown by a vertical line in the representation below.

$$\uparrow\uparrow\uparrow\uparrow\uparrow\uparrow|\downarrow\downarrow\downarrow\downarrow|\uparrow\uparrow\uparrow\uparrow\uparrow|\downarrow\downarrow|\uparrow\uparrow\uparrow\uparrow|\downarrow\downarrow\downarrow\downarrow\downarrow\downarrow$$

How many ways can n domain walls be arranged? Calculate the entropy, $S(U)$, and hence show that the energy is related to the temperature as

$$U = \frac{N\epsilon}{\exp(\epsilon/k_{\mathrm{B}}T) + 1}.$$

Sketch the energy and the heat capacity as a function of temperature, paying particular attention to the asymptotic behaviour for low and for high temperatures.

9. N atoms are arranged to lie on a simple cubic crystal lattice. Then M of these atoms are moved from their lattice sites to lie at the interstices of the lattice, that is points which lie centrally between the lattice sites. Assume that the atoms are placed in the interstices in a way which is completely independent of the positions of the vacancies. Show that the number of ways of taking M atoms from lattice sites and placing them on interstices is $W = (N!/M!(N-M)!)^2$ if there are N interstitial sites where displaced atoms can sit.

Suppose that the energy required to move an atom away from its lattice site into any interstitial site is ϵ. The energy is $U = M\epsilon$ if there are M interstitial atoms. Use the formula for S in terms of W and the formula which defines the temperature to obtain

$$\frac{M}{N} = \frac{1}{e^{\epsilon/2k_{\mathrm{B}}T} + 1}.$$

10. The entropy of black body radiation is given by the formula $S = \frac{4}{3}\sigma V^{1/4}U^{3/4}$ where σ is a constant. Determine the temperature of the radiation and show that $PV = U/3$.

11. The entropy of a two-dimensional gas of particles in an area A is given by the expression

$$S = Nk_{\mathrm{B}}\left\{\ln\left(\frac{A}{N}\right) + \ln\left(\frac{mU}{2\pi\hbar^2 N}\right) + 2\right\}$$

where N is the number of particles and U is the energy of the gas. Calculate the temperature of the gas and the chemical potential. Refer to eqn (2.9.7) for the definition of the chemical potential.

12. Consider a system of N non-interacting particles, each fixed in position and carrying a magnetic moment. The system is immersed in a magnetic

induction field **B**. Each particle may then exist in one of the two energy states: a spin-down state of energy $-\mu B$ and a spin-up state of energy μB. Treat the particles as being distinguishable.

Derive a formula for the entropy in terms of the number of particles in the upper state, n. Sketch $S(n)$. Find the value of n for which $S(n)$ is a maximum. What is the relation between S and the temperature? By treating the energy as a continuous variable, show that the system can have negative temperatures. If a system with a negative temperature makes thermal contact with one with a positive temperature, which way will the heat flow?

13. A system is made up of N oscillators (with N very large) each with energy levels n with $n = 0, 1, 2, 3, \ldots$. The total energy of the system is U so there can be a division of the energy into $U/\hbar\omega$ quanta of energy. ($U/\hbar\omega$ is an integer.) These quanta of energy must be divided amongst the oscillators somehow. The number of arrangements is

$$W = \frac{(N - 1 + U/\hbar\omega)!}{(N - 1)!(U/\hbar\omega)!}.$$

From this expression obtain a formula for the temperature of the system as a function of U, and hence calculate the average energy per oscillator as a function of temperature.

14. A system of N distinguishable particles is arranged such that each particle can exist in one of two states: one has energy ε_1, the other has energy ε_2. The populations of these states are n_1 and n_2 respectively ($N = n_1 + n_2$). The system is placed in contact with a heat bath at temperature T. A simple quantum process occurs in which the populations change: $n_2 \rightarrow n_2 - 1$ and $n_1 \rightarrow n_1 + 1$ with the energy released going into the heat bath. Calculate the change in the entropy of the two-level system, and the change in the entropy of the heat bath. If the process is reversible, what is the ratio of n_2 to n_1?

5

The canonical ensemble

On the distribution in phase called canonical, in which the index of probability is a linear function of energy.
                                                                    J. Willard Gibbs

5.1 A system in contact with a heat bath

In the last chapter we invited you to imagine an ensemble of systems all prepared in the same way: they all have the same number of particles, energy, volume, shape, magnetic field, electric field, et cetera. The probability that a system is in a particular quantum state is given by the fraction of the ensemble in this state. We assume that the fraction equals the classical probability so that all the quantum states are equally probable. This ensemble is called the *microcanonical ensemble*, following the nomenclature introduced by Willard Gibbs.

In this chapter we ask you to imagine a similar ensemble of systems all prepared in the same way with the same number of particles, volume, shape, magnetic field, electric field, and so on, but this time *the energy of each system is not constant*. Energy can be passed from one system to its neighbours. Therefore the energy of each system fluctuates. Each system is in thermal contact with the remainder which acts as a heat bath for the system. This ensemble is called the *canonical ensemble*.

In many experiments involving condensed matter the system is placed in contact with a heat bath. The heat bath is any system with a large heat capacity so that its temperature does not change when heat energy is added to it. The system and heat bath eventually reach thermal equilibrium at the temperature of the heat bath. The temperature of the system is stabilized, which is why many experiments are done this way. When a system is in contact with a heat bath it is said to be in its canonical state. According to *Chambers 20th Century Dictionary*: canon: n. a law or rule, esp. in ecclesiastical matters: a general rule or principle: a standard or criterion: ... . Canonical, in this context, means standard.

The combined system, the system being studied together with the heat bath, is thermally isolated so that its total energy, U_T, is constant. The combined system can be treated using the techniques developed in Chapter 4. The entropy of the combined system is given by $k_B \ln(W)$ where W is the number of accessible quantum states of the combined system. The system we want to study, let us call it system A, has energy U_A and the heat bath has energy

$$U_R = U_T - U_A.$$

The number of accessible quantum states of the combined system, $W(U_A)$, is a function of U_A, the energy of system A. The volume and number of particles in system A are constant but the energy is allowed to vary. We write

$$W(U_A) = W_A(U_A) \times W_R(U_T - U_A) \tag{5.1.1}$$

where $W_R(U_T - U_A)$ is the number of accessible quantum states of the heat bath and $W_A(U_A)$ is the number of accessible quantum states of system A when it has energy U_A.

Suppose that, out of all the quantum states of system A, a particular quantum state is chosen, ψ_i. This quantum state is an eigenstate of the energy operator with eigenvalue E_i. Hence $U_A = E_i$ and $W_A(E_i) = 1$ since one quantum state is chosen. With system A in quantum state ψ_i the total number of quantum states for the combined system is

$$W(E_i) = 1 \times W_R(U_T - E_i). \tag{5.1.2}$$

The temperature of the heat bath is defined by the equation

$$\frac{1}{k_B T} = \left(\frac{\partial \ln(W_R)}{\partial U_R} \right)_V. \tag{5.1.3}$$

When energy is added to the heat bath the temperature does not change; this is the property which defines a heat bath. Integration of eqn (5.1.3) gives

$$W_R = \gamma e^{U_R/k_B T} \tag{5.1.4}$$

where γ is an integration constant. It follows that

$$W(E_i) = W_R(U_T - E_i) = \gamma e^{(U_T - E_i)/k_B T}. \tag{5.1.5}$$

The basic postulate of statistical mechanics is that all accessible quantum states are equally probable. The total number of accessible quantum states of the combined system is

$$N_{total} = \sum_j W(E_j) = \gamma e^{U_T/k_B T} \sum_j e^{-E_j/k_B T} \tag{5.1.6}$$

where the sum is over all quantum states of system A. The classical probability that the system is in the quantum state ψ_i is equal to the number of quantum states associated with ψ_i divided by the total number of states:

$$p_i = \frac{W(E_i)}{\sum_j W(E_j)} = \frac{e^{-E_i/k_B T}}{\sum_j e^{-E_j/k_B T}} = \frac{e^{-E_i/k_B T}}{Z}. \tag{5.1.7}$$

Equation (5.1.7) is called the *Boltzmann probability distribution*; it tells us the probability that a particular quantum state of the system is occupied. The quan-

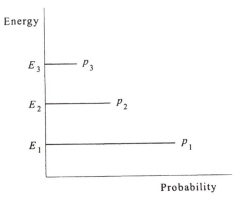

Fig. 5.1 The Boltzmann probability distribution with the energy plotted on the vertical axis and the probability on the horizontal axis.

tum states of the heat bath have dropped out completely; the heat bath defines the temperature of the system but has no other effect.

At low temperatures only the quantum states of low energy are occupied; as the temperature is increased, quantum states of higher energy start to have a significant probability of being occupied. Figure 5.1 illustrates a Boltzmann probability distribution with the probabilities drawn along the horizontal axis and the energies along the vertical. For infinitely high temperatures all the probabilities are the same. The distribution of probability of system A being in its different quantum states changes with temperature.

The Boltzmann probability distribution can be used for any system in contact with a heat bath provided the number of particles in the system is constant. The system could be a single particle such as an electron in an atom with a set of single-particle energy states, or it could be a large object containing a lot of particles, such as block of gold.

5.2   The partition function

The denominator in eqn (5.1.7) is called the *partition function*, and is denoted by the symbol Z. The partition function is given by the equation

$$Z = \sum_j e^{-E_j/k_B T} \tag{5.2.1}$$

where the sum is taken over all the different quantum states of the system. Z is an abbreviation of the German word 'Zustandssumme' which means sum-over-states. When we work out the partition function that is precisely what we do: sum over quantum states.

As an example, consider a very simple system which has three energy levels, each of which is *non-degenerate* (the word non-degenerate implies that there is only one quantum state for each level). The energy levels have energies $E_1 = 0$, $E_2 = 1.4 \times 10^{-23}$ J, $E_3 = 2.8 \times 10^{-23}$ J. If the heat bath has a temperature of 2 K the partition function is

$$Z = e^0 + e^{-1/2} + e^{-1} = 1.9744$$

since $E_2/k_BT = 0.5$ and $E_3/k_BT = 1$. The probability of being in the state of energy E_1 is 0.506, of energy E_2 is 0.307, and of energy E_3 is 0.186.

In eqn (5.2.1) the sum in the equation for Z is over the different quantum states of system A. An alternative way of writing Z is to sum over energy levels, E_n. If there are g_n quantum states all with the energy E_n, then Z can be written as

$$Z = \sum_n g_n e^{-E_n/k_BT} \tag{5.2.2}$$

where the sum is over all the different energy levels of system A. The quantity g_n is called the *degeneracy* of the energy level.

As an example, consider a system which has two energy levels, one of energy $E_0 = 0$ with a degeneracy of two, and the other of energy $E_1 = 1.3806 \times 10^{-23}$ J with a degeneracy of three. The partition function at a temperature of 1 K is

$$Z = 2 e^0 + 3 e^{-1} = 3.104$$

since $E_1/k_BT = 1$.

5.3 Definition of the entropy in the canonical ensemble

The system under study, system A, does not have a constant energy when it is in contact with a heat bath. The energy of the system fluctuates. Successive measurements of the energy will find it in different energy states. The relative probability of making a measurement on the system and finding the system in one of the many quantum states is given by the Boltzmann probability distribution. The probabilities are not all equal.

We can calculate the entropy of the system by using the notion of an ensemble of systems. Imagine $(M - 1)$ replica systems in contact with each other and with the original system; these replicas act as a heat bath if we take M to be huge. Imagine a block of gold as the system, and the block of gold and its replicas are all stacked together in a pile, as illustrated in Fig. 5.2. We will call this the *Gibbs ensemble*. Because of the contact between them, energy can be transferred from one system to its neighbours, which in turn transfer energy to their neighbours, and so on. The collection of systems is thermally isolated from the rest of the universe, so that the collection can be treated using the techniques of the microcanonical ensemble.

Each replica system is identical to the rest, but the replicas are taken to be objects which can easily be distinguished from each other by their position. Perhaps the systems are large classical objects containing lots of particles. We

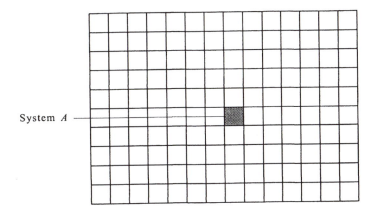

Fig. 5.2 System A is in contact with $(M - 1)$ replica systems which act as a heat bath. System A can gain or lose heat to its neighbours which in turn can gain or lose heat to theirs.

shall describe the state of each system quantum mechanically by saying that it is in a particular quantum mechanical state, ψ_i.

Let the number of systems in the quantum state ψ_i be n_i out of the total number of systems M. There are n_1 systems in quantum state ψ_1, n_2 systems in quantum state ψ_2, and so on. The number of ways of arranging n_1 systems to be in quantum state ψ_1, n_2 in quantum state ψ_2, $\ldots$, is

$$W = \frac{M!}{n_1! n_2! n_3! \ldots}. \tag{5.3.1}$$

Let us take M to be so enormous that all the n_i are huge. The entropy for the M systems is given by

$$S_M = k_{\mathrm{B}} \ln(W) = k_{\mathrm{B}} \left(M \ln(M) - \sum_i n_i \ln(n_i) \right) \tag{5.3.2}$$

where we have used Stirling's approximation, as well as the relation $M = \sum_i n_i$. By using the identity $M \ln(M) = \sum_i n_i \ln(M)$, the equation for S_M can be rewritten as

$$S_M = -k_{\mathrm{B}} M \left(\sum_i \left(\frac{n_i}{M} \right) \ln \left(\frac{n_i}{M} \right) \right). \tag{5.3.3}$$

As M tends to infinity so the ratio n_i/M tends to the probability of finding the system in the quantum state ψ_i. Hence the entropy per system can be written as

$$S = \frac{S_M}{M} = -k_{\mathrm{B}} \sum_i p_i \ln(p_i) \tag{5.3.4}$$

the formula for the average entropy of a system in the canonical ensemble, that is when it is contact with a heat bath.

This expression for the entropy simplifies when all the probabilities are equal. If there are W quantum states, the probabilities are $p_i = W^{-1}$ and so

$$
\begin{aligned}
S &= -k_{\mathrm{B}} \sum_{i=1}^{W} p_i \ln(p_i) \\
&= -k_{\mathrm{B}} \sum_{i=1}^{W} \frac{1}{W} \ln\left(\frac{1}{W}\right) = k_{\mathrm{B}} \ln(W),
\end{aligned}
$$

which is Boltzmann's formula for the entropy.

5.4   The bridge to thermodynamics through Z

It may seem that the partition function is just the normalization constant needed to get the sum of the probabilities equal to one. But it is much more important than that. Its importance arises because it enables us to make a direct connection between the quantum states of the system and its thermodynamic properties, properties such as the free energy, the pressure, and the entropy. We can calculate all the thermodynamic properties of a system from a knowledge of the partition function.

Let us prove this connection. In the canonical ensemble the probability of being in the quantum state ψ_i is

$$
p_i = \frac{\mathrm{e}^{-E_i/k_{\mathrm{B}}T}}{Z}. \tag{5.1.7}
$$

When the probabilities are not equal the entropy is given by the formula

$$
S = -k_{\mathrm{B}} \sum_i p_i \ln(p_i). \tag{5.3.4}
$$

The following manipulation—or is it legerdemain?— is worth careful study. No trickery is used although the result appears as if by magic. By writing

$$
\ln(p_i) = -E_i/k_{\mathrm{B}}T - \ln(Z)
$$

and substituting $\ln(p_i)$ in eqn (5.3.4) we get

$$
S = k_{\mathrm{B}} \sum_i p_i \left(\frac{E_i}{k_{\mathrm{B}}T} + \ln(Z)\right). \tag{5.4.1}
$$

The sum over quantum states in eqn (5.4.1) has a part due to $\ln(Z)$ and a part containing $E_i/k_{\mathrm{B}}T$ which involves the mean or average energy

$$
\bar{U} = \sum_i p_i E_i. \tag{5.4.2}
$$

Because the sum of the probabilities is one we get from eqns (5.4.1) and (5.4.2)

$$S = \frac{\bar{U}}{T} + k_B \ln(Z)$$

which can be rewritten as

$$\bar{U} - TS = -k_B T \ln(Z). \tag{5.4.3}$$

The quantity $(\bar{U} - TS)$ is the average value of the Helmholtz free energy, F, a function of state. Equation (5.4.3) tells us that we can relate the free energy to the partition function Z as

$$F = -k_B T \ln(Z). \tag{5.4.4}$$

This equation is one of the foundation stones of statistical mechanics; engrave it on your memory.

Once we know all the quantum states of a system and their energy eigenvalues, we can calculate Z and hence obtain the free energy of the system. Z acts as a bridge connecting the microscopic energy states of the system to the free energy and thence to all the large-scale properties of a system.

Before proceeding, you should re-read this section and try to digest it thoroughly for it is of central importance to the rest of the subject. In particular you should remember these equations:

$$Z = \sum_j e^{-E_j/k_B T} \tag{5.2.1}$$

$$F = -k_B T \ln(Z). \tag{5.4.4}$$

We can measure the energy levels and their degeneracies using the techniques of spectroscopy, and hence calculate Z for any temperature. The free energy can be obtained by lots of measurements of the heat capacity: start at the lowest temperature possible and keep track of the temperature as energy is added slowly to the system. In this way we can measure the heat capacity and the internal energy as a function of temperature, and hence obtain entropy and then the free energy. We can test to see whether our understanding of the energy levels of the system is consistent with the measurement of the free energy. Whenever it has been thoroughly tested it is found to work; the theory is not just a figment of our imagination.

5.5 The condition for thermal equilibrium

The Boltzmann probability distribution occurs when a system is in contact with a heat bath. The requirement that the probability distribution satisfies eqn (5.1.7) places a severe constraint on a system. If the distribution is not the Boltzmann probability distribution, then the system cannot be in thermal equilibrium with the heat bath. How do we tell if this is the case?

Suppose the system of N atoms has a series of energy eigenvalues with the ground state having an energy $E_0 = 0$. The system is in contact with a heat bath at temperature T so that the probability of finding it in quantum state ψ_i is given by the Boltzmann probability distribution:

$$p_i = \frac{e^{-E_i/k_B T}}{Z}. \tag{5.1.7}$$

The ratio of the probability of being in quantum state ψ_i to that of being in the ground state is

$$\frac{p_i}{p_0} = e^{-E_i/k_B T}. \tag{5.5.1}$$

This equation must be correct for all the quantum states, for otherwise the distribution would not be the Boltzmann distribution. It follows that the temperature of the heat bath can be found from the equation

$$T = -\frac{E_i}{k_B \ln(p_i/p_0)} \tag{5.5.2}$$

provided that the value of T so obtained is the same for all states ψ_i. We can use eqn (5.5.2) to calculate the temperature from the probabilities.

As an example, suppose there are four non-degenerate energy levels with energies $0, 1.9 \times 10^{-20}$ J, 3.6×10^{-20} J, and 5.2×10^{-20} J. The system is observed repeatedly and it is found that the probabilities of being in these levels are $p_0 = 0.498$, $p_1 = 0.264$, $p_2 = 0.150$, and $p_3 = 0.088$. Is the system in thermal equilibrium and if so what is the temperature?

If the energy levels are non-degenerate we can use eqn (5.5.2) to estimate the temperature. This gives values of 2168 (using p_1/p_0), 2173 (using p_2/p_0), and 2173 (using p_3/p_0). Since the numbers are the same, within the accuracy of the probabilities that are given, the system can be assumed to be in thermal equilibrium at a temperature of about 2172 K.

5.6 Thermodynamic quantities from ln(Z)

Consider a small reversible change in F in which the number of particles is kept constant. Such a change can be written as

$$dF = -P\,dV - S\,dT. \tag{2.6.9}$$

Hence

$$P = -\left(\frac{\partial F}{\partial V}\right)_T \tag{5.6.1}$$

and

$$S = -\left(\frac{\partial F}{\partial T}\right)_V. \tag{5.6.2}$$

These are straightforward thermodynamic relations. By using the expression for F given by eqn (5.4.4) we can obtain both the entropy and the pressure from $\ln(Z)$:

$$P = k_{\mathrm{B}} \left(\frac{\partial (T \ln(Z))}{\partial V} \right)_T, \tag{5.6.3}$$

$$S = k_{\mathrm{B}} \left(\frac{\partial (T \ln(Z))}{\partial T} \right)_V. \tag{5.6.4}$$

Provided we can work out the energies of the quantum states, and how they vary with the volume of the system, we can calculate the pressure and the entropy of the system. This piece of mathematics is very important, for it shows that there is an intimate connection between the energy levels of the system and thermodynamic quantities such as the entropy and the pressure.

From eqn (5.6.1) we can also get the *isothermal compressibility*, K, a measure of how the volume of a system changes when the pressure is altered at constant temperature:

$$K^{-1} = -V \left(\frac{\partial P}{\partial V} \right)_T = V \left(\frac{\partial^2 F}{\partial V^2} \right)_T. \tag{5.6.5}$$

The inverse of the compressibility is proportional to the second derivative of the free energy with respect to volume. We can also obtain the heat capacity at constant volume from eqns (2.5.11) and (5.6.2):

$$C_V = T \left(\frac{\partial S}{\partial T} \right)_V = -T \left(\frac{\partial^2 F}{\partial T^2} \right)_V. \tag{5.6.6}$$

The heat capacity at constant volume is proportional to the second derivative of the free energy with respect to temperature. Quantities which are proportional to the second derivative of the free energy with respect to some variable are called *response functions*. They can easily be calculated from $k_{\mathrm{B}} T \ln(Z)$. The pressure, entropy, heat capacity, and isothermal compressibility all can be calculated once we know $\ln(Z)$.

Another quantity that we want to calculate is the average internal energy of the system, $\bar{U} = TS + F$. This can be calculated using eqns (5.4.4) and (5.6.4). We get

$$\bar{U} = k_{\mathrm{B}} T^2 \left(\frac{\partial \ln(Z)}{\partial T} \right)_V. \tag{5.6.7}$$

At this stage it is a good idea to see whether we can calculate all the thermodynamic properties of a system starting from a set of energy levels. For this purpose we shall study two examples. First we shall calculate the thermal properties of a two-level system, something that was studied in Chapter 4, and obtain the same expressions for the thermodynamic functions. Then we shall give an elementary quantum mechanical description of a single particle in a box, calculate its partition function, and then obtain the equation of state for a gas of particles, the basis of the gas laws.

These examples are rather mathematical, for they involve intricate differentiation of complicated mathematical expressions. But the most important thing is not the mathematics; it is the fact that there is a systematic way of calculating the thermodynamic properties of a system. The steps are simple: select a model for the system; calculate the partition function from the energy levels of the model system; deduce the free energy; differentiate to get the entropy and pressure, and, if desired, the heat capacity and compressibility; fit to the experimental data to see if the model is valid.

5.7 Two-level system

The simplest possible quantum system is one in which there are just two non-degenerate quantum states with energies ε and $-\varepsilon$. An example is a single particle of spin $1/2$ localized on a lattice site and placed in a magnetic induction field B. The energy eigenvalues of the particle are $+\mu B$ and $-\mu B$ where μ is the magnetic moment of the particle. Imagine the system to be a single particle with spin $1/2$. The particle is replicated M times to build up an ensemble of systems. We assume that each particle with its spin is distinguishable from the rest.

The partition function of a single two-level system is just

$$Z = e^{\varepsilon/k_B T} + e^{-\varepsilon/k_B T} \tag{5.7.1}$$

so the free energy is

$$F = -\varepsilon - k_B T \ln\left(1 + e^{-2\varepsilon/k_B T}\right). \tag{5.7.2}$$

The entropy is then obtained from F

$$
\begin{aligned}
S &= -\left(\frac{\partial F}{\partial T}\right)_V \\
&= k_B \ln\left(1 + e^{-2\varepsilon/k_B T}\right) + \frac{2\varepsilon/T}{e^{2\varepsilon/k_B T} + 1}.
\end{aligned} \tag{5.7.3}
$$

The heat capacity at constant volume is

$$C_V = T\left(\frac{\partial S}{\partial T}\right)_V = \frac{\varepsilon^2}{k_B T^2 \cosh^2\left(\varepsilon/k_B T\right)}. \tag{5.7.4}$$

As the temperature increases the quantity $x = \varepsilon/k_B T$ becomes small, and $\cosh(x)$ tends to one. The heat capacity for high temperatures dies away as T^{-2}. In contrast, for low temperatures the heat capacity dies away to zero as $e^{-2\varepsilon/k_B T}$. In between the heat capacity has a maximum. In Fig. 5.3 we show the temperature dependence of the heat capacity of a single two-level system. This is commonly referred to as the *Schottky heat capacity*.

The average internal energy, $(TS + F)$, of a two-level system is

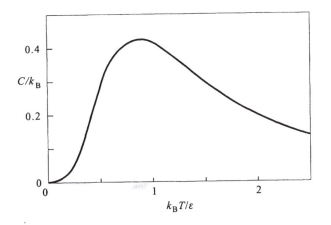

Fig. 5.3 The diagram shows the heat capacity of a single two-level system in units of k_B, as a function of $k_B T/\varepsilon$. For low temperatures C dies away to zero; at high temperatures it dies away as T^{-2}; in between the heat capacity has a maximum. The shape of this curve is characteristic of a Schottky heat capacity of a two-level system.

$$\bar{U} = -\varepsilon \tanh\left(\varepsilon/k_B T\right) \tag{5.7.5}$$

so for N two-level systems we get

$$\bar{U} = -N\varepsilon \tanh\left(\varepsilon/k_B T\right). \tag{5.7.6}$$

This is almost the same expression as we obtained in Chapter 4 (see eqn (4.4.5)). The difference is that in the canonical ensemble we are concerned with the average energy; in the microcanonical ensemble the energy of the system is fixed.

5.8 Single particle in a one-dimensional box

Imagine a system which is simply a box with a single particle in it. The box is replicated $M - 1$ times to build up the Gibbs ensemble. We assume that each particle is localized in its box and that each box is distinguishable from the rest. To determine the thermodynamics of each particle we need to list all the independent quantum states of the system, determine their energies, and calculate the partition function.

Let us start this programme by considering the quantum states of a particle of mass m moving along the x direction. *Schrödinger's wave equation* in one dimension is

$$-\frac{\hbar^2}{2m}\frac{\partial^2 \phi(x)}{\partial x^2} + V(x)\phi(x) = \varepsilon\phi(x) \tag{5.8.1}$$

where $V(x)$ is the potential energy, $\phi(x)$ is the single-particle wavefunction, and ε is the energy. The particle is confined in a one-dimensional box of length L.

The potential energy is taken to be zero in the range $0 < x < L$ and infinite elsewhere. As a result, the wavefunction has to satisfy the boundary conditions $\phi(x = L) = 0$ and $\phi(x = 0) = 0$. The solution of Schrödinger's equation which satisfies these boundary conditions is a *standing* wave of the form

$$\phi_n(x) = A \sin\left(\frac{n\pi x}{L}\right). \tag{5.8.2}$$

The first boundary condition is automatically satisfied by such a wavefunction; the second boundary condition is satisfied provided n is a positive, non-zero integer. Why do we exclude negative integers?

We want *independent solutions* of Schrödinger's equation corresponding to independent quantum states. If n is chosen to be a negative integer then the wavefunction is linearly dependent on the wavefunction for the corresponding positive integer. It is not an independent solution. For example the wavefunction

$$
\begin{aligned}
\phi_{-3}(x) &= A \sin\left(\frac{-3\pi x}{L}\right) \\
&= -A \sin\left(\frac{3\pi x}{L}\right) = -\phi_3(x)
\end{aligned}
$$

is the same as the wavefunction $\phi_3(x)$ apart from a minus sign. It is not an independent solution. We restrict the set of quantum states to those with n a positive integer so as to ensure that we only consider independent quantum states.

The energy eigenvalue for wavefunction $\phi_n(x)$ is

$$\varepsilon_n = \left(\frac{\hbar^2 \pi^2 n^2}{2mL^2}\right) \equiv \alpha n^2. \tag{5.8.3}$$

The quantity α sets the scale of energy. For an electron in a length of one centimetre the value of α is

$$\alpha = \frac{\left(1.054 \times 10^{-34}\right)^2 \pi^2}{2 \times 9.11 \times 10^{-31} \times 10^{-4}} \, J = 6.0 \times 10^{-34} \, J$$

or in terms of temperature

$$\frac{\alpha}{k_B} = 4.4 \times 10^{-11} \, K$$

which is very small. If we do experiments at a temperature of 300 K, the quantity $\gamma = \alpha/k_B T$ is about 1.5×10^{-13}. Changing L to a micron would increase γ by 10^8 to 1.5×10^{-5}. It is still small. γ is comparable to one only for much smaller lengths or for much lower temperatures (or both).

Our immediate objectives are to work out the partition function and then the thermodynamic properties. The partition function for translational motion in one dimension is

$$Z_{trans} = \sum_{n=1}^{\infty} e^{-\varepsilon_n / k_B T} = \sum_{n=1}^{\infty} e^{-\gamma n^2} \qquad (5.8.4)$$

where the factor γ is minute. Successive values of $e^{-\gamma n^2}$ are smaller than their predecessor by a tiny amount, so that the series converges very slowly indeed.

How are we to do the sum? The trick is to pretend the sum is an integral over a continuous variable, n (see Appendix B). Turning the sum into an integral is a very good approximation provided the length L is sufficiently large to make γ very small. Thus

$$Z_{trans} = \sum_{n=1}^{\infty} e^{-\gamma n^2} \approx \int_0^{\infty} e^{-\gamma n^2} dn$$

$$= \left(\frac{\pi}{4\gamma}\right)^{\frac{1}{2}} = \left(\frac{L^2 m k_B T}{2\pi \hbar^2}\right)^{\frac{1}{2}}. \qquad (5.8.5)$$

Here we have used the standard integral

$$\int_0^{\infty} e^{-x^2} dx = \frac{\sqrt{\pi}}{2}$$

which is proved in Appendix B, and used the variable $x = \sqrt{\gamma} n$. The final result can be written as $Z_{trans} = L/\lambda_D$ where

$$\lambda_D = \left(\frac{2\pi \hbar^2}{m k_B T}\right)^{\frac{1}{2}}. \qquad (5.8.6)$$

The quantity λ_D is called the *thermal de Broglie wavelength*. According to de Broglie, the wavelength, λ, of a particle is related to the momentum, p, as $\lambda = h/p$. There is a distribution of momentum of the particle due to thermal excitation, and therefore a distribution of wavelengths. The thermal de Broglie wavelength is the thermal average wavelength of a free particle of momentum p in a three-dimensional system.

Notice that the partition function becomes infinite in the mathematical limit that $\hbar$ tends to zero—taking $\hbar$ as a variable which can tend to zero is nothing more than a mathematical trick; physically $\hbar$ is a constant. If we take this limit, $\hbar \to 0$, first then Z_{trans} is infinite and the theory makes no sense. It is more sensible to keep $\hbar$ finite, to calculate the thermodynamics through the free energy, evaluate say a difference in entropy, and then to take the mathematical limit that $\hbar$ tends to zero at the end of the calculation. That was the way that physicists managed to get sensible answers before the theory of quantum mechanics was created.

The free energy of a single gas particle in a length L is

$$F = -k_B T \ln \left(\frac{L^2 m k_B T}{2\pi \hbar^2}\right)^{\frac{1}{2}}, \qquad (5.8.7)$$

the entropy is

$$S = -\left(\frac{\partial F}{\partial T}\right)_V = \frac{k_B}{2}\left(\ln\left(\frac{L^2 m k_B T}{2\pi\hbar^2}\right) + 1\right), \tag{5.8.8}$$

and so the contribution to the heat capacity is

$$C_V = T\left(\frac{\partial S}{\partial T}\right)_V = \frac{k_B}{2}. \tag{5.8.9}$$

Translational motion of a single particle in one dimension gives a contribution to the heat capacity of $k_B/2$.

5.9 Single particle in a three-dimensional box

The same method can be used to describe the quantum states of a particle in a cube of volume L^3. The wavefunction for a particle in a cube of length L is

$$\phi_i(x, y, z) = A\sin\left(\frac{n_1\pi x}{L}\right)\sin\left(\frac{n_2\pi y}{L}\right)\sin\left(\frac{n_3\pi z}{L}\right). \tag{5.9.1}$$

This wavefunction describes a *standing* wave in all three dimensions; a wave which vanishes when x, y, or $z = 0$ and where x, y, or $z = L$. Here the suffix i on the wavefunction $\phi_i(x, y, z)$ stands for the set of quantum numbers (n_1, n_2, n_3). In fact we could denote the quantum state by

$$\phi_i = |n_1, n_2, n_3\rangle \tag{5.9.2}$$

instead of using the wavefunction given by eqn (5.9.1)

The energy of a free particle of mass m involves these three quantum numbers n_1, n_2, and n_3. The quantum state is given by the solution of

$$-\frac{\hbar^2}{2m}\nabla^2\phi_i(x, y, z) + V(x, y, z)\phi_i(x, y, z) = \varepsilon_i\phi_i(x, y, z). \tag{5.9.3}$$

The potential energy is zero inside the box and infinite elsewhere. It follows that the three quantum numbers n_1, n_2, and n_3 are all positive integers, and that the energy eigenvalues are

$$\varepsilon_i = \frac{\hbar^2\pi^2}{2mL^2}\left(n_1^2 + n_2^2 + n_3^2\right). \tag{5.9.4}$$

The partition function for translational motion is

$$Z_{\text{trans}} = \sum_{n_1=1}^{\infty}\sum_{n_2=1}^{\infty}\sum_{n_3=1}^{\infty} e^{-\gamma(n_1^2 + n_2^2 + n_3^2)}. \tag{5.9.5}$$

The factor γ is the same as the one introduced in section 5.8. We can rewrite eqn (5.9.5) as

$$Z_{\text{trans}} = \left(\sum_{n_1=1}^{\infty} e^{-\gamma n_1^2} \right) \left(\sum_{n_2=1}^{\infty} e^{-\gamma n_2^2} \right) \left(\sum_{n_3=1}^{\infty} e^{-\gamma n_3^2} \right)$$

where each sum is the same as that in the one-dimensional calculation. Hence the partition function for translational motion of a single particle in a cube in three dimensions is

$$Z_{\text{trans}} = L^3 \left(\frac{mk_{\text{B}}T}{2\pi\hbar^2} \right)^{\frac{3}{2}}. \tag{5.9.6}$$

From Z_{trans} we can get all the thermodynamics of a single particle in a box of volume $V = L^3$. For example, the free energy of a single particle in a volume V is

$$
\begin{aligned}
F &= -k_{\text{B}}T \ln\left(Z_{\text{trans}} \right) \\
&= -k_{\text{B}}T \left\{ \ln\left(V \right) + \frac{3}{2} \ln\left(\frac{mk_{\text{B}}T}{2\pi\hbar^2} \right) \right\}.
\end{aligned} \tag{5.9.7}
$$

The pressure is

$$P = -\left(\frac{\partial F}{\partial V} \right)_T = \frac{k_{\text{B}}T}{V} \tag{5.9.8}$$

which can be rewritten as $PV = k_{\text{B}}T$. This is the formula for the contribution to the pressure of a single particle in a gas. For N particles the equation would be $PV = Nk_{\text{B}}T$ provided the particles do not interact with each other. When N is Avogadro's number, N_A, then we get

$$PV = N_A k_{\text{B}}T \tag{5.9.9}$$

the equation of state for an ideal gas as long as we put $R = N_A k_{\text{B}}$.

We can also calculate the contribution to the heat capacity and entropy of a single gas particle in a volume V:

$$S = -\left(\frac{\partial F}{\partial T} \right)_V = k_{\text{B}} \left(\ln\left(V \right) + \frac{3}{2} \ln\left(\frac{mk_{\text{B}}T}{2\pi\hbar^2} \right) + \frac{3}{2} \right). \tag{5.9.10}$$

The contribution to the heat capacity of a single particle at constant volume is

$$C_V = T \left(\frac{\partial S}{\partial T} \right)_V = \frac{3k_{\text{B}}}{2}. \tag{5.9.11}$$

In three dimensions a single particle gives a contribution of $3k_{\text{B}}/2$ to the heat capacity. For N non-interacting particles we expect the heat capacity to be $3Nk_{\text{B}}/2$.

Although this calculation for a single particle in a box is correct, it needs careful handling if we want to use it as the basis of the statistical mechanics of a gas of N particles. Some of the formulae just need to be multiplied by N:

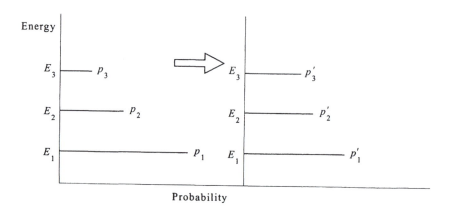

Fig. 5.4 The change in the Boltzmann probability distribution when heat is added to a system but no work is done on it. In this process the energy levels remain fixed but the probabilities change.

for example the heat capacity is given by $3Nk_B/2$. This is the not the case for the free energy and for the entropy. To calculate the thermodynamics of the gas properly we really need to evaluate the partition function for the N-particle system. We shall deal with this topic in Chapter 6.

5.10 Expressions for heat and work

Heat and work are very simple concepts in statistical mechanics: the addition of a small amount of heat corresponds to changing the probabilities slightly, keeping the energy levels fixed (Fig. 5.4); the addition of a small amount of work corresponds to changing the energy levels slightly keeping the probabilities fixed (Fig. 5.5). Any process which causes a shift in the energy levels can give rise to work. Examples of such processes include changing the volume, the magnetic induction field, or the electric field, for all these processes can change the energy levels of the system.

The average internal energy is defined to be

$$\bar{U} = \sum_i p_i E_i. \tag{5.4.2}$$

Let us suppose the system starts in its canonical state because it is in equilibrium with a heat bath at temperature T. The probabilities of the system being in each quantum state are given by

$$p_i = \frac{e^{-E_i/k_B T}}{Z} \tag{5.1.7}$$

where E_i is the energy eigenvalue for quantum state ψ_i.

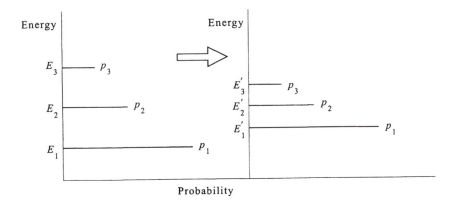

Fig. 5.5 The change in the Boltzmann probability distribution when work is done on a system but no heat is added to it. In this process the probabilities remain constant but the energy levels change.

An infinitesimal process carried out quasistatically leads to changes in the average energy of the system of

$$d\bar{U} = \sum_i dp_i\, E_i + \sum_i p_i\, dE_i \qquad (5.10.1)$$

for we can neglect the term $\sum_i dp_i\, dE_i$ as being second order in small quantities. The change in the average internal energy is the sum of these two terms when the changes are small.

Now we know that a small change in the internal energy for reversible processes can be written as $d\bar{U} = đ Q_{\text{rev}} + đ W_{\text{rev}}$ so it seems likely that one of the terms in eqn (5.10.1) can be identified as $đ Q_{\text{rev}}$ and the other as $đ W_{\text{rev}}$. We shall first identify $đ Q_{\text{rev}}$. In what follows we shall drop the subscript 'rev'.

Imagine that the *external parameters* are constant: the volume is kept constant, and the magnetic and electric fields are kept constant: all the constraints we can make on the system are imposed. The energy levels are functions of the external parameters alone. We cannot alter the external parameters which means that the energy levels are fixed, that is $dE_i = 0$. The only way we can increase the internal energy of the system is to add heat to the system. Therefore

$$đ Q = \sum_i dp_i\, E_i \qquad (5.10.2)$$

is the heat added in an infinitesimal process. The addition of heat occurs when the energy levels are fixed and the probabilities are altered somehow. Notice that in this case there is no restriction that dp_i is small.

It follows that the work done on the system is given by

$$\text{d}W = \sum_i p_i \, \text{d}E_i. \tag{5.10.3}$$

Reversible work is given by the weighted average of changes in the energy levels for constant probabilities. This statement gives us a very simple picture of the work done on a system. Notice that the change in the energy levels does not have to be small.

In general, both the energy levels and the probabilities are altered when a system undergoes a change. If the changes in the energy levels and the probabilities are not small then we cannot identify heat and work separately. The identification of heat and work can only be made for infinitesimal changes in the energy levels and probabilities.

We can use this idea to calculate the work done when the volume of a system changes. As an example, consider the change in the energy levels of gas atoms in a cube of volume $V = L^3$. The single-particle energy levels are

$$
\begin{aligned}
\varepsilon_i &= \frac{\hbar^2 \pi^2}{2mL^2} \left(n_1^2 + n_2^2 + n_3^2 \right) \\
&= \frac{\hbar^2 \pi^2}{2mV^{2/3}} \left(n_1^2 + n_2^2 + n_3^2 \right).
\end{aligned} \tag{5.9.4}
$$

We can change the energy level by changing the volume. Let us shrink the cube symmetrically in each dimension. This changes the energy levels, since

$$\frac{\partial \varepsilon_i}{\partial V} = -\frac{2}{3} \left(\frac{\hbar^2 \pi^2}{2mV^{5/3}} \right) \left(n_1^2 + n_2^2 + n_3^2 \right) = -\frac{2\varepsilon_i}{3V}.$$

By changing the volume of the system each of the single-particle energy levels changes. Since the total energy, E_i, of the quantum state is made up of a sum of these single-particle energy levels we find that

$$\frac{\partial E_i}{\partial V} = -\frac{2E_i}{3V}.$$

We can do work by changing the volume of the gas. The work is given by

$$
\begin{aligned}
-P \, \text{d}V &= \sum_i p_i \, \text{d}E_i = \sum_i p_i \left(\frac{\partial E_i}{\partial V} \right) \text{d}V \\
&= -\frac{2}{3V} \left(\sum_i p_i E_i \right) \text{d}V = -\frac{2\bar{U}}{3V} \text{d}V
\end{aligned}
$$

from which it follows that $PV = 2\bar{U}/3$.

Work can be done in other ways. Any quantity which couples to the energy levels can be used to do work. Suppose we have a gas of atoms in a magnetic induction field. The electrons in each atom couple to the magnetic field through their orbital motion and their spin. The coupling of the magnetic induction field to the energy levels of gas atoms is called the *Zeeman effect*. Such a system can do magnetic work. When the magnetic induction field changes, all the energy levels move, and work is done.

We could also apply electric fields to a gas of atoms. The shift in the energy of gas atoms with electric field is called the *Stark effect*. Any interaction which will couple to the energy levels can give rise to work. We could apply both an electric and a magnetic field to the system and do both electrical and magnetic work. The changes in energy levels induced by fields can be used to obtain expressions for different sorts of work; from these expressions we can derive equations of state for the system. Let us illustrate the technique with a simple example.

Suppose we have a system in which the energy levels depend on the volume of the sample, V, and on a uniform magnetic induction field which acts only along the z direction, $\mathbf{B} = \hat{\mathbf{k}} B_z$. (Here $\hat{\mathbf{k}}$ is a unit vector along the z direction.) Infinitesimal changes in the energy level, $E_i(V, B_z)$, arise from two processes: we can change the volume keeping the magnetic induction constant, or we can change the magnetic induction and keep the volume constant. For small changes in both quantities

$$dE_i = \left(\frac{\partial E_i}{\partial V}\right) dV + \left(\frac{\partial E_i}{\partial B_z}\right) dB_z. \tag{5.10.4}$$

The work now has two terms

$$đW = \sum_i p_i \left\{ \left(\frac{\partial E_i}{\partial V}\right)_{B_z} dV + \left(\frac{\partial E_i}{\partial B_z}\right)_V dB_z \right\}. \tag{5.10.5}$$

Since the first term involves a change in volume it must be just $-P\,dV$; the second term is the magnetic work

$$đW = \int \mathbf{H}.d\mathbf{B}\, d^3r = \sum_i p_i \left(\frac{\partial E_i}{\partial B_z}\right)_V dB_z. \tag{5.10.6}$$

The magnetic field can be written as $\mathbf{H} = (\mathbf{B}/\mu_0) - \mathbf{M}$ where $\mathbf{M}$ is the magnetic moment per unit volume, or magnetization. Suppose, for simplicity, that the magnetic field $\mathbf{H}$ as well as $\mathbf{B}$ is uniform over space, and lies along the z direction. Then

$$\sum_i p_i \left(\frac{\partial E_i}{\partial B_z}\right)_V dB_z = \frac{1}{\mu_0} \int B_z\, dB_z\, d^3r - \int M_z\, dB_z\, d^3r. \tag{5.10.7}$$

This equation can be used to calculate the magnetization of a simple system. It is useful to exclude from the energy levels E_i the energy associated with the magnetic induction field, that is the first term on the right-hand side of eqn (5.10.7). This energy would be present even if the magnetic material were absent. Thus we get

$$M_z = -\frac{1}{V}\sum_i p_i \left(\frac{\partial E_i}{\partial B_z}\right)_V.$$ (5.10.8)

For a single spin $1/2$ system with energy levels μB_z and $-\mu B_z$ a simple calculation using the Boltzmann probability distribution gives the magnetization from one spin as

$$M_z = \frac{\mu}{V}\tanh\left(\frac{\mu B_z}{k_B T}\right).$$

For N spin $1/2$ particles we recover the same formula for the magnetization as we did in Chapter 4.

The great advantage of using the canonical ensemble is seen when we calculate the magnetization of a spin $3/2$ particle with energy levels of say $3g\mu B_z/2$, $g\mu B_z/2$, $-g\mu B_z/2$, and $-3g\mu B_z/2$. Equation (5.10.8) gives

$$\begin{aligned}
M_z &= \frac{3g\mu}{2ZV}e^{3g\mu B_z/2k_B T} + \frac{g\mu}{2ZV}e^{g\mu B_z/2k_B T}\\
&\quad - \frac{g\mu}{2ZV}e^{-g\mu B_z/2k_B T} - \frac{3g\mu}{2ZV}e^{-3g\mu B_z/2k_B T}
\end{aligned}$$

with Z the partition function

$$Z = e^{3g\mu B_z/2k_B T} + e^{g\mu B_z/2k_B T} + e^{-g\mu B_z/2k_B T} + e^{-3g\mu B_z/2k_B T}.$$

It is simple doing the calculation using eqn (5.10.8), whereas the problem is much harder using the techniques of the microcanonical ensemble. (If you doubt this statement try it!)

5.11   Rotational energy levels for diatomic molecules

Consider a molecule, such as carbon monoxide, made up of two different atoms, one carbon and one oxygen separated by a distance d. Such a molecule can exist in quantum states of different orbital angular momentum with each state having energy

$$\varepsilon_l = \frac{\hbar^2 l(l+1)}{2I}$$ (5.11.1)

as shown in Appendix C. Here $I(=\mu d^2)$ is the moment of inertia of the molecule about an axis through its centre of mass, and $l = 0, 1, 2, 3, \dots$ is the quantum number associated with the orbital angular momentum. Each energy level of the rotating molecule has a degeneracy $g_l = (2l+1)$. This means for example that there is one $l = 0$ state, three $l = 1$ states, five $l = 2$ states, et cetera. The partition function is given by

$$Z = \sum_{l=0}^{\infty}(2l+1)\,e^{-\hbar^2 l(l+1)/2I k_B T}.$$ (5.11.2)

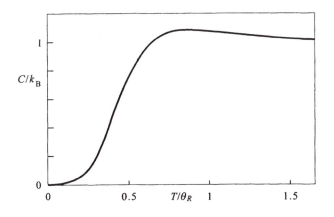

Fig. 5.6 The contribution to the heat capacity in units of k_B from the rotational states of a single diatomic molecule as a function of T/θ_R.

For large values of $\hbar^2/Ik_BT$ only the first few terms in the summation are important; we can truncate the series after the first two terms

$$Z = 1 + 3\,e^{-\hbar^2/Ik_BT} + \cdots.$$

By expanding

$$\ln(Z) = \ln\left(1 + 3\,e^{-\hbar^2/Ik_BT}\right) = 3\,e^{-\hbar^2/Ik_BT} + \cdots$$

we get

$$F = -3k_BT\,e^{-\hbar^2/Ik_BT} + \cdots.$$

It is then straightforward to calculate the entropy and the heat capacity, both of which die away exponentially as the temperature decreases. The rotational states with l greater than 0 are said to be frozen out as the temperature is lowered.

For very small values of $\hbar^2/Ik_BT$ (that is for high temperatures) the sum can be replaced with little error by an integral. We put $y = l(l+1)$ and treat y as a continuous variable. Since $dy = (2l+1)dl$, we find

$$Z \approx \int_0^\infty e^{-\hbar^2 y/2Ik_BT}\,dy = \frac{2Ik_BT}{\hbar^2}.$$

Consequently the free energy is

$$F = -k_BT\ln\left(\frac{2Ik_BT}{\hbar^2}\right).$$

and the entropy is just

Table 5.1 The characteristic rotational frequency, expressed as a wavenumber or as a temperature, for several gases

	CO_2	I_2	HI	HCl	H_2
B cm^{-1}	0.39	0.037	6.5	10.6	60.9
θ_R K	0.56	0.053	9.4	15.26	88

$$S = -\left(\frac{\partial F}{\partial T}\right)_V = k_B \left\{ \ln\left(\frac{2Ik_BT}{\hbar^2}\right) + 1 \right\}.$$

It follows that the contribution to the heat capacity at constant volume for high temperatures is k_B.

In Fig. 5.6 we show the variation of the heat capacity with temperature for one molecule: it is very small for low temperatures but, as the temperature increases, it rises to a maximum and then settles down to a value of k_B.

A common notation is to write $\hbar^2/2I = hcB$ where c is the velocity of light and so B has the dimensions of an inverse length; hcB can also be written as $k_B\theta_R$ where θ_R has dimensions of temperature and indicates the demarcation between high and low temperatures. Values of these parameters for simple gases are given in Table 5.1.

5.12   Vibrational energy levels for diatomic molecules

To a good approximation the vibrational states of a diatomic molecule are those of a harmonic oscillator. There is an attractive potential $V(r)$ holding the atoms together, a potential which has a minimum at a separation of d. For a small displacement from the minimum, the potential is of the form

$$V(r) \approx V(d) + \frac{(r-d)^2}{2}\left(\frac{d^2V(r)}{dr^2}\right)_{r=d} + \cdots$$

that is the potential energy is that of a simple harmonic oscillator. The quantum treatment of a diatomic molecule is discussed in more detail in Appendix C. The energy levels of a one-dimensional harmonic oscillator are given by

$$\varepsilon_n = \hbar\omega\left(n + 1/2\right) \tag{5.12.1}$$

where ω is the angular frequency of vibration and n is a positive integer or zero. The partition function is a geometric series which can be summed exactly. It is

$$
\begin{aligned}
Z &= \sum_{n=0}^{\infty} e^{-\hbar\omega(n+1/2)/k_BT} \\
&= e^{-\hbar\omega/2k_BT} + e^{-3\hbar\omega/2k_BT} + e^{-5\hbar\omega/2k_BT} + \cdots \\
&= \frac{e^{-\hbar\omega/2k_BT}}{1 - e^{-\hbar\omega/k_BT}}.
\end{aligned} \tag{5.12.2}
$$

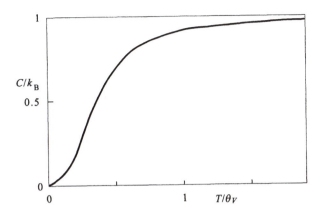

Fig. 5.7 The contribution to the heat capacity in units of k_B from the vibrational states of a simple harmonic oscillator as a function of T/θ_V.

For high temperatures, $T \gg \hbar\omega/k_B$, the partition function is approximately $k_B T/\hbar\omega$ and this leads to a similar formula for the entropy as for the rotation of a diatomic molecule:

$$S \approx k_B \left\{ \ln\left(\frac{k_B T}{\hbar\omega}\right) + 1 \right\}.$$

The contribution to the heat capacity is k_B at high temperatures.

When we calculate the entropy of a harmonic oscillator at low temperatures we must use the exact expression for Z. The free energy can be written as

$$F = \frac{\hbar\omega}{2} + k_B T \ln\left(1 - e^{-\hbar\omega/k_B T}\right) \tag{5.12.3}$$

and the entropy is

$$S = -k_B \ln\left(1 - e^{-\hbar\omega/k_B T}\right) + \frac{\hbar\omega/T}{e^{\hbar\omega/k_B T} - 1} \tag{5.12.4}$$

from which it is easy to find the heat capacity as

$$C_V = \frac{\hbar^2\omega^2}{k_B T^2} \frac{e^{\hbar\omega/k_B T}}{\left(e^{\hbar\omega/k_B T} - 1\right)^2}. \tag{5.12.5}$$

Again, the heat capacity dies away exponentially at low temperature, as is illustrated in Fig. 5.7.

The vibrational frequencies can be expressed either in terms of wave numbers, $\nu = \omega/2\pi c$, which have dimensions of an inverse length, or in terms of a characteristic temperature, θ_V, defined as $hc\nu/k_B$. Typical values are given in Table 5.2.

Table 5.2 The characteristic vibrational frequency, expressed as a wavenumber or a temperature, for several gases

	I_2	F_2	HCl	H_2
ν cm^{-1}	215	892	2990	4400
θ_V K	309	1280	4300	6330

By comparing the numbers in Tables 5.1 and 5.2 it can be seen that characteristic vibration energies are much larger than characteristic rotation energies. Always the vibrational energy is larger than the rotational one, so that as the temperature is lowered, first the excited vibrational states are frozen out and then the excited rotational states.

5.13 Factorizing the partition function

Sometimes there are situations where the system has single-particle energy levels which can be written as a sum of functions, each of which depends on one quantum number only. Thus for quantum numbers n_i

$$\varepsilon_i = g(n_1) + f(n_2) + h(n_3) + \cdots \tag{5.13.1}$$

that is a sum of energies which depends on a single quantum number. For example, a diatomic molecule such as carbon monoxide vibrates at frequency ω to a good approximation with energy levels

$$g(n) = \varepsilon_n = \hbar\omega \left(n + 1/2\right).$$

The energy levels of rotation of a diatomic molecule are approximately

$$f(l) = \varepsilon_l = \frac{\hbar^2 l(l+1)}{2I}.$$

The carbon monoxide molecule can also move freely about in space with kinetic energy for translational motion of

$$\varepsilon_i = \frac{\hbar^2 \pi^2}{2mL^2} \left(n_1^2 + n_2^2 + n_3^2\right) = h(n_1) + h(n_2) + h(n_3).$$

The kinetic energy of translational motion is completely unaffected by the vibration and rotation of the molecule. The total energy is

$$\varepsilon_j = \frac{\hbar^2 \pi^2}{2mL^2} \left(n_1^2 + n_2^2 + n_3^2\right) + \hbar\omega \left(n + 1/2\right) + \frac{\hbar^2 l(l+1)}{2I}.$$

In this approximation the energy of the molecule is the sum of the energy levels of vibration, rotation, and translation each with their quantum numbers n_1, n_2, n_3, n, and l. In making this statement it is assumed that the vibrational frequency is independent of the rotational state of the system, labelled by the quantum number l. In fact that is not exactly true: the vibrational energy is shifted a little bit by the rotational motion. If the molecule is rotating, there is a centrifugal potential which slightly alters the vibrational frequency and shifts the vibrational energy levels. However, in practice this is a very small correction to the energy, and it can be neglected.

The partition function for the diatomic molecule is

$$Z = \sum_{n_1=0}^{\infty} \sum_{n_2=0}^{\infty} \sum_{n_3=0}^{\infty} \sum_{n=0}^{\infty} \sum_{l=0}^{\infty} (2l+1) e^{-\varepsilon_j/k_B T}.$$

Because the energy eigenvalue is the sum of terms each with its own quantum number we get

$$Z = Z_{\text{trans}} \times Z_{\text{vib}} \times Z_{\text{rot}}. \tag{5.13.2}$$

The partition function is just the product of the partition functions for the translational, rotational, and vibrational states of the molecule. When the energy can be written as the sum of subenergies, with each subenergy involving an independent quantum number, the partition function is the product of partition functions for each source of energy.

There may be yet further quantum states of the molecule with extra energy levels. If the electrons in the atoms have a set of spin states, then we must include the partition function for spin of the electrons; if the nuclei have a set of spin states we must include their partition function. Of course if the separation in energy of the spin states is large compared to $k_B T$ then these spin states are effectively frozen out. Usually the opposite is the case for both electronic and nuclear spins. We shall ignore spin states for the present in the interests of simplicity.

The free energy is

$$\begin{aligned} F &= -k_B T \ln(Z) \\ &= -k_B T \ln(Z_{\text{trans}}) - k_B T \ln(Z_{\text{vib}}) - k_B T \ln(Z_{\text{rot}}). \end{aligned} \tag{5.13.3}$$

We add the free energy of rotation to that of vibration and translation. The free energies can be added, provided that the energy levels of rotation and vibration are independent of the other quantum numbers. For most molecules this is a good approximation.

The conclusion is this: when we add together energy levels, each with their own set of quantum numbers, we multiply partition functions and we *add* the free energies. It follows that for such systems we add the entropies and heat capacities of the subsystems. Consequently we expect the heat capacities to increase as the temperature rises as first the rotational and then the vibrational modes get excited. This is illustrated in Figs 5.8 and 5.9 where we plot C_P against T for hydrogen and for carbon monoxide. The heat capacity of normal molecular hydrogen rises to a peak at $21\,\text{K}$ where the liquid boils, and then

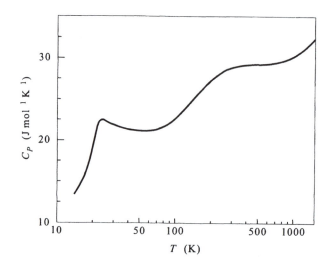

Fig. 5.8 The molar heat capacity of normal molecular hydrogen at a pressure of 1 bar. The heat capacity for temperatures around 30 K is about 22 J mol^{-1} K^{-1}, that is about $5R/2$. Remember that $C_P = C_V + R$, so this means C_V is about $3R/2$. As the temperature increases above 100 K, the rotational states become excited and the heat capacity, C_P, rises to $7R/2$. Above 1000 K the heat capacity rises again as the vibrational states become excited. The liquid boils at 21 K where there is a peak in the heat capacity. There is always a rise in the heat capacity near a phase transition. Data taken from the tables of Vargaftik (1975).

the heat capacity drops back to a value of about 20 J mol^{-1} K^{-1}, that is about $5R/2$. As the temperature increases so the rotational modes become excited above 100 K and the heat capacity rises to about $7R/2$. About 1000 K the heat capacity rises again as the vibrational states become excited. In Fig. 5.9 we show the heat capacity of carbon monoxide for temperatures ranging from 200 to 2000 K. As the temperature rises the vibrational modes begin to get excited and the heat capacity increases. The solid curve is the fit of the simple theory to the heat capacity with one adjustable parameter of $\theta_V = 3000$ K. The rotational energy levels are always excited. Spectroscopy shows the rotational spectrum to be consistent with $cB = 115.2$ GHz which corresponds to a temperature $\theta_R = 5.5$ K, way below the temperature range shown in Fig. 5.9.

5.14 Equipartition theorem

Let us consider a gas which is enclosed in a box. Try to imagine that quantum mechanics has not been invented, so that the particles are described classically. The particles can be thought of as perfectly elastic balls which bounce off each other and the walls of the container. You can imagine the atoms as if they were golf balls in a tiled bathroom, bouncing off the walls and rarely colliding with

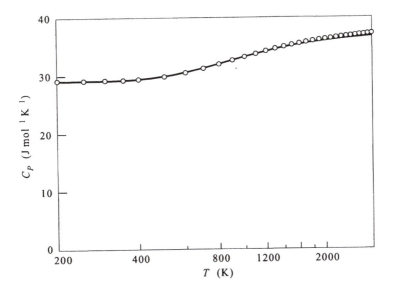

Fig. 5.9 The molar heat capacity of carbon monoxide at a constant pressure of 0.1 bar as a function of temperature. The data are taken from the tables compiled by Vargaftik (1975). The only adjustable parameter is θ_V which we took to be 3000 K.

each other. At each instant every ball is at a particular place moving with a well-defined velocity. Our aim is to get as complete a description as possible of the particles in the gas using this mental picture.

To a classical physicist momentum is a continuous variable, with components p_x, p_y, and p_z, each of which is continuous. Let us mentally slice up the x, y, and z components of momentum into tiny intervals of width dp_x, dp_y, and dp_z and make up little boxes in momentum space. We can allocate the momentum of each particle to one of these boxes. Each such box is labelled according to some scheme. Perhaps a particular box is labelled as 83 so we can denote its momentum as $\mathbf{p}_{83}$.

Similarly we can slice up the x, y, and z components of the position vector, $\mathbf{r}$, into tiny intervals of width dx, dy, and dz, make up little boxes in real space, and then allocate the position vector of each particle to one of these numbered boxes. In this way we can describe the microscopic state of the gas of particles as, for example, the collection of numbers

$$\psi_i = |(\mathbf{r}_{57}, \mathbf{p}_{83}) ; (\mathbf{r}_{812}, \mathbf{p}_{72}) ; \ldots\rangle$$

where the quantities in the first bracket are the position and momentum of the first particle, the next bracket for the second particle, and so forth. That seems to be a very precise description of the microscopic state of the system. It is particularly precise if we can make the boxes in real space and momentum space smaller and smaller without limit.

Can we do this? The answer is no: we cannot measure both the position and momentum at the same instant. Nature frustrates our attempt to do this. In the language of quantum mechanics this is known as the *uncertainty principle* which says that there is a fundamental lack of precision in the simultaneous measurement of position and momentum of a particle. This is one of the central ideas of quantum mechanics. Of course this restriction was unknown to classical physicists like Boltzmann and Gibbs.

Even so, there was a fundamental problem with the classical picture which gave a foretaste of quantum mechanics. Suppose a particle is trapped in a box and we try to specify exactly where it is inside the volume. We slice the volume of the box into tiny cells so that the position vector can be labelled by a discrete index j. As the cells get smaller and smaller the number of cells increases. In the limit that the cells become infinitely small, there is an infinite number of them because there is an infinite number of places where each particle could be. The degeneracy of the microscopic state is infinite, for the particle could be in any one of an infinite number of places. This makes the entropy, $k_B \ln(W)$, infinite. The calculation of thermodynamic quantities becomes nonsense unless we stop dividing the cells at some stage.

To avoid this difficulty, we can treat the momentum and position vectors as discrete quantities as described above. Each particle is associated with a little box in real space and a box in momentum space. The momentum vector is mentally divided up into tiny cells in momentum space, that is into little boxes with momentum having components in the range p_x to $(p_x + dp_x)$, p_y to $(p_y + dp_y)$, and p_z to $(p_z + dp_z)$. In the same way we divide the position vector into tiny boxes in real space with position vectors having components in the range x to $(x + dx)$, y to $(y + dy)$, and z to $(z + dz)$. These tiny cells in real space and in momentum space, when combined, are called *phase space*. The particle exists in a box in phase space.

For example, in a one-dimensional world a cell in phase space for one particle corresponds to a region with coordinates p_x to $(p_x + dp_x)$ and x to $(x + dx)$. The particle is smeared out in phase space over an area h, where h is a constant. Each area in phase space can be labelled separately by its position and momentum value as $\{x, p_x\}$. We then sum over all the possible values of $\{x, p_x\}$ in phase space and this gives a finite value to W.

From $\ln(W)$ we can calculate the entropy. The absolute value of the entropy depends on the choice of h. It is possible to measure changes in the entropy, but one cannot measure the absolute value of the entropy. For example the heat capacity involves the small change in entropy when we add a bit of heat to a system causing the temperature to rise. Entropy changes are independent of h, so we can calculate the heat capacity and get sensible results. That is what classical physicists did; it is one of the techniques of classical statistical mechanics.

A more sophisticated way of calculating the thermodynamics is through the partition function. For example, in a one-dimensional world, the partition function for one particle is

$$Z = \frac{1}{h} \int_{-\infty}^{\infty} dx \int_{-\infty}^{\infty} dp_x \, e^{-E(p_x,x)/k_B T} \qquad (5.14.1)$$

where $E(p_x, x)$ is the energy of the particle in terms of its momentum and position. This expression can be easily be extended to systems with more than one dimension or with more than one particle.

Consider the various terms which can appear in the expression for the energy of the particle. The particles have kinetic energy of translation, of rotation, and energy associated with vibration. Let us consider a general coordinate q together with its associated (conjugate) momentum, p. If the part of the energy of the system associated with these variables is of the form

$$E(p,q) = \alpha q^2 + \beta p^2, \qquad (5.14.2)$$

then the classical prescription for calculating Z is

$$Z = \frac{1}{h} \int_{-\infty}^{\infty} dq \int_{-\infty}^{\infty} dp \, e^{-E(p,q)/k_B T}. \qquad (5.14.3)$$

Let us assume that the quantities α and β are greater than zero. Then

$$Z = \left(\frac{\pi k_B T}{h\alpha}\right)^{\frac{1}{2}} \left(\frac{\pi k_B T}{h\beta}\right)^{\frac{1}{2}}. \qquad (5.14.4)$$

The free energy is $F = -k_B T \ln(Z)$, and we can obtain the entropy from F. It is straightforward to show that such a term in the energy produces a contribution to the heat capacity at constant volume of k_B.

In contrast, if say α is zero then the integral over q must be over a finite range, that is it is bounded, for otherwise Z is infinite. Let us take the q integral to extend from $-\Delta q/2$ to $\Delta q/2$. Then

$$
\begin{aligned}
Z &= \frac{1}{h} \int_{-\Delta q/2}^{\Delta q/2} dq \int_{-\infty}^{\infty} dp \, e^{-\beta p^2/k_B T} \\
&= \frac{\Delta q}{h} \left(\frac{\pi k_B T}{\beta}\right)^{\frac{1}{2}},
\end{aligned}
$$

and the same calculation gives a contribution to the heat capacity at constant volume of $k_B/2$.

The *equipartition theorem* can be stated thus: every degree of freedom of a body which contributes a quadratic term of a coordinate or momentum to the total energy has an average energy $k_B T/2$ and gives a contribution to the heat capacity of $k_B/2$. A *degree of freedom* means one of the quadratic terms in the energy. For example a vibrational mode has an energy

$$E(p_x, x) = \frac{p_x^2}{2m} + \frac{1}{2}kx^2.$$

It has terms quadratic in the momentum and in the coordinate and so it has two degrees of freedom. For high temperatures the heat capacity of gases can be understood simply by counting the number terms in the energy which are quadratic in either the coordinate or the momentum.

If the particle is monatomic then the kinetic energy in three dimensions is

$$E\left(p_x, p_y, p_z\right) = \frac{p_x^2 + p_y^2 + p_z^2}{2m}.$$

There are three quadratic terms, so three degrees of freedom. The contribution to the heat capacity at constant volume per atom is $3k_B/2$. Remember that $C_P = C_V + R$, so that the contribution to the heat capacity at constant pressure is $5k_B/2$.

If the particle is diatomic there is one mode of vibration with two degrees of freedom and two rotational modes corresponding to two degrees of freedom. Why are there two rotational modes? To specify the orientation of the molecule needs only two angles and so there are two conjugate momenta. The energy associated with rotation is quadratic in these momenta; hence there are two rotational degrees of freedom. There are three degrees of freedom for translational motion. Consequently at high temperatures the contribution to the heat capacity at constant volume per molecule is $7k_B/2$. Since $C_P = C_V + R$ for diatomic gases as well, the contribution to the heat capacity at constant pressure is $9k_B/2$.

If the particle is triatomic the situation is more complicated. The centre of mass of the molecule can move in three spatial directions each with a conjugate momentum. The kinetic energy of translational motion of the centre of mass is quadratic in these momenta and so there are three degrees of freedom for translational motion. If there is no axis of symmetry, we need three angles to specify the orientation of the molecule and consequently there are three conjugate momenta. The energy does not depend on the angles, but it is quadratic in each of the three conjugate momenta so there are three rotational degrees of freedom. A triatomic molecule has three atoms so there must be nine kinetic energy terms which are quadratic in momenta. We have accounted for six of them so there must be three vibrational modes of vibration each with a potential and a kinetic energy term; hence there are six degrees of freedom for vibrations. The total of 12 degrees of freedom gives a contribution per molecule to the heat capacity at constant volume of $6k_B$.

If the molecule possesses an axis of symmetry the situation is different. There are three translational degrees of freedom for the motion of the centre of mass. We now need two angles to specify the orientation of the molecule; hence there are two rotational degrees of freedom coming from the kinetic energy of rotation. It follows that there must be four vibrational modes of the molecule corresponding to eight degrees of freedom. The 13 degrees of freedom give a contribution per molecule to the heat capacity at constant volume of $13k_B/2$. Notice that the symmetry of the molecule affects its heat capacity!

These simple sums reflect the classical picture. In reality the modes get frozen out as the gas is cooled as shown in Fig 5.8. As the temperature is lowered first the modes of vibration get frozen out, then the rotational motion stops, and finally the translational motion slows down as the system liquefies and stops when the liquid freezes.

5.15 Minimizing the free energy

One of the simplest and most powerful techniques for calculating the equilibrium properties of a system comes from the observation that the free energy is a minimum when the system is in equilibrium. All we have to do is obtain an expression for the free energy as a function of a parameter (or set of parameters) and then locate the minimum in the free energy. This procedure generates an equation of state. The value of this approach lies not only in the range of problems to which it can successfully be applied, but also in the generation of physical intuition in guessing the solution to very difficult problems. We concentrate here on easy ones.

5.15.1 Minimizing the Helmholtz free energy

First we prove the idea that a system of constant volume which is displaced from thermal equilibrium evolves so that the Helmholtz free energy tends to a minimum. In other words the *Helmholtz free energy is a minimum for a system held at constant volume and temperature.* Consider system A to be in thermal contact with a large heat bath or reservoir, which is at temperature T given by

$$\frac{1}{k_B T} = \left(\frac{\partial \ln(W_R)}{\partial U_R} \right)_V \tag{5.1.3}$$

where U_R is the internal energy of the reservoir. This equation can be rewritten as

$$W_R = \gamma e^{U_R / k_B T} \tag{5.1.4}$$

where γ is independent of U_R. When system A is placed in thermal contact with the heat bath, the joint number of accessible states is

$$
\begin{aligned}
W &= W_A \gamma e^{U_R / k_B T} \\
&= \gamma e^{(U_T - U_A + k_B T \ln(W_A)) / k_B T}
\end{aligned}
\tag{5.15.1}
$$

where $U_T \, (= U_A + U_R)$ is the total energy of the joint system. The combined system is in equilibrium when $k_B \ln(W)$ is a maximum. The idea is to maximize $\ln(W)$ by varying U_A; to do this we must minimize the quantity

$$\frac{U_A - k_B T \ln(W_A)}{k_B T} = \frac{U_A - T S_A}{k_B T}. \tag{5.15.2}$$

The temperature of system A is the same as that of the heat bath because they are in thermal equilibrium. Hence we can replace T by T_A in eqn (5.15.2); it follows that we minimize the Helmholtz free energy, $F_A = U_A - T_A S_A$. The system is in equilibrium when the total entropy is a maximum; it follows that the system is in equilibrium when the free energy of system A is a minimum.

We can derive an *equation of state* for the system from the condition that the free energy is a minimum when the system is in equilibrium with a heat bath. By way of illustration, consider n vacancies in a solid with each vacancy raising the energy by an amount ε. The solid is composed of N atoms on $N + n$ lattice sites. The number of ways of arranging the vacancies on the lattice sites is $(N + n)!/N!n!$, and the entropy is

$$S = k_B \ln \left(\frac{(N+n)!}{N!n!} \right).$$

Therefore the Helmholtz free energy is

$$
\begin{aligned}
F &= U - TS \\
&= n\varepsilon - k_B T \ln \left(\frac{(N+n)!}{N!n!} \right).
\end{aligned}
$$

We minimize F by varying n to find the equilibrium number of vacancies, which gives the equation

$$\frac{\varepsilon}{k_B T} = \ln \left(\frac{N+n}{n} \right)$$

from which it is easy to obtain n as

$$n = \frac{N}{e^{\varepsilon/k_B T} - 1}.$$

This is the result we obtained in Chapter 4 (eqn 4.4.9) by using the techniques of the microcanonical ensemble. We have obtained eqn (4.4.9) more directly by minimizing the free energy.

5.15.2   Minimizing the Gibbs free energy

The same technique can be used to find the condition for equilibrium for a system which is kept at a constant *pressure*, not at constant volume. We show that the *Gibbs free energy is a minimum for a system held at constant pressure and temperature*. The definition of pressure is $P = k_B T (\partial \ln(W)/\partial V)_T$ (eqn 5.6.1). For a heat bath of energy U_R, volume V_R, pressure P_R, and at temperature T, the number of accessible states of the heat bath is

$$W_R = \gamma e^{(U_R + P_R V_R)/k_B T}. \tag{5.15.3}$$

Suppose that the system attached to the heat bath has volume V_A and energy U_A. The total energy $U_T = U_A + U_R$ is fixed, as is the total volume $V_T = V_A + V_R$.

The total number of states of the combined system (A plus heat bath) is given by

$$
\begin{aligned}
W &= W_A(U_A, V_A) \times \gamma e^{(U_T - U_A + P_R(V_T - V_A))/k_B T} \\
&= \gamma e^{(U_T - U_A + P_R(V_T - V_A) + k_B T \ln(W_A))/k_B T}.
\end{aligned}
\tag{5.15.4}
$$

The system is in equilibrium when the entropy is a maximum. The idea is to maximize $\ln(W)$ for constant pressure P_R, and to do this the quantity $(U_A + P_R V_A - T S_A)$ must be minimized. The system is in thermal and mechanical equilibrium so that $T_A = T$ and $P_A = P_R$. In order to maximize W we must minimize the Gibbs free energy, $G = U_A + P_A V_A - T_A S_A$, of system A.

To illustrate the method let us reconsider vacancies in a solid but this time allow the presence of vacancies to increase the volume of the solid. For simplicity let us suppose that both atoms and vacancies occupy the same volume. As the number of vacancies increases so the volume of solid expands. Suppose there are n vacancies in the lattice with N atoms on $N + n$ lattice sites. The energy of n vacancies is $n\varepsilon$ and the volume occupied is $(N + n)v$, where v is the volume occupied by each atom or vacancy. The Gibbs free energy is $F + PV$ so

$$
G = n\varepsilon - k_B T \ln\left(\frac{(N+n)!}{N! n!}\right) + Pv(N + n).
$$

The aim is to minimize G by varying n which gives

$$
\frac{\varepsilon + Pv}{k_B T} = \ln\left(\frac{N + n}{n}\right)
$$

from which it is easy to obtain

$$
n = \frac{N}{e^{(\varepsilon + Pv)/k_B T} - 1}.
$$

From the definition of G as $F + PV$ and from the relation $dF = -S\, dT - P\, dV$ (eqn (2.6.9)) it follows that $dG = V\, dP - S\, dT$. Hence the average volume of the solid is

$$
V = \left(\frac{\partial G}{\partial P}\right)_{T,n} = Nv\left\{1 + \frac{1}{e^{(\varepsilon + Pv)/k_B T} - 1}\right\}.
$$

The presence of thermally excited vacancies makes the solid expand.

5.16 Problems

1. A system consists of three energy levels which are non-degenerate (there is only one quantum state for each level). The energy levels are $E_1 = 0$, $E_2 = 1.4 \times 10^{-23}$ J, $E_3 = 2.8 \times 10^{-23}$ J. Given that the system is at a temperature of $1\,\mathrm{K}$ determine the partition function and calculate the probability that the system is in each level. (Take $k_B = 1.4 \times 10^{-23}\,\mathrm{J\ K^{-1}}$.)

2. A system has four non-degenerate energy levels. The energy levels are $E_1 = 0, E_2 = 1.4 \times 10^{-23}$ J, $E_3 = 4.2 \times 10^{-23}$ J, $E_4 = 8.4 \times 10^{-23}$ J. Given that the system is at a temperature of 5 K, what is the probability that the system is in the $E_1 = 0$ level?

3. A system has two non-degenerate energy levels with an energy gap of 0.1 eV $= 1.6 \times 10^{-20}$ J. What is the probability of the system being in the upper level if it is in thermal contact with a heat bath at a temperature of 300 K? At what temperature would the probability be 0.25?

4. A system has non-degenerate energy levels with energy $\varepsilon = (n + \frac{1}{2})\hbar w$ where $\hbar w = 1.4 \times 10^{-23}$ J and n a positive integer or zero. What is the probability that the system is in the $n = 1$ state if it is in contact with a heat bath of temperature 1 K?

5. A system has three energy levels of energy 0, 100 k_B, and 200 k_B, with degeneracies of 1, 3, and 5 respectively. Calculate the partition function, the relative population of each level, and the average energy at a temperature of 100 K.

6. A system has two modes of oscillation: one with frequency w, the other with frequency $2w$. The ground state energy of the system has an energy eigenvalue of $\frac{3}{2}\hbar w$. If the system is in contact with a heat bath at temperature T, what is the probability that the system has an energy less than $4\hbar w$?

7. The partition function of a system is given by the equation

$$Z = e^{aT^3V}$$

where a is a constant. Calculate the pressure, the entropy, and the internal energy of the system.

8. Calculate the probability that a harmonic oscillator $\varepsilon_n = (n + 1/2)\hbar w$ is a state with n an odd number if the oscillator is in contact with a heat bath at temperature T.

9. The average kinetic energy $(= 3k_B T/2)$ of hydrogen atoms in a stellar gas is 1 eV. What is the ratio of the number of atoms in the second excited state $(n = 3)$ to the number in the ground state $(n = 1)$? The energy levels of the hydrogen atom are $\varepsilon_n = -\alpha/n^2$ where $\alpha = 13.6$ eV, and the degeneracy of the nth level is $2n^2$.

10. A system has two energy levels with an energy gap of 3.2×10^{-21} J; the upper level is twofold degenerate, the lower level is non-degenerate. What is the probability that the lower level is occupied if the system is in thermal contact with a heat bath at a temperature of 150 K?

11. A system of particles occupying single-particle states and obeying Boltzmann statistics is in thermal contact with a heat bath at temperature T. The populations are: 3.1% for 0.0281 eV; 8.5% for 0.0195 eV; 23% for 0.0109 eV; 63% for 0.0023 eV. What is the temperature of the system?

12. System A is placed in contact with a reservoir which exerts a constant pressure P_R and maintains a constant temperature T_R. In subsection 5.15.2

we showed that the entropy of the joint system is a maximum when

$$G = U_A + P_R V_A - T_R S_A$$

is a minimum. Show that this condition is satisfied when

$$T_A = T_R; \qquad P_A = P_R.$$

13. The lowest energy level of O_2 is threefold degenerate. The next level is doubly degenerate and lies $0.97\,\text{eV}$ above the lowest level. Take the lowest level to have an energy of 0. Calculate the partition function at $1000\,\text{K}$ and at $3000\,\text{K}$.

14. Consider a lattice gas model of a system in which there are N sites, each of which can be empty or occupied by one particle, the energy being 0 for no particles on a site and ϵ for one particle. Each particle has a magnetic moment μ which in the presence of a magnetic induction field leads to a shift in energy into either μB or $-\mu B$. What is the partition function of the system? Evaluate the mean energy and the magnetization of the lattice gas.

15. A three-dimensional isotropic harmonic oscillator has energy levels

$$\varepsilon_{n_1, n_2, n_3} = \hbar \omega (n_1 + n_2 + n_3 + 3/2),$$

where each of n_1, n_2, and n_3 can be $0, 1, 2, 3$, et cetera. Find the degeneracies of the levels of energy $7\hbar\omega/2$ and $9\hbar\omega/2$.

Given that the system is in thermal equilibrium with a heat bath at a temperature T, show that the $9\hbar\omega/2$ level is more populated than the $7\hbar\omega/2$ level if $k_B T$ is larger than $\hbar\omega/\ln(5/3)$.

16. Show that the Helmholtz free energy, F, of a set of N localized particles, each of which can exist in levels of energy 0, ϵ, 2ϵ, and 3ϵ having degeneracies $1, 3, 3$, and 1 respectively, is

$$F = -3N k_B T \ln\left(1 + \exp(-\epsilon/k_B T)\right).$$

17. The partition function for an interacting gas is assumed to be

$$Z = \left(\frac{V - Nb}{N}\right)^N \left(\frac{m k_B T}{2\pi\hbar^2}\right)^{3N/2} e^{N^2 a^2 / V k_B T}$$

where a and b are constants. Show that the pressure is of the same form as van der Waals equation. Calculate the internal energy.

18. If the chlorine molecule at 290 K were to rotate at the angular frequency predicted by the equipartition theorem what would be the average centripetal force? (The atoms of Cl are 2×10^{-10} m apart and the mass of the chlorine atom 35.45 a.m.u.)

19. Consider an electron of mass 9.11×10^{-31} kg which is localized in a cube of volume 10^{-6} m^3. The wavefunction of the particle vanishes at the faces

of the cube. What is the energy eigenvalue of the single-particle state with quantum numbers up to $n_1^2 + n_2^2 + n_3^2 = 14$? What is the degeneracy of each energy level ignoring the spin of the electron?

20. Pluto is believed to have a radius of 1500 km, a mass of 1.5×10^{22} kg, and a surface temperature of 55 K. Could Pluto support an atmosphere of methane, CH_4?

21. By putting $\beta = 1/k_B T$, write $Z = \sum_i e^{-\beta E_i}$ and show that

$$\left(\frac{\partial \ln(Z)}{\partial \beta} \right) = \sum_i p_i E_i = \bar{U}.$$

Consider the second derivative of $\ln(Z)$ with respect to β. Show that

$$\left(\frac{\partial^2 \ln(Z)}{\partial \beta^2} \right) = \sum_i E_i^2 \frac{e^{-\beta E_i}}{Z} - \left(\sum_i E_i \frac{e^{-\beta E_i}}{Z} \right)^2 = \sum_i E_i^2 p_i - \bar{U}^2.$$

Prove that standard deviation in energy is given by

$$\Delta U^2 = \left(\frac{\partial^2 \ln(Z)}{\partial \beta^2} \right).$$

Show that the average internal energy of a harmonic oscillator is

$$\bar{U} = \frac{\hbar \omega}{2} + \frac{\hbar \omega}{e^{\hbar \omega / k_B T} - 1}.$$

Calculate the average energy and the standard deviation in energy for a two-level system with energies ε and $-\varepsilon$.

22. In 1931 Kappler measured Boltzmann's constant by studying the fluctuations of a mirror suspended vertically from a torsion wire. The mirror was surrounded by a gas at a constant temperature of 278.1 K. He measured over a long time the fluctuations in the angle made by the normal to the mirror to a line perpendicular to the torsion wire. The mean square angular displacement of the mirror from its equilibrium position was $\overline{\theta^2} = 4.178 \times 10^{-6}$ with the angle measured in radians. The restoring torque on the wire when it was displaced by an angle θ from equilibrium was $\tau = -k\theta$. The torsion constant of the wire k was 9.428×10^{-16} kg m^2 s^{-2}. Use the equipartition theorem to obtain a value of Boltzmann's constant.

23. An electrical circuit consisting of an inductor and a capacitor is in thermal equilibrium with its surroundings at temperature T. Find the high-temperature formula for the root mean square current through the inductor, L, treating the motion of charge in the circuit as that of a harmonic oscillator.

24. A non-linear triatomic molecule has nine modes: three for translation, three for rotation, and three for vibration. What is the expected heat capacity per

molecule at high temperatures? If the triatomic molecule is linear, identify the number of the modes for translation, for rotation, and for vibration. Hence calculate the heat capacity per molecule at high temperatures.

25. How many degrees of freedom are there for a diatomic molecule in a two-dimensional world?

26. Consider a glass in which some fraction of atoms may occupy one of two sites in the unit cell. These two positions give rise to two energy levels, one with energy Δ_i and the other with energy $-\Delta_i$. The suffix i is a label which denotes the number of the atom.

(a) If each atom has energy $\pm\Delta$, calculate the contribution of these atoms to the heat capacity of the solid.

(b) If the glass is such that there is a spread in energies with values of Δ_i ranging uniformly from zero to Δ_0 find the temperature dependence of the heat capacity when $\Delta_0 \gg k_B T$.

6

Identical particles

Symmetry, as wide or as narrow as you may define it, is one idea by which man through the ages has tried to comprehend order, beauty, and perfection.   *Hermann Weyl*

The partition function involves a sum over independent quantum states of the N-particle system. When the particles which make up a system are indistinguishable it is more difficult to identify the independent quantum states of the system. When N is one there is no difficulty in identifying the independent quantum states; when N is two or larger problems arise.

Suppose we have a system with two identical particles which can be in single-particle states $\phi_i(x)$ of energy ε_i. A possible quantum state of the two-particle system has energy $\varepsilon_i + \varepsilon_j$ with one particle in $\phi_i(x)$ and the other in $\phi_j(x)$. This seems at first sight to be a situation in which the energy is the sum of sub-energies and so the partition function factorizes into a product of partition functions. One might expect that $Z_2 = Z_1 \times Z_1$. For N particles the same argument gives $Z_N = Z_1^N$. If the particles are distinguishable from each other then this argument is correct; if they are indistinguishable then by using this expression we are including in the summation more quantum states than are present.

In the last chapter the entropy due to translational motion of a single gas particle in a volume V was calculated to be

$$S = k_\mathrm{B} \left(\ln(V) + \frac{3}{2} \ln \left(\frac{m k_\mathrm{B} T}{2 \pi \hbar^2} \right) + \frac{3}{2} \right). \tag{5.9.10}$$

If we were to assume that the partition function due to translational motion of N particles is given by Z_1^N then the entropy would be just N times eqn (5.9.10). The volume is an extensive quantity, so the entropy would contain a term varying as $N \ln(V) = N \ln(N) + N \ln (V/N)$. When the ratio N/V is held fixed, the entropy would contain a term of the form $N \ln(N)$. Consequently it is not extensive as it should be. Clearly something is wrong. This chapter is concerned with the reasons for this apparent paradox.

6.1 Identical particles

One of these men is Genius to the other; And so of these: which is the natural man,
And which the spirit? Who deciphers them?

William Shakespeare, Comedy of Errors

All matter in the universe is made up of particles, each with its own set of
attributes. The central problem is distinguishability: how do we tell one electron
from another? All electrons are identical replicas of each other; they are all the
same. There is no feature, there is no label which will enable us to tell the
difference between them. Once we have studied one electron and found all its
properties then all electrons are the same. They cannot be told apart. They just
repeat endlessly the same set of attributes.

Identical particles are absolutely indistinguishable from each other. There
is no observable change in the system when any two particles are exchanged.
This observation leads to complications in the quantum mechanical description
of the microscopic state of the system, complications which have no counterpart
in a classical description. Classically, each particle can be identified by its precise position. All the particles can be numbered according to their position at
an instant. Subsequently, any particular particle can be observed continuously.
When, for example, the particle scatters off other identical particles it can be
identified by its trajectory during the collision. In principle, a classical particle
can be distinguished by its position for all time.

The uncertainty principle spoils all this for us. We cannot specify the position
of a particle accurately at each stage, nor can we follow the precise trajectory of a
particle. When two identical particles collide and get very close to each other we
cannot tell which one is which. Suppose the position of each particle is uncertain
by an amount Δx during a collision process. If this uncertainty is greater than
the minimum separation of the particles in the collision then the two particles
cannot be identified separately. They can exchange places during a collision and
nobody can be sure whether or not it has happened. Do the particles scatter as
shown in Fig. 6.1*a* or as in Fig. 6.1*b*?

There is no experiment which we can devise that will discriminate between
the two possibilities for identical particles.

6.2 Symmetric and antisymmetric wavefunctions

All particles that appear in Nature have a symmetry property: their quantum
state is either completely symmetric or antisymmetric with respect to interchange
of any two particles. We call this *exchange symmetry*.

Consider two identical quantum particles such as electrons with particle 1
placed at x_1 and particle 2 placed at x_2. The quantum state of the system
is described by a real space wavefunction $\psi(x_1, x_2)$ which, for the present, is
assumed to be just a function of the two coordinates. The first coordinate in the
bracket tells us where particle 1 is (it is at x_1), the second coordinate tells us
where particle 2 is (it is at x_2).

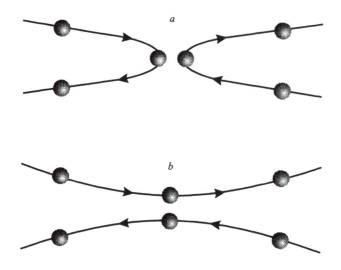

Fig. 6.1 The scattering of two identical particles in two possible collisions, a and b, shown as classical trajectories. Initially particles enter from opposite directions, collide, and move apart. Both collision processes have exactly the same initial and final trajectories, the difference being that the particles have exchanged places in b as compared to a. If we now treat the particles as quantum objects, they have an uncertainty in their position. If the uncertainty is comparable to the minimum separation of the trajectories, then the exchange of particles during the collision is possible, and we cannot tell which process has occurred.

Now we exchange the particles so that particle 1 is at x_2 and particle 2 is at x_1. The quantum state is described by the wavefunction $\psi(x_2, x_1)$. The quantity which can be related to experiment is the modulus of the wavefunction squared, $|\psi(x_1, x_2)|^2$: the quantity $|\psi(x_1, x_2)|^2 \, d^3x_1 \, d^3x_2$ is equal to the probability that the particles are localized in volume elements d^3x_1 and d^3x_2 around the points x_1 and x_2; these probabilities could (notionally) be measured experimentally. The indistinguishability of the particles means that the probability is unaltered when the particles are exchanged; if this statement were false then we could use the difference in the probabilities to distinguish the particles. Thus we must have

$$|\psi(x_1, x_2)|^2 = |\psi(x_2, x_1)|^2 . \tag{6.2.1}$$

It follows that the two wavefunctions must obey the equation

$$\psi(x_1, x_2) = e^{i\alpha} \, \psi(x_2, x_1) \tag{6.2.2}$$

where α is a phase factor. But what is the phase factor?

If we insist that two exchanges lead back to the original wavefunction then the only possibilities are $\alpha = 0$ and $\alpha = \pi$. For the choice $\alpha = 0$ we get

$$\psi(x_1, x_2) = \psi(x_2, x_1) \qquad (6.2.3)$$

so that the wavefunction keeps the same sign on interchange of particles. In contrast when $\alpha = \pi$ we have

$$\psi(x_1, x_2) = -\psi(x_2, x_1) \qquad (6.2.4)$$

so that the wavefunction changes sign on interchange of particles. A wavefunction is said to be *symmetrical* if it is unchanged on interchange of any two particles. It is called *antisymmetrical* if its sign is changed on interchange of any two particles.

The Hamiltonian of the system must be a symmetrical function of the coordinates of the particles if they are identical, for otherwise we could tell the particles apart. It follows that if the system starts in a symmetrical state it must always remain in a symmetrical state, for there is nothing to disturb the symmetry. In the same way an antisymmetrical state always remains in an antisymmetrical state. As a result it could be that only particles in symmetrical or antisymmetrical states appear in Nature. This appears to be always the case, even though other states are mathematically possible. In other words all the particles that appear in Nature are described by many-particle wavefunctions which are either completely symmetrical or completely antisymmetrical on interchange of identical particles.

6.3 Bose particles or bosons

For these particles the wavefunction keeps the same sign when the particles are swapped over. Particles with this property are named after the Indian physicist Satyendra Nath Bose (1924). The statistical mechanics of these particles was analysed by Einstein (1924, 1925). Examples of Bose particles are photons, mesons, and ^{4}He atoms, that is helium atoms with two protons and two neutrons in the nucleus and two electrons orbiting the nucleus. Some of these particles, such as the photon, are truly elementary; others are bound states of elementary particles which act like Bose particles.

Here is a simple example of an unnormalized two-particle wavefunction that satisfies the symmetry requirement:

$$\psi_{\text{Bose}}(x_1, x_2) = \phi_i(x_1)\phi_j(x_2) + \phi_i(x_2)\phi_j(x_1) \qquad (6.3.1)$$

where $\phi_i(x_1)$ is a single-particle wavefunction with energy eigenvalues ε_i. The two-particle wavefunction in eqn (6.3.1) is unaltered when we interchange x_1 and x_2. There is an internal symmetry to the wavefunction which means that it is unchanged when the particles are swapped over.

For non-interacting particles the energy eigenvalue for this wavefunction is $E_{ij} = \varepsilon_i + \varepsilon_j$. There is only one quantum state with this set of labels i, j. If we were to ignore the exchange symmetry, then we could construct two wavefunctions $\chi_1(x_1, x_2) = \phi_i(x_1)\phi_j(x_2)$ and $\chi_2(x_1, x_2) = \phi_i(x_2)\phi_j(x_1)$, both of which have energy $\varepsilon_i + \varepsilon_j$. At first sight it seems that we should sum over these two quantum states; but these two wavefunctions do not satisfy the exchange symmetry condition on the wavefunction. There is only one quantum state that does that properly. If we included the two quantum states, then we would double-count when calculating the partition function.

The same problem of overcounting occurs if we have a three-particle wavefunction describing bosons. The appropriate unnormalized wavefunction when i, j, and k are all different is

$$
\begin{aligned}
\psi_{\text{Bose}}(x_1, x_2, x_3) \; = \; & \phi_i(x_1)\phi_j(x_2)\phi_k(x_3) + \phi_i(x_2)\phi_j(x_1)\phi_k(x_3) \\
& + \phi_i(x_2)\phi_j(x_3)\phi_k(x_1) + \phi_i(x_3)\phi_j(x_2)\phi_k(x_1) \\
& + \phi_i(x_3)\phi_j(x_1)\phi_k(x_2) + \phi_i(x_1)\phi_j(x_3)\phi_k(x_2).
\end{aligned} \tag{6.3.2}
$$

This wavefunction is completely symmetric on interchange of any two particles. There is only one such wavefunction which satisfies this requirement.

Notice that there is no difficulty in constructing a wavefunction which allows two particle to exist in the same single-particle state. For example, for two particles we can have the wavefunction $\psi_1(x_1, x_2) = \phi_i(x_1)\phi_i(x_2)$. In fact we can construct quantum states for bosons in which there are any number of particles in each single-particle state. The number of particles in the single-particle state can be zero or any positive integer.

We can label the quantum state of a system of non-interacting identical particles by how many particles are in each single-particle state. In general, the quantum state can be written as

$$
\psi_i = |n_1, n_2, n_3, \cdots\rangle \tag{6.3.3}
$$

where n_i is the number of particles in the single-particle state $\phi_i(x)$. The set of numbers, n_i, which can be zero or any positive integer for Bose particles, gives a complete description of the quantum state of the system. The energy of this quantum state is given by the expression

$$
E_i = n_1\varepsilon_1 + n_2\varepsilon_2 + n_3\varepsilon_3 + \cdots \tag{6.3.4}
$$

for non-interacting particles.

6.4 Fermi particles or fermions

For this type of particle the wavefunction changes sign when the particles are swapped over. This kind of particle was first considered by Fermi (1926a,b) and independently by Dirac (1926). Examples of fermions include electrons, neutrinos, protons, and ^{3}He atoms, that is helium atoms with two protons and one neutron in the nucleus. Some of these particles, such as electrons and neutrinos, are believed to be truly elementary; others, such as protons and ^{3}He, are bound states of several elementary particles, the bound state acting like a Fermi particle. For example the proton is made of three Fermi particles which are called quarks.

Swapping any two particles over changes the sign of the wavefunction if the wavefunction is completely antisymmetric. Here is a simple example of an un-normalized two-particle wavefunction for non-interacting identical particles that satisfies the exchange symmetry requirement:

$$\psi_{\text{Fermi}}(x_1, x_2) = \phi_i(x_1)\phi_j(x_2) - \phi_i(x_2)\phi_j(x_1). \qquad (6.4.1)$$

The wavefunction changes sign if we interchange x_1 and x_2. There is an internal symmetry to the wavefunction which means that it reverses sign when particles are exchanged. Notice something quite surprising: if $i = j$ the wavefunction vanishes. We cannot put the two particles into the same single-particle state because the wavefunction for such a state is identically zero. This is an interference effect of two waves. At most there can be one particle in each single-particle state.

This form of wavefunction can be written as a determinant

$$\psi_{\text{Fermi}}(x_1, x_2) = \begin{vmatrix} \phi_i(x_1) & \phi_j(x_1) \\ \phi_i(x_2) & \phi_j(x_2) \end{vmatrix}. \qquad (6.4.2)$$

Writing the wavefunction as a determinant can easily be extended to states with three or more particles. For example for three particles we need a 3×3 determinant

$$\psi_{\text{Fermi}}(x_1, x_2, x_3) = \begin{vmatrix} \phi_i(x_1) & \phi_j(x_1) & \phi_k(x_1) \\ \phi_i(x_2) & \phi_j(x_2) & \phi_k(x_2) \\ \phi_i(x_3) & \phi_j(x_3) & \phi_k(x_3) \end{vmatrix}. \qquad (6.4.3)$$

The determinant vanishes if any two columns are the same, which happens if we try to put two particles in the same single-particle state.

We can label the quantum state of a system of identical particles by how many particles are in each single-particle state. For both Fermi and Bose particles the quantum state can be written as

$$\psi_i = |n_1, n_2, n_3, \cdots\rangle \qquad (6.3.3)$$

where n_i is the number of particles in the single-particle state $\phi_i(x)$. For Fermi particles the numbers, n_i, are either zero or one. The energy of this quantum state is given by the expression

$$E_i = n_1\varepsilon_1 + n_2\varepsilon_2 + n_3\varepsilon_3 + \cdots \qquad (6.3.4)$$

for non-interacting particles. This equation is valid for both Fermi and Bose particles, the only difference being that n_i is zero or one for Fermi particles, whereas it can be any positive integer or zero for Bose particles.

The restriction on the number of Fermi particles in any single-particle state is called the *Pauli exclusion principle*. It was proposed by Pauli (1925) in order to explain the structure of Mendeleev's periodic table of elements in terms of the

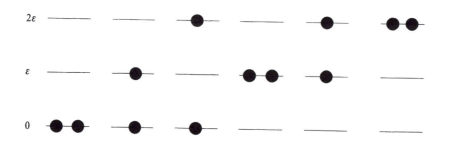

Fig. 6.2 The possible arrangements of two identical Bose particles in three non-degenerate energy levels of energy 0, ε, and 2ε.

single-particle energy levels of the electron (a fermion) in an atom. The Pauli principle is very important, for it restricts the number of possible structures that can occur, and generates regularity among those that do. In the case of electrons orbiting atomic nuclei it generates the periodicity among the atomic elements as revealed by Mendeleev's periodic table. It plays a similar role in the structures made up of three quarks, the baryons of elementary particle physics.

6.5 Calculating the partition function for identical particles

As an illustration of the different statistics, consider a very simple problem of two identical mobile atoms with each atom having non-degenerate quantum states of energy 0, ε, and 2ε. Let one particle be in a single-particle state of energy ε_i (one of these three energies), the other in the state of energy ε_j. The total energy is $E_{i,j} = \varepsilon_i + \varepsilon_j$. These two atoms are in contact with a heat bath at temperature T. What is the partition function if the atoms are indistinguishable?

6.5.1 Bosons

Let the atoms be bosons. For Bose particles there is no restriction on the number of particles in any single-particle state. We show the possible arrangement of the particles in Fig. 6.2. For the first state shown in Fig. 6.2, both particles are in the single-particle state of zero energy. We can label the quantum states by the number of particles in each single-particle state. Thus $|2,0,0\rangle$ indicates the quantum state with two particles in the single-particle state of zero energy. The possible energy levels and their degeneracies are: $E = 0$ (one state $= |2,0,0\rangle$), $E = \varepsilon$ (one state $= |1,1,0\rangle$), $E = 2\varepsilon$ (states $|1,0,1\rangle$ and $|0,2,0\rangle$), $E = 3\varepsilon$ (one state $=|0,1,1\rangle$), and $E = 4\varepsilon$ (one state $= |0,0,2\rangle$). The partition function for the two Bose particles is

$$Z_{\text{Bose}} = 1 + e^{-\varepsilon/k_{\text{B}}T} + 2\,e^{-2\varepsilon/k_{\text{B}}T} + e^{-3\varepsilon/k_{\text{B}}T} + e^{-4\varepsilon/k_{\text{B}}T}.$$

6.5.2 Fermions

Let the atoms be fermions. We cannot put both fermions in the same single-particle state so there are three possible energy levels. They have energies ε, 2ε, and 3ε corresponding to the states $|1,1,0\rangle$, $|1,0,1\rangle$, and $|0,1,1\rangle$. These energy levels are non-degenerate so the partition function is

$$Z_{\text{Fermi}} = e^{-\varepsilon/k_{\text{B}}T} + e^{-2\varepsilon/k_{\text{B}}T} + e^{-3\varepsilon/k_{\text{B}}T}.$$

The partition function is quite different in the two cases.

6.5.3 A large number of energy levels

Now let us consider two identical particles, either fermions or bosons, with a more extensive set of single-particle quantum states. Perhaps there are a thousand single-particle states instead of three. The total energy of a quantum state is $E_{ij} = \varepsilon_i + \varepsilon_j$. There is just one quantum state with this energy if i is not equal to j, that is the state

$$\psi_\alpha = (\phi_i(x_1)\phi_j(x_2) \pm \phi_i(x_2)\phi_j(x_1))$$

with a plus sign for bosons and a minus sign for fermions. This quantum state can be written as

$$\psi_\alpha = |0, 0, \ldots, 1, 0, 0, \ldots 1, 0, 0, \ldots\rangle$$

with one particle in the single-particle state $\phi_i(x_1)$, the other in $\phi_j(x_2)$. If the two particles are bosons there could also be the quantum state with two particles in the same single-particle state. A possible quantum state is

$$\psi_\beta = \phi_i(x_1)\phi_i(x_2)$$

which could be written as

$$\psi_\beta = |0, 0, \ldots, 2, 0, 0, \ldots 0, 0, 0, \ldots\rangle.$$

This quantum state does not exist for fermions.

The partition function for bosons is given by

$$
\begin{aligned}
Z_{\text{Bose}} = \ & e^{-2\varepsilon_1/k_{\text{B}}T} + e^{-2\varepsilon_2/k_{\text{B}}T} + e^{-2\varepsilon_3/k_{\text{B}}T} + \cdots \\
& + e^{-(\varepsilon_1+\varepsilon_2)/k_{\text{B}}T} + e^{-(\varepsilon_1+\varepsilon_3)/k_{\text{B}}T} + e^{-(\varepsilon_1+\varepsilon_4)/k_{\text{B}}T} + \cdots \\
& + e^{-(\varepsilon_2+\varepsilon_3)/k_{\text{B}}T} + e^{-(\varepsilon_2+\varepsilon_4)/k_{\text{B}}T} + e^{-(\varepsilon_2+\varepsilon_5)/k_{\text{B}}T} + \cdots
\end{aligned}
$$

whereas for fermions it is

$$
\begin{aligned}
Z_{\text{Fermi}} = \ & e^{-(\varepsilon_1+\varepsilon_2)/k_{\text{B}}T} + e^{-(\varepsilon_1+\varepsilon_3)/k_{\text{B}}T} + e^{-(\varepsilon_1+\varepsilon_4)/k_{\text{B}}T} + \cdots \\
& + e^{-(\varepsilon_2+\varepsilon_3)/k_{\text{B}}T} + e^{-(\varepsilon_2+\varepsilon_4)/k_{\text{B}}T} + e^{-(\varepsilon_2+\varepsilon_5)/k_{\text{B}}T} + \cdots.
\end{aligned}
$$

There are no terms of the form $e^{-2\varepsilon_1/k_{\text{B}}T}$ in the partition function of fermions.

Suppose we were to pretend the particles were distinguishable. The energy is a simple sum of terms involving different quantum numbers which implies that we can factorize the partition function. If we were to assume that Z_N for two particles is simply the product

$$Z_2 = \left(\sum_{i=1}^{M} e^{-\varepsilon_i/k_BT} \right) \left(\sum_{j=1}^{M} e^{-\varepsilon_j/k_BT} \right)$$

and we were to multiply this expression out, then we would get terms

$$\begin{aligned} Z_2 &= e^{-2\varepsilon_1/k_BT} + e^{-2\varepsilon_2/k_BT} + e^{-2\varepsilon_3/k_BT} + \cdots \\ &+ 2e^{-(\varepsilon_1+\varepsilon_2)/k_BT} + 2e^{-(\varepsilon_1+\varepsilon_3)/k_BT} + \cdots. \end{aligned}$$

If we were to believe this expression to be correct, then there would have to be two quantum states with energy $\varepsilon_1 + \varepsilon_2$. The quantum state with this energy is counted twice, not once. This is wrong for indistinguishable particles, be they fermions or bosons: for both types of particle there is only one quantum state which has energy $\varepsilon_1 + \varepsilon_2$.

The quick (and crude) way around this problem is to divide by 2!; for the majority of the terms in the partition function this fixes things. If we do this we get

$$\begin{aligned} Z_2 &= \tfrac{1}{2} e^{-2\varepsilon_1/k_BT} + \tfrac{1}{2} e^{-2\varepsilon_2/k_BT} + \tfrac{1}{2} e^{-2\varepsilon_3/k_BT} + \cdots \\ &+ e^{-(\varepsilon_1+\varepsilon_2)/k_BT} + e^{-(\varepsilon_1+\varepsilon_3)/k_BT} + \cdots. \end{aligned}$$

Dividing by 2! would be correct but for those terms in the sum where $i = j$; these terms do not need to be divided by 2!. If the particles are fermions, there is no such state with $i = j$, and there should be no contribution to the partition function from such terms. The approximation given above includes them. If the particles are bosons, then the doubly occupied state is allowed and it contributes equally with all the rest. But the contribution to Z_2 from doubly occupied states is incorrect in the above expression.

How are we to get out of this difficulty? It cannot be resolved in a simple way, but we can argue as follows. Suppose there are so many single-particle states of low energy that the probability of finding the two particles in the same single-particle state is negligible. Under these circumstances we can ignore the double-counting problem, and just divide by 2!. For example for two particles in a large box at high temperatures the probability of double occupancy of a single-particle state is extremely small.

The same approach can be used for an N-particle system. Provided we only count quantum states in which there is no double occupancy of a single-particle state then we just need to divide by $N!$. For a dilute gas with N particles we assume that the partition function, Z_N, is given to a fair approximation by

$$Z_N \simeq \frac{Z_1^N}{N!} \tag{6.5.1}$$

(or, to be more precise, $\ln(Z_N) \simeq \ln\left(Z_1^N/N!\right)$). Let us apply the theory to a monatomic gas of spinless particles so that there are no vibrational and rotational states and the spin degeneracy is zero. The single-particle partition function for a monatomic spinless particle is given by $Z_1 = Z_{\text{trans}}$, the partition function for translation. The partition function for N such particles is

$$Z_N = \frac{(Z_{\text{trans}})^N}{N!}. \tag{6.5.2}$$

Quantum mechanics enters in this subtle way to introduce the factor $N!$ which at first sight should not be there. It is a consequence of the indistinguishability of identical particles.

The free energy, using Stirling's approximation, is

$$F = -Nk_BT \left\{ \ln(Z_{\text{trans}}) - \ln(N) + 1 \right\}. \tag{6.5.3}$$

Using eqn (5.9.6) we get

$$F = -Nk_BT \left\{ \ln\left(\frac{V}{N}\right) + \frac{3}{2}\ln\left(\frac{mk_BT}{2\pi\hbar^2}\right) + 1 \right\}. \tag{6.5.4}$$

From the free energy it is simple to derive an expression for the entropy as

$$S = Nk_B \left\{ \ln\left(\frac{V}{N}\right) + \frac{3}{2}\ln\left(\frac{mk_BT}{2\pi\hbar^2}\right) + \frac{5}{2} \right\}. \tag{6.5.5}$$

This equation, the *Sackur–Tetrode formula* for the entropy of an ideal gas, is valid for all dilute gases at temperatures such that all vibrational or rotational motion has been frozen out.

It is instructive to compare eqns (6.5.4) and (6.5.5) with eqns (5.9.7) and (5.9.10) respectively. Notice that neither the entropy nor the free energy for the N-particle system is given by N times the value of the entropy or free energy for a single particle, a somewhat surprising result. However, when the number density, N/V, is kept constant, the entropy is proportional to the number of particles; that is, it is extensive as it should be. If the $N!$ term were omitted, the thermodynamics would be wrong: the formulae we would get for the entropy and the free energy would not be extensive.

Confirmation that the Sackur–Tetrode formula is correct can be obtained from experiment; all that is needed is careful measurement of the heat capacity of the system as a function of temperature and measurements of the latent heats as explained earlier. The Sackur–Tetrode formula gives excellent agreement with the experimentally determined entropy for a dilute gas at high temperatures. For example the entropy of krypton at its boiling point is found experimentally to be $144.56 \, \text{J mol}^{-1}\,\text{K}^{-1}$; the theoretical value, using the measured value of Planck's constant, is $145.06 \, \text{J mol}^{-1}\,\text{K}^{-1}$. Similar agreement is found for other monatomic spinless gases.

6.6 Spin

The analysis we have given so far has not given the complete story: most particles possess an intrinsic angular momentum called *spin* and we have ignored this in our treatment of identical particles. According to quantum mechanics the angular momentum quantum number of a system can only have the following values: $0, \hbar/2, \hbar, 3\hbar/2, 2\hbar, \ldots$. The spin of a particle, being an angular momentum, can only have these values. Experimentally we find that bosons always have an integer spin, that is $0, \hbar, 2\hbar, \ldots$, whereas fermions have a half-integer spin, that is $\hbar/2, 3\hbar/2, 5\hbar/2, \ldots$. Electrons and protons have a spin of $\hbar/2$; they are fermions. Photons have a spin of $\hbar$; a photon is a boson. The statistics of a particle depends on whether the particle is a fermion or a boson.

There is a remarkable connection between the statistics of a particle and its spin, but the proof of this relationship requires knowledge of relativistic quantum field theory. (See for example Mandl and Shaw 1993, pp. 72–3.) It is much simpler to accept the connection between spin and statistics as an experimental fact.

In the description which we have given so far we have assumed that the particles only have a space coordinate; the spin has been neglected. This is correct for spinless particles which are bosons. It is not correct for spin 1/2 particles which are fermions. For such particles we must deal with the part of the wavefunction associated with spin.

To simplify the discussion consider two identical fermions which are interchanged. The total wavefunction of the system must change sign on interchange of particles. By the term 'total' we mean the product of the spin and the real space (or orbital) wavefunction

$$\Psi_{\text{total}} = \psi_{\text{space}} \chi_{\text{spin}}. \tag{6.6.1}$$

If the spin wavefunction changes sign on interchange of particles, the space wavefunction must keep the same sign; and vice versa.

There are four possible spin wavefunctions for the two fermions. One state is $\chi_1 = |\uparrow, \uparrow\rangle$ which means that both particles have spin up. Another state is $\chi_\alpha = |\uparrow, \downarrow\rangle$ which means that particle 1 has spin up, particle 2 has spin down. A third state is $\chi_\beta = |\downarrow, \uparrow\rangle$ which means that particle 1 has spin down and particle 2 has spin up. The fourth state is $\chi_3 = |\downarrow, \downarrow\rangle$ which means that both particles have spin down. We can also create the combinations $\chi_2 = |\uparrow, \downarrow\rangle + |\downarrow, \uparrow\rangle$ and $\chi_4 = |\uparrow, \downarrow\rangle - |\downarrow, \uparrow\rangle$. The spin functions χ_1, χ_2, and χ_3 are all symmetric on interchange of the two particles. They form a spin triplet of states with total spin of one. In contrast the state χ_4 changes sign on interchanging the particles so it is antisymmetric. It forms a spin singlet corresponding to a total spin of zero.

Let us return to the Pauli principle. The total wavefunction must be antisymmetric on interchange of identical Fermi particles. If the spin wavefunction is symmetric on interchange, the real space wavefunction must be antisymmetric. For two particles a possible wavefunction is

$$\Psi(1,2) = \{\phi_i(x_1)\phi_j(x_2) - \phi_j(x_1)\phi_i(x_2)\} \, (|\uparrow,\downarrow\rangle + |\downarrow,\uparrow\rangle)$$

for the spin state χ_2. There are similar states involving χ_1 and χ_3.

If the spin wavefunction is antisymmetric on interchange, the real space wavefunction must be symmetric. For two particles a possible wavefunction is

$$\Psi(1,2) = \{\phi_i(x_1)\phi_j(x_2) + \phi_j(x_1)\phi_i(x_2)\} \, (|\uparrow,\downarrow\rangle - |\downarrow,\uparrow\rangle).$$

For this state we can have $i = j$. There can be two particles in the same orbital state, but one must be spin up, the other spin down. Out of the four possible spin states, it is only this one which allows there to be two particles in the same single-particle orbital state.

6.7   Identical particles localized on lattice sites

When the particles form a solid, there is kinetic energy associated with the localization of the particles about their lattice sites. For heavy particles the wavefunction is very sharply peaked at each lattice site. The wavefunction dies away rapidly from the lattice site so that the probability of finding a particle near a neighbouring site is incredibly small. The chance of any two particles exchanging places by a tunnelling motion is very small.

The effects arising from Bose or Fermi statistics only make themselves felt if the particles can exchange places. In a gas or liquid this is relatively easy, and so statistics play a role in the properties of these systems. In solids the chance of the particles exchanging places is extremely small and so statistics plays no role.

Let us consider solid and liquid ^{4}He as an example of a system made of Bose particles, and solid and liquid ^{3}He as an example of a system made of Fermi particles. These particles have a relatively light mass so the wavefunctions describing the localization of the particles will be more spread out than for more massive particles. In the liquid phase exchange of particles is relatively easy. The restriction of the wavefunction on interchange of particles is very important for these liquids. It dominates the behaviour of the system; as a result liquid ^{4}He has very different properties compared to those of liquid ^{3}He.

In contrast, in the solid phase the exchange of particles is very difficult. The effects of Bose or Fermi statistics play no role directly on the statistical mechanics of the system, although there is an indirect effect. Remember that bosons have an integral spin and fermions have a half-integral spin. This leads to a difference in the spin degeneracy of Bose and Fermi particles. For example a boson of spin zero has no spin degeneracy, a boson of spin one has a spin degeneracy of three. A fermion of spin one-half has a spin degeneracy of two. The spin degeneracy is different for Bose and Fermi particles. The statistical mechanics of the solid is affected by the spin states of the particles and so the statistics of the particles, Bose or Fermi, has an indirect effect on the thermodynamics of the solid.

Of course a small amount of exchange does arise, either from a tunnelling rotation of several atoms in the cage which surrounds them or from a movement of vacancies. The energy involved in these processes is very small for ^{3}He, of order 10^{-26} J or less (Roger *et al.* 1983). It is a fairly good approximation to ignore exchange effects provided the system is at a much higher temperature, but at temperatures of a few mK and below these exchange processes cannot be ignored. We are then faced with the problem of constructing a realistic Hamiltonian of the system and then evaluating the partition function of the system. We defer further discussion until Chapter 11.

How do we treat the statistical mechanics of the spins which are situated on each lattice site? Imagine solid ^{3}He with each atom localized on its lattice site. The solid is placed in a magnetic induction field **B**. Due to the spin of the ^{3}He atom there is a splitting of the energies of the spin states into $+\mu_N B$ or $-\mu_N B$ where μ_N is the nuclear magnetic moment. There are N such two-level systems, and so the total partition function is

$$Z_N = Z_1 \times Z_1 \times Z_1 \times Z_1 \times \cdots = Z_1^N \qquad (6.7.1)$$

where Z_1 is the single-particle partition function $Z_1 = e^{\mu_N B/k_B T} + e^{-\mu_N B/k_B T}$. There is no factor of $N!$ in this case, for we assume that there is no possibility of exchange of particles between neighbouring lattice sites. Each particle is distinguishable by the lattice site on which it sits.

Since the total partition function is just Z_1^N, the free energy is

$$F = -k_B T \ln(Z_N) = -N k_B T \ln(Z_1). \qquad (6.7.2)$$

Therefore the free energy is proportional to N, the number of two-level systems, which makes perfect sense. It follows that the entropy and heat capacity are extensive as well.

If we were to repeat the calculation with each particle having M levels, the resulting partition function would have the same form, Z_1^N, but the single-particle partition function would be

$$Z_1 = \sum_{i=1}^{M} e^{-\varepsilon_i/k_B T}.$$

Everything is relatively straightforward: as long as we consider distinguishable particles on particular sites, like these two-level systems, we can work out the total partition function just by multiplying together the partition functions for each system in turn. The free energy is just the sum of all the individual free energies. Once we get the free energy it is straightforward to get all the thermodynamics.

6.8   Identical particles in a molecule

Consider a molecule made of identical particles which is rotated by 180 degrees so that the identical particles are exchanged. Under rotation by 180 degrees the wavefunction is altered by a factor of $(-1)^l$ where l is the angular momentum

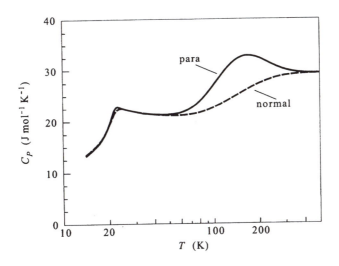

Fig. 6.3 The molar heat capacity of para- and normal hydrogen as a function of temperature from the data tabulated by Vargaftik (1975). Below 50 K the two heat capacities are identical and arise from translational motion; above 50 K the heat capacities are different because the rotational parts of the partition function are not the same: the partition function for para-hydrogen is given by eqn (6.8.1) whereas for normal hydrogen it is given by eqn (6.8.2).

quantum number as discussed in Appendix C. An 's' orbital is unchanged by such a rotation, a 'p' orbital changes sign.

As an example, consider a CO_2 molecule with two ^{16}O nuclei which is rotated by 180 degrees. The ^{16}O nuclei are bosons with spin zero and hence the spin wavefunction is symmetric. The total wavefunction must keep the same sign on rotation by 180 degrees. It follows that only even values of l are allowed by the Pauli exclusion principle. The rotational partition function is

$$Z = \sum_{l=\text{even}}^{\infty} (2l + 1)\, e^{-\hbar^2 l(l+1)/2Ik_BT}. \tag{6.8.1}$$

Another example is a hydrogen molecule. The nucleus of each atom is a proton, a fermion of spin $1/2$. The total wavefunction of the protons must change sign on rotation by 180 degrees. The spin state of the protons with total spin zero is $\chi_4 = |\uparrow\downarrow\rangle - |\downarrow\uparrow\rangle$ which is antisymmetric on interchange of particles. The orbital wavefunction of the protons must keep the same sign on interchange of particles and so l must be 0, 2, 4, ... , an even integer. This type of hydrogen is called *para-hydrogen*.

The three spin states χ_1, χ_2, and χ_3 form a triplet with a total spin of one. They are all symmetric on interchange of particles. The orbital wavefunction of the protons must change sign on interchange of particles and so l must be 1, 3, 5, . . . , an odd integer. This type of hydrogen is called *ortho-hydrogen*.

Ordinary hydrogen consists of a mixture of para- and ortho-hydrogen. The total partition function is

$$
Z = \sum_{l=\text{even}}^{\infty} (2l+1)\, e^{-\hbar^2 l(l+1)/2Ik_B T}
$$
$$
+ 3 \sum_{l=\text{odd}}^{\infty} (2l+1)\, e^{-\hbar^2 l(l+1)/2Ik_B T}. \tag{6.8.2}
$$

The factor of three in the last term in eqn (6.8.2) is for the three spin states. The adjustment to chemical equilibrium of these two species is very slow except in the presence of a catalyst. We can think of them as two different species of molecule each with its own heat capacity.

The data shown in Fig. 6.3 are the heat capacity of normal hydrogen, which consists of an equilibrium mixture of ortho- and para-hydrogen, together with the heat capacity of para-hydrogen.

6.9   Problems

1. Calculate the free energy of a system with spin one on each site, given that the levels associated with the three spin states have energies ε, 0, and $-\varepsilon$.

2. Calculate the free energy of a system with N particles, each with spin 3/2 with one particle per site, given that the levels associated with the four spin states have energies $\frac{3}{2}\varepsilon$, $\frac{1}{2}\varepsilon$, $-\frac{1}{2}\varepsilon$, and $-\frac{3}{2}\varepsilon$ and degeneracies 1, 3, 3, and 1 respectively.

3. Consider two identical particles which are to be placed in four single-particle states. Two of these states have energy 0, one has energy ε, the last has energy 2ε. Calculate the partition function given that the particles are (a) fermions and (b) bosons.

4. Suppose there are single-particle energy eigenvalues of 0, ε, 2ε, and 3ε which are non-degenerate. A total of 6ε is to be shared between four particles. List the configuration of the particles and their degeneracies for: distinguishable particles; indistinguishable Bose particles; indistinguishable Fermi particles.

5. Consider a system of two atoms, each having only four single-particle states of energies 0, ε, 2ε, and 3ε. The system is in contact with a heat bath at temperature T. Write down the energy levels and the partition function given that the particles obey:
   (a) classical statistics because the particles are distinguishable;
   (b) Fermi–Dirac statistics because they are indistinguishable Fermi particles, which implies that two atoms have to be in different single-particle states;
   (c) Bose–Einstein statistics because they are indistinguishable Bose particles, which implies that the two atoms can be in the same single-particle states. You may assume that the particles have no spin.

6. A molecule of hydrogen in its ground state can exist in two forms: ortho-hydrogen where the nuclear spins are parallel, resulting in a net spin; and para-hydrogen where the nuclear spins are antiparallel, resulting in no net spin. The ortho form of hydrogen has three spin states all of the same energy $\varepsilon > 0$, the para form has one state of zero energy. The hydrogen molecules form a solid made up of N molecules, with each molecule distinguishable from the rest by being localized on a lattice site. Assume that the spins of neighbouring molecules couple very weakly. The solid is thermally isolated and has a fixed energy $U = n\varepsilon$ where n is the number of molecules in the ortho state. Show that the entropy of the system in the microcanonical ensemble is

$$W = \frac{N!}{n!(N-n)!}3^n,$$

and obtain an expression for the temperature of the system. How does the number of molecules in the para state vary with temperature?

7. Consider a system of N identical, non-interacting, magnetic ions of spin $1/2$, magnetic moment μ, in a crystal at temperature T in a uniform magnetic induction field B. Calculate the partition function, Z_N, for this system and from it obtain a formula for the entropy, S.
   The crystal is initially in thermal equilibrium with a reservoir at $T = 1$ K, in a magnetic induction field of 10 tesla. The crystal is thermally isolated and the field is reduced adiabatically (so the entropy is constant) until the field is 10^{-2} tesla. What temperature do the spins reach at the end of the process?

8. Two identical particles of mass m interact with an external potential which is harmonic, but they do not interact with each other. The Hamiltonian of the system is

$$\hat{H} = \frac{p_1^2}{2m} + \frac{p_2^2}{2m} + \frac{m\omega^2}{2}(x_1^2 + x_2^2).$$

   What are the energy levels of the system? Given that the particles are bosons, determine the partition function. Repeat the calculation given that the two particles are fermions.

9. Consider a mixture of two ideal monatomic gases. The partition function of the mixture can be written as $Z_N = (Z_a^{N_a} Z_b^{N_b})/(N_a! N_b!)$ where Z_a is the partition function of a single particle of gas a of which there are N_a particles, and Z_b is the partition function of a single particle of gas b of which there are N_b particles. The total number of particles is $N = N_a + N_b$. If the particles were identical the partition function would be $Z_N = Z_a^N/N!$. Calculate the free energy in the two cases, assuming that $Z_a = Z_b$. Show that the factor $1/N_a! N_b!$ leads to an extra *mixing* entropy of the form

$$S_{\text{mixing}} = -k_B \left(N_a \ln \left(\frac{N_a}{N} \right) + N_b \ln \left(\frac{N_b}{N} \right) \right).$$

7

Maxwell distribution of molecular speeds

And here I wish to point out that, in adopting this statistical method of considering the average number of groups of molecules selected according to their velocities, we have abandoned the strict kinetic method of tracing the exact circumstances of each individual molecule in all its encounters. It is therefore possible that we arrive at results which, though they fairly represent the facts as long as we are supposed to deal with gas in mass, would cease to be applicable if our faculties and instruments were so sharpened that we could detect and lay hold of each molecule and trace it through its course.

James Clerk Maxwell

In 1859 Maxwell published an expression for the distribution of the speeds of particles in a gas which he obtained by analysing collision processes between the particles. Previously, people had speculated that the particles of the gas all moved with the same speed; he showed that there could be a distribution of speeds, and calculated the distribution. Subsequently, Boltzmann (1877) treated the kinetic theory of gases from a microscopic point of view using notions of probability. He was able to derive Maxwell's distribution of speeds and to relate it to the temperature of the particles. In this chapter we treat the particles using quantum mechanics and derive the Maxwell distribution of speeds.

7.1 The probability that a particle is in a quantum state

Suppose a single particle is confined in a rectangular box of volume $V = L_x L_y L_z$ where L_x, for example, is the length of the box along the x direction. The energy eigenfunctions depend on the boundary conditions which are applied to the wavefunction. We choose the boundary condition that the wavefunction vanishes on the surface of the box, which yields standing wave solutions. These boundary conditions are appropriate when the wavefunction must vanish outside the box, as happens when the particle is confined inside. For travelling waves we use periodic boundary conditions, as discussed in Appendix D.

For a particle in a box, the eigenfunction describing standing waves is

$$\phi_i(x, y, z) = A \sin\left(\frac{n_1 \pi x}{L_x}\right) \sin\left(\frac{n_2 \pi y}{L_y}\right) \sin\left(\frac{n_3 \pi z}{L_z}\right) \qquad (7.1.1)$$

a generalization of eqn (5.9.1). The suffix i stands for the set of quantum numbers (n_1, n_2, n_3) which are positive integers. The corresponding energy eigenvalue is

$$\varepsilon_i = \frac{\hbar^2 \pi^2}{2m} \left(\frac{n_1^2}{L_x^2} + \frac{n_2^2}{L_y^2} + \frac{n_3^2}{L_z^2} \right) \tag{7.1.2}$$

the generalization of eqn (5.9.4). The partition function for translational motion is

$$Z_{\text{trans}} = L_x L_y L_z \left(\frac{m k_B T}{2 \pi \hbar^2} \right)^{\frac{3}{2}}. \tag{7.1.3}$$

(We drop the subscript trans from now on.)

Armed with this partition function we can calculate the thermodynamics of the system. If the thermodynamics were all that we could calculate, then statistical mechanics would only give us information about the large-scale properties of matter. We want more. By studying collision processes Maxwell showed that the particles in a gas have a distribution of speeds which is a new sort of information, something beyond thermodynamics. The question is: can we calculate the distribution of speeds of particles using the techniques of statistical mechanics? The answer is that we can, and for low-density gases at high temperatures the theory works perfectly.

According to Boltzmann, the probability of finding a given particle in a particular single-particle state of energy ε_i is

$$p_i = \frac{e^{-\varepsilon_i / k_B T}}{Z}. \tag{5.1.7}$$

If there are N particles in the box then the probability that this particular single-particle state is occupied by one of the N particles is roughly

$$n_i = N p_i = N \frac{e^{-\varepsilon_i / k_B T}}{Z} \tag{7.1.4}$$

that is N times the probability that a single particle fills the state. This is the result that we would get if we ignored the possibility that two particles could ever be in the same single-particle state, for then there would be no problems arising from the indistinguishability of atoms, that is whether they are bosons or fermions. We call this way of calculating n_i the method of Maxwell and Boltzmann—it is commonly referred to as *Maxwell–Boltzmann statistics*. We will call n_i as given by eqn (7.1.4) the *Maxwell–Boltzmann distribution of particles* in a single-particle state.

Equation (7.1.4) is an approximation; it is not exact. As an analogy, assume that the probability of a single entry winning the jackpot in a lottery is 1 in 10^4. If someone makes an entry each week for a hundred weeks, then a simple-minded calculation gives his chance of winning as 1 in 10^2; we multiply the probability for a single entry by the number of entries. This is what was done in eqn (7.1.4) when we multiplied by N. However, if he made 10^4 entries week after week (for 200 years), clearly something goes wrong with this calculation, for he cannot be certain that he will win even though the calculation tells us that the probability is one. It is clear that if he made 10^5 entries the probability cannot be $10^5 \times 10^{-4}$ since probabilities cannot be greater than one.

In the same way the probability of finding a particle in the single-particle state is given by $N\,e^{-\varepsilon_i/k_B T}/Z$, only if this number is much less than one as is the case for low-density gases at high temperatures. The approach goes wrong for high-density gases at low temperatures, even when the particles do not interact with each other, because the average number of particles in a single-particle state becomes larger than one. The proper treatment of high-density gases at low temperatures is given in Chapter 10.

7.2   Density of states in k space

The single-particle states describing standing waves are not eigenfunctions of the momentum operator. In fact if we were to measure the x component of momentum we would get either $\hbar k_x$ or $-\hbar k_x$ where

$$k_x = \frac{\pi n_1}{L_x} \tag{7.2.1a}$$

with n_1 a positive integer. Similarly for measurements of the y and z components of momentum we could get $\pm \hbar k_y$ and $\pm \hbar k_z$ respectively where

$$k_y = \frac{\pi n_2}{L_y}; \tag{7.2.1b}$$

$$k_z = \frac{\pi n_3}{L_z} \tag{7.2.1c}$$

with n_2 and n_3 positive integers. We cannot say for certain which value we would get. Nevertheless, we can construct a 'wave vector' **k** as

$$\mathbf{k} = \hat{\mathbf{i}}k_x + \hat{\mathbf{j}}k_y + \hat{\mathbf{k}}k_z \tag{7.2.2}$$

where $\hat{\mathbf{i}}$, $\hat{\mathbf{j}}$, and $\hat{\mathbf{k}}$ are unit vectors along the x, y, and z directions respectively.

Suppose we measured the x, the y, and the z components of the momentum and got a positive value for each component. In this case the measured wave vector of the particle would be given by eqn (7.2.2). In this sense the standing wave states can be represented by this vector, **k**, with n_1, n_2, and n_3 positive integers.

Each of the allowed single-particle states can be represented as a point in an abstract space as shown in Fig. 7.1. The axes of the space, called *k space*, are k_x, k_y, and k_z. The set of points form a lattice in **k** space. Only if the wave vector **k** lies exactly on the lattice point does it correspond to a standing wave state for a single particle.

The unit cell in this lattice has a volume of $\pi^3/L_x L_y L_z$. Therefore the density of points in **k** space is $L_x L_y L_z/\pi^3$. When the volume of the box, $V = L_x L_y L_z$, becomes very large the density of points in **k** space becomes enormous. If we want to evaluate the single-particle partition function we need a simple way of

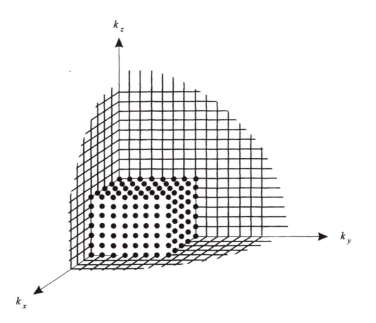

Fig. 7.1 The lattice of points in **k** space. Each point represents a standing wave state for a single particle. For reasons of clarity we do not show all the points.

summing over states, that is over the points in **k** space. The trick we use is to turn the sum into an integral. This trick will be used repeatedly to evaluate the partition function for a variety of systems.

We can represent the wave vector **k** using spherical polar coordinates defined by analogy with the representation of the position vector in spherical polar coordinates

$$k = \sqrt{k_x^2 + k_y^2 + k_z^2} \qquad\qquad (7.2.3a)$$

$$\theta = \cos^{-1}\left(\frac{k_z}{k}\right) \qquad\qquad (7.2.3b)$$

$$\phi = \tan^{-1}\left(\frac{k_y}{k_x}\right). \qquad\qquad (7.2.3c)$$

Here θ is the angle **k** makes with the k_z axis, and the projection of **k** on the $k_x k_y$ plane makes an angle ϕ to the k_x axis. The angles are shown in Fig. 7.2. Because the wave vector **k** has k_x, k_y, and k_z positive, both the angles θ and ϕ are restricted to the range 0 to $\pi/2$.

Suppose we want to study a small region of **k** space with volume element $dV_k = k^2 \sin(\theta) dk\, d\theta\, d\phi$. The number of single-particle states in this volume element of **k** space is given by

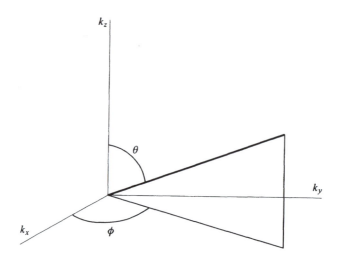

Fig. 7.2 The wave vector **k** in spherical coordinates. The wave vector makes an angle θ with the k_z axis, and the projection on the horizontal plane makes an angle ϕ with the k_x axis.

$$\mathrm{d}N = k^2 \sin(\theta)\mathrm{d}k\,\mathrm{d}\theta\,\mathrm{d}\phi \times \left(\frac{L_x L_y L_z}{\pi^3}\right)$$

since the density of points is $L_x L_y L_z/\pi^3$. If we integrate over all angles the total number of states in this small region of **k** space is

$$\mathrm{d}N = \left(\frac{V}{2\pi^2}\right) k^2\,\mathrm{d}k = D(k)\,\mathrm{d}k \qquad (7.2.4)$$

where $D(k)\,\mathrm{d}k = Vk^2\,\mathrm{d}k/2\pi^2$ is the number of single-particle states in **k** space in a thin shell lying between k and $k+\mathrm{d}k$. The quantity $D(k)$ is called the *density of states*. Thus the density of states in three dimensions is $Vk^2/2\pi^2$. Exactly the same expression for $D(k)$ for a three-dimensional system is obtained for travelling waves as is shown in Appendix D. The density of states is independent of the boundary conditions we impose on the waves.

Once we have calculated the density of states it is easy to obtain the single-particle partition function which is given by

$$Z = \int_0^\infty D(k)\,\mathrm{e}^{-\varepsilon(k)/k_\mathrm{B}T}\mathrm{d}k. \qquad (7.2.5)$$

This is a very simple formula. However, before we use it we want to examine the density of states in a little more detail.

The density of states depends on the number of spatial dimensions of the system. The expression for $D(k)$ in one or two dimensions is quite different from

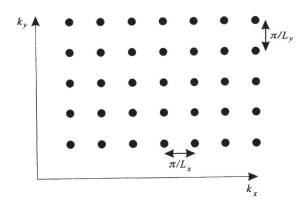

Fig. 7.3 The lattice of points in **k** space for a two-dimensional system. Each point represents a standing wave state for a single particle.

that in three. To illustrate the point we calculate the density of states in two dimensions. The eigenfunction describing standing waves for a particle in a box in two dimensions is of the form

$$\phi_i(x, y) = A \sin\left(\frac{n_1 \pi x}{L_x}\right) \sin\left(\frac{n_2 \pi x}{L_y}\right). \tag{7.2.6}$$

We can introduce a two-dimensional wave vector

$$\mathbf{k} = \hat{\mathbf{i}} k_x + \hat{\mathbf{j}} k_y \tag{7.2.7}$$

where

$$k_x = \frac{\pi n_1}{L_x}; \qquad k_y = \frac{\pi n_2}{L_y}. \tag{7.2.8}$$

The single-particle quantum states exist on a two-dimensional lattice of points in **k** space as shown in Fig. 7.3. We plot k_x and k_y as orthogonal coordinates along the x and y directions of real space. Each point at integer values of n_1 and n_2 corresponds to a particular value of k_x and k_y and gives a point in the plane. Since n_1 and n_2 must be positive integers, the points in **k** space are restricted to positive values of k_x and k_y. The area corresponding to each point is $\pi^2/L_x L_y$, so the density of states is $L_x L_y/\pi^2$.

Instead of using Cartesian coordinates (k_x, k_y), we convert to polar coordinates (k, θ) where

$$k = \sqrt{k_x^2 + k_y^2}, \tag{7.2.9a}$$

$$\theta = \tan^{-1}\left(\frac{k_y}{k_x}\right). \tag{7.2.9b}$$

The number of states in a small area of $\mathbf{k}$ space, $dA_k = k\,dk\,d\theta$ is

$$dN = k\,dk\,d\theta \left(\frac{L_x L_y}{\pi^2}\right).$$

By integrating this expression over θ we get the number of states in the range k to $k + dk$ as

$$dN = \left(\frac{L_x L_y}{2\pi}\right) k\,dk = D(k)\,dk. \tag{7.2.10}$$

Here $D(k)$ is density of states in two dimensions given by

$$D(k)\,dk = \frac{A}{2\pi} k\,dk \tag{7.2.11}$$

where $A = L_x L_y$. The density of states $D(k)$ clearly differs from the corresponding expression in three dimensions, $D(k) = Vk^2/2\pi^2$. In one dimension the number of states in a range dk is

$$D(k)\,dk = \frac{L}{\pi}dk. \tag{7.2.12}$$

Hence the density of states in one dimension is L/π.

7.3   Single-particle density of states in energy

Often it is convenient to use the energy as a variable, and to express the density of single-particle states in terms of energy instead of wave vector. This is very widespread practice in solid state physics. We write the *single-particle density of states* in energy as $D(\varepsilon)$ so that in a range from to $\varepsilon + d\varepsilon$ there are $D(\varepsilon)\,d\varepsilon$ single-particle states. The density of states in energy depends on the dispersion, that is the dependence of $\varepsilon(k)$ on k as well as the number of dimensions of the system.

In one dimension

$$D(\varepsilon)\,d\varepsilon = \frac{L}{\pi}\left(\frac{dk}{d\varepsilon(k)}\right)d\varepsilon, \tag{7.3.1}$$

in two dimensions the density of states in energy is

$$D(\varepsilon)\,d\varepsilon = \frac{Ak}{2\pi}\left(\frac{dk}{d\varepsilon(k)}\right)d\varepsilon, \tag{7.3.2}$$

and in three dimensions it is

$$D(\varepsilon)\,d\varepsilon = \frac{Vk^2}{2\pi^2}\left(\frac{dk}{d\varepsilon(k)}\right)d\varepsilon. \tag{7.3.3}$$

As an example, let us consider a free particle of mass m with energy $\varepsilon(k) = \hbar^2 k^2/2m$. The density in states in energy in three dimensions is

$$D(\varepsilon) = \frac{Vkm}{2\pi^2\hbar^2} = \frac{Vm\,(2m\varepsilon)^{1/2}}{2\pi^2\hbar^3}.$$

However, if the dispersion relation is different from that of a free particle, the density of states in energy changes. For a highly relativistic particle the dispersion relation is $\varepsilon(k) = \hbar c k$. In this case

$$D(\varepsilon) = \frac{Vk^2}{2\pi^2\hbar c} = \frac{V\varepsilon^2}{2\pi^2\hbar^3 c^3}.$$

The density of states in energy depends on the dispersion relation as well as on the number of dimensions.

As a final example let consider an excitation whose dispersion is given by $\varepsilon(k) = \alpha k^{3/2}$ with α a constant. This is the dispersion relation of a ripple on the surface of a liquid, a ripple which we are treating as if it were made up of particles. The system is two-dimensional since it is a surface. One could rewrite the dispersion relation as $k = (\varepsilon(k)/\alpha)^{2/3}$ and so

$$\frac{dk}{d\varepsilon(k)} = \frac{2}{3\alpha^{2/3}\varepsilon^{1/3}}.$$

By using eqn (7.3.2) we find the density of states in energy in two dimensions is

$$D(\varepsilon) = \frac{Ak}{2\pi}\frac{2}{3\alpha^{2/3}\varepsilon^{1/3}} = \frac{A\varepsilon^{1/3}}{3\pi\alpha^{4/3}}.$$

7.4 The distribution of speeds of particles in a classical gas

The Boltzmann probability distribution tells us the probability that a particle is in the single-particle state φ_k with energy $\varepsilon(k)$ is

$$p_k = \frac{e^{-\varepsilon(k)/k_B T}}{Z}. \tag{5.1.7}$$

If there are N particles in a gas, the probability that one of these particle is in the single-particle state is just

$$n_k = Np_k = \frac{Ne^{-\varepsilon(k)/k_B T}}{Z}, \tag{7.1.4}$$

which is N times the probability that a single particle fills the state. This equation is valid provided n_k is much less than one, as was discussed in section 7.1. When this is the case n_k is the *average number of particles in the state*.

Now let us ask how many particles are in states whose wave vectors lie between k and $k+dk$. There are $D(k)\,dk$ such states and each state contains on average n_k particles. Hence the average number of particles, $f(k)\,dk$, in the range between k and $k+dk$ is

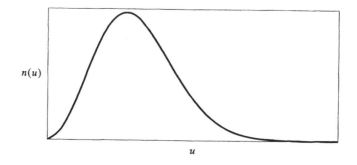

$n(u)$

u

Fig. 7.4 The Maxwell distribution of speeds for particles in a three-dimensional system. For low speeds the distribution increases as u^2 because the density of states varies as k^2; for high speeds the distribution dies away because few single-particle states of high speed are occupied.

$$f(k)\,\mathrm{d}k = \frac{Ne^{-\varepsilon(k)/k_B T}}{Z}D(k)\,\mathrm{d}k. \qquad (7.4.1)$$

The particles are distributed into states with different wave vectors. The distribution depends on the number of spatial dimensions through the density of states, $D(k)$, as well as on the Boltzmann probability factor. For a free particle in three dimensions we get, using $Z = V/\lambda_D^3$ (eqn (5.9.6)),

$$f(k)\,\mathrm{d}k = \left(\frac{N\lambda_D^3}{2\pi^2}\right)k^2\,e^{-\varepsilon(k)/k_B T}\,\mathrm{d}k, \qquad (7.4.2)$$

where $\lambda_D = (2\pi\hbar^2/mk_B T)^{1/2}$ is the thermal de Broglie wavelength. Notice that the volume cancels out completely. The distribution varies as k^2 for small k but it decreases to zero due to the exponential term for large k.

The speed of a non-relativistic particle of wave vector k is $u = \hbar k/m$. Consequently we can convert from wave vector k to speed u to get the Maxwell distribution of speeds. Let there be $n(u)\,\mathrm{d}u$ particles with speeds between u and $u + \mathrm{d}u$. For a gas in three dimensions we get

$$n(u)\,\mathrm{d}u = \left(\frac{N\lambda_D^3 m^3}{2\pi^2\hbar^3}\right)u^2\,e^{-mu^2/2k_B T}\,\mathrm{d}u. \qquad (7.4.3)$$

The distribution is proportional to $u^2\,e^{-mu^2/2k_B T}$ for particles in three dimensions. This distribution is shown in Fig. 7.4. This is called the *Maxwell distribution of molecular speeds*.

Once we have derived the distribution with wave vector, it is simple to calculate the average value of any quantity by using the formula

$$\bar{A} = \frac{\int_0^\infty f(k) A(k) \, dk}{\int_0^\infty f(k) \, dk},$$ (7.4.4)

where $A(k)$ is the value of the quantity corresponding to wave vector k and the bar denotes the average value. For example, the average kinetic energy is

$$\overline{\frac{\hbar^2 k^2}{2m}} = \frac{\int_0^\infty f(k) \, (\hbar^2 k^2/2m) \, dk}{\int_0^\infty f(k) dk},$$

with $A(k) = \hbar^2 k^2/2m$.

We prove eqn (7.4.4) by starting with the expression

$$\bar{A} = \sum_k p_k A(k).$$ (3.7.2)

The probability of a given particle being in the single-particle state ϕ_k is given by the Boltzmann distribution. The sum over states is turned into an integral over k weighted by the density of states:

$$\bar{A} = \int_0^\infty D(k) A(k) \frac{e^{-\varepsilon(k)/k_B T}}{Z} \, dk.$$ (7.4.5)

This expression can be used directly, but a useful trick is to take $A(k) = 1$ for all k. Then

$$1 = \int_0^\infty D(k) \frac{e^{-\varepsilon(k)/k_B T}}{Z} \, dk.$$ (7.4.6)

By dividing eqn (7.4.5) by eqn (7.4.6) we get

$$
\begin{aligned}
\bar{A} &= \frac{\int_0^\infty D(k) A(k) \, e^{-\varepsilon(k)/k_B T} dk}{\int_0^\infty D(k) e^{-\varepsilon(k)/k_B T} dk} \\
&= \frac{\int_0^\infty f(k) A(k) \, dk}{\int_0^\infty f(k) \, dk}
\end{aligned}
$$ (7.4.4)

by using eqn (7.4.1) for $f(k)$.

As an example of the usefulness of eqn (7.4.4), let us calculate the average value of the wave vector. This is given by

$$\bar{k} = \frac{\int_0^\infty f(k) k \, dk}{\int_0^\infty f(k) \, dk} = \frac{\int_0^\infty k^3 \, e^{-\hbar^2 k^2/2mk_B T} dk}{\int_0^\infty k^2 \, e^{-\hbar^2 k^2/2mk_B T} dk}.$$

To evaluate this expression we convert to a variable $x = \hbar k/(2mk_B T)^{1/2}$. Then

$$\bar{k} = \sqrt{\frac{2mk_B T}{\hbar^2}} \frac{\int_0^\infty x^3 \, e^{-x^2} dx}{\int_0^\infty x^2 \, e^{-x^2} dx}.$$

The integrals which are involved are given in Appendix B. We get finally

$$\bar{k} = \sqrt{\frac{8mk_{\rm B}T}{\pi\hbar^2}}.$$

The average speed of the particle is then

$$\bar{u} = \frac{\hbar\bar{k}}{m} = \sqrt{\frac{8k_{\rm B}T}{\pi m}}.$$

The same technique can be used to get the average value of k^2: we get

$$\overline{k^2} = \frac{\int_0^\infty k^4\, e^{-\hbar^2 k^2/2mk_{\rm B}T}\mathrm{d}k}{\int_0^\infty k^2\, e^{-\hbar^2 k^2/2mk_{\rm B}T}\mathrm{d}k}.$$

The integrals can be calculated using those given in Appendix B; the final answer is

$$\overline{k^2} = \left(\frac{3mk_{\rm B}T}{\hbar^2}\right).$$

The average value of the square of the speed is

$$\overline{u^2} = \left(\frac{\hbar}{m}\right)^2 \overline{k^2} = \frac{3k_{\rm B}T}{m}.$$

It follows from this that the average kinetic energy of each particle is

$$\frac{1}{2}m\overline{u^2} = \frac{3}{2}k_{\rm B}T.$$

In an ideal classical gas the average kinetic energy of the particles is proportional to the temperature.

Another quantity of interest is the root mean square speed $\sqrt{\overline{u^2}}$. For a particle of mass m in three dimensions it is $\sqrt{3k_{\rm B}T/m}$. For example, a gas of nitrogen molecules of mass 4.65×10^{-26} kg at a temperature of 273 K has a root mean square speed of

$$\sqrt{\overline{u^2}} = \sqrt{\frac{3 \times 1.3806 \times 10^{-23} \times 273}{4.65 \times 10^{-26}}} = 493\,\mathrm{m\,s^{-1}}.$$

In contrast, the average speed of such a molecule is

$$\bar{u} = \sqrt{\frac{8 \times 1.3806 \times 10^{-23} \times 273}{\pi \times 4.65 \times 10^{-26}}} = 454\,\mathrm{m\,s^{-1}}.$$

7.5   Molecular beams

Many modern engineering devices use narrow beams of particles. *Molecular beams* can be produced by allowing molecules of a gas to escape from a small opening in the walls of the container into a region where the vapour pressure is kept low by continuous pumping. The beam of molecules is collimated by a series of baffles which only allows molecules travelling in a particular direction to emerge. If the molecules are of a metallic substance such as silver, then it is necessary to heat the molecules in an oven or electric furnace. If the hole in the oven is sufficiently small then the silver molecules rattle around in the oven for a long time before emerging. In this time the molecules come into thermal equilibrium with the walls of the oven. The emerging beam can then be characterized by the temperature of the oven.

The distribution of molecular speeds of the molecules emerging from an orifice in the oven is not the Maxwell distribution, $n(u)$. Molecules with a high speed approach the hole more rapidly than those of low speed and so the distribution of emerging molecules is weighted towards those of higher speeds.

Suppose we consider particles with speed in the range u to $u + du$ which cross an area A at an angle θ to the normal to the area. In a time t they travel a distance ut and sweep out a volume $Aut \cos(\theta)$. The number of particles in this volume with speeds in the range u to $u + du$ and whose direction of motion lies in the range θ to $\theta + d\theta$ and ϕ to $\phi + d\phi$ is

$$Aut \cos(\theta) \frac{n(u)\, du}{V} \frac{d\theta \sin(\theta)\, d\phi}{4\pi}.$$

When we integrate over angles with ϕ going from 0 to 2π and θ going from 0 to $\pi/2$ we get the flux of particle with speed in the range u to $u + du$ per unit area

$$
\begin{aligned}
f(u)\, du &= \frac{un(u)\, du}{4\pi V} \int_0^{\frac{\pi}{2}} d\theta \cos(\theta) \sin(\theta) \int_0^{2\pi} d\phi \\
&= \frac{un(u)\, du}{4V}. \qquad (7.5.1)
\end{aligned}
$$

The *flux of particles leaving through the hole* with speed u is proportional to the product of the Maxwell distribution and the speed, that is to $un(u)$. The total flux per unit area is found by integrating over all speeds; it is $\frac{1}{4}n\bar{u}$ where $n = N/V$ is the number density and

$$\bar{u} = \frac{\int_0^\infty u\, n(u)\, du}{\int_0^\infty n(u)\, du}$$

is the average speed. By measuring the distribution of speeds of emerging particles we expect to find the function

$$f(u) = \frac{N\lambda_D^3 m^3}{8\pi^2 V \hbar^3} u^3\, e^{-mu^2/2k_B T}. \qquad (7.5.2)$$

The distribution of speeds of emerging particles can be measured experimentally. One way of doing this is to create a pulsed beam of molecules from an oven

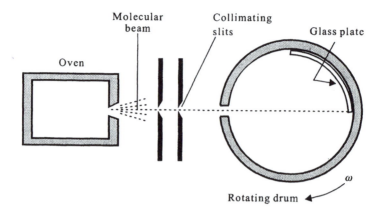

Fig. 7.5 A schematic view of the apparatus used to study the distribution of speeds of particles emerging from a hole in an oven. The beam of particles passes through collimated slits before passing through a hole in a rotating cylinder. The particles take a finite time to cross the cylinder. The fastest particles hit the glass plate first, the slowest ones arrive later and are deposited on a different part of the glass plate. By studying the spatial distribution of the particles deposited on the plate we can infer the distribution of speeds of the particles.

at temperature T, as illustrated in Fig. 7.5. The beam is collimated and then passes through a tiny slit in a rotating drum. Typically the drum rotates at a frequency, f, of 5000 r.p.m. If the drum were at rest, all the molecules in the collimated beam would strike the face opposite the slit. As the drum is rotating, the faster molecules in the beam arrive at a point slightly shifted from the point opposite the slit, the slower ones arrive at a point which is shifted much more.

Suppose the drum has a diameter d. The time for a molecule of speed u to cross the drum is $t = d/u$. In this time the drum rotates through an angle $\theta = 2\pi f d/u$. The faster molecules are shifted by a small angle, the slower ones by a large angle. By measuring the distribution of particles as a function of angle, we can infer the distribution of emerging particles as a function of speed. Experiments that have been done in this way have confirmed that the particles do follow the predicted distribution. In Fig. 7.6 we show some of the data collected by Miller and Kusch (1955) for thallium atoms. The data collected at two temperatures fall on the same curve provided the amplitude of the peak is scaled to be two, and the speed is divided by the speed at the peak. The solid line shown in Fig. 7.6 is the theoretical curve, eqn (7.5.2), scaled the same way.

In many situations only the molecules that travel at high speed really matter. For example, evaporation from a liquid involves molecules crossing the surface of the liquid. The faster molecules are more able to leave than the slower. The rate at which a chemical reaction proceeds often varies with temperature. The high speed molecules which are present at high temperatures have enough energy to

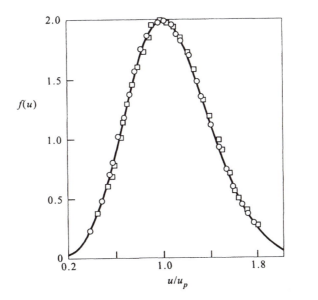

Fig. 7.6 The data of Miller and Kusch (1955) for the distribution of speeds of thallium atoms for two temperatures: circles for $T = 870 \pm 4\,\mathrm{K}$, squares for $T = 944 \pm 5\,\mathrm{K}$. The data have been scaled so that the maximum amplitude is 2 and the peak in the distribution is at u_p. The speed at the maximum can be calculated either from fitting the distribution or from the temperature. For $T = 944 \pm 5\,\mathrm{K}$ this gives $u_p = 395 \pm 2\,\mathrm{m\,s^{-1}}$ from fitting the distribution and $392 \pm 1\,\mathrm{m\,s^{-1}}$ from the temperature. For $T = 870 \pm 4\,\mathrm{K}$ the values are $u_p = 376 \pm 2\,\mathrm{m\,s^{-1}}$ from fitting the distribution and $376 \pm 1\,\mathrm{m\,s^{-1}}$ from the temperature.

activate the reaction.

The molecular composition of the atmosphere of a star or a planet depends on the *escape velocity*. For the Earth (of radius R_E) the escape velocity $(2gR_E)^{1/2}$, is $11\,000\,\mathrm{m\,s^{-1}}$. The average speed of a hydrogen molecule in the Earth's atmosphere (taken to be at an average temperature of $250\,\mathrm{K}$) is about $1800\,\mathrm{m\,s^{-1}}$. These values seem to indicate that hydrogen will not be able to escape from the Earth. But about 2 in 10^9 molecules travel at about six times the average speed, so there is always a finite fraction of molecules travelling faster than $11\,000\,\mathrm{m\,s^{-1}}$. They can escape from the Earth if they travel vertically upwards. Given a long enough time virtually all of them will escape; this is the reason why the atmosphere contains very few hydrogen molecules.

The corresponding numbers for oxygen and nitrogen are different; the typical molecular velocity is about four times smaller, and this means that the fraction of molecules which can escape are those which are more than 24 standard deviations from the mean. Very few of these molecules escape each year, and that is why we can breathe easily.

7.6 Problems

1. Calculate the single-particle density of states in energy of a free particle in two dimensions.

2. Estimate the root mean square velocity and corresponding wavelength for neutrons in thermal equilibrium with a reservoir at 77 K. Estimate the most probable neutron velocity.

3. Calculate the single-particle density of states in energy for an excitation of energy $\varepsilon(k) = \alpha k^{3/2}$ in three dimensions.

4. Calculate the root mean square speed of a ^{4}He atom in a gas at temperature 20°C.

5. Calculate the root mean square speed and the mean speed of hydrogen molecules of mass 3.32×10^{-27} kg at 300 K.

6. According to the theory of relativity, the energy, ε, of a particle of rest mass, m, and momentum, $\mathbf{p} = \hbar \mathbf{k}$, is

$$\varepsilon^2 = p^2 c^2 + m^2 c^4.$$

(a) What is the density of states in energy of a relativistic particle ($p^2 \gg m^2 c^2$) enclosed in a volume V?

(b) Give an expression for the partition function of an ideal gas of N relativistic particles at high temperature.

7. Calculate the partition function for a relativistic particle ($p \gg mc$) in one dimension.

8. The oven in a simple molecular beam apparatus contains H_2 molecules at a temperature of 300 K and at a pressure of 1 mm of mercury. The hole in the oven has a diameter of 100 μm which is much smaller than the molecular mean free path. Calculate:

(a) the distribution in speeds of the molecules in the beam;
(b) the mean speed of the molecules in the beam;
(c) the most probable speed of molecules in the beam;
(d) the average rotational energy of the molecules;
(e) the flux of molecules through the hole. (A pressure of 760 mm of mercury is equivalent to 1.013×10^5 N m^{-2}.)

9. Calculate the partition function of a relativistic particle ($p \gg mc$) in two dimensions. You will need to use polar coordinates to describe $\mathbf{k}$ space.

10. What fraction of H_2 gas molecules at sea level and at $T = 300$ K has sufficient speed to escape the Earth's gravitational field? Leave your answer in integral form.

11. Calculate the partition function in three dimensions for a particle whose energy varies with wave vector as $\varepsilon(k) = \alpha k^3$.

12. Consider the emission and absorption of light by molecules of a hot gas. Derive an expression for the frequency distribution, $F(\nu)$, expected for a

spectral line of central frequency ν_0 due to Doppler broadening in an ideal gas at temperature T.

13. Calculate the Maxwell–Boltzmann distribution and average speed of free particles in two dimensions.

14. Calculate the average reciprocal speed, $\overline{u^{-1}}$, of a Maxwell–Boltzmann distribution for a particle of mass m in three dimensions.

15. The oven represented in Fig. 7.5 contains bismuth at a temperature of 830 °C. The drum has a radius of 0.1 m and rotates at 5000 r.p.m. For molecules travelling at the root mean square speed find the displacement from the point opposite the slit for both Bi and Bi_2 molecules. (The mass of a bismuth atom is 209 a.m.u.)

16. The energy distribution of neutrons from a reactor is investigated with a velocity selector. The length of the neutron path through the selector is 10 m; the velocity of the neutron is defined by the time taken for the neutron to travel along the path. The table gives the relative number of neutrons

t ms	Number	t ms	Number
30	1.2	2.5	18.8
15	2.7	2.14	15.0
7.5	8.2	1.87	11.3
5	13.4	1.67	7.2
3.75	19.0	1.5	4.1
3	20.5	1.2	2.1

per unit velocity range observed for different flight times in milliseconds.

(a) Plot a graph of the distribution of particles with speed.

(b) Show that the distribution of speeds of the particles is of the form

$$f(u) = K u^3 \, e^{-mu^2/2k_B T}$$

if we assume that the particles are in thermal equilibrium at temperature T.

(c) From the maximum in the distribution, estimate the temperature of the reactor.

(d) Plot a suitable graph as a function of u^2 which enables you to find a more reliable estimate of the temperature. Comment on the shape of the observed distribution compared to that expected if the neutrons in the reactor formed a Maxwellian distribution of speeds.

17. A monatomic gas is contained in a vessel from which it leaks through a fine hole. Show that the average kinetic energy of the molecules leaving through the hole is $2k_B T$.

8
Planck's distribution

[We consider] the distribution of the energy U among N oscillators of frequency ν. If U is viewed as divisible without limit, then an infinite number of distributions are possible. We consider however—and this is the essential point of the whole calculation—U as made up of an entirely determined number of finite equal parts, and we make use of the natural constant $h = 6.55 \times 10^{-27}$ erg-sec. This constant when multiplied by the common frequency of the oscillators gives the element of energy in ergs ...

Max Planck

8.1   Black-body radiation

The modern theory of *black-body radiation*, which was developed by Max Planck in 1900, gave birth to the subject of quantum physics. Planck's analysis (1900, 1901) changed the course of science. What, then, is a black body and how does it radiate?

A *black body* is defined as one which absorbs all the radiation which is incident upon it, none is reflected. When it is cold it looks black: any radiation which is incident on the body is absorbed. The same body, when heated, will glow. If you were to look through a tiny hole in a furnace, then the reddish glow you would see would be a fair approximation to the visible portion of black-body radiation. The visible part of the spectrum is for wavelengths ranging from 400 nm (violet) to 750 nm (red). Radiation of wavelength outside this range cannot be seen by the naked eye. What you see is only a small part of the total spectrum of radiation.

All objects emit electromagnetic radiation at a rate which varies with their temperature. All of us are emitting electromagnetic radiation at the moment, just because our bodies are at a temperature of about 310 K, but we cannot see the radiation because it is at the wrong wavelength for our eyes. Instead we can take pictures of people by using an infrared camera which detects this radiant energy. An infrared camera detects radiation of wavelength greater than 750 nm.

When a body is in thermal equilibrium with its surroundings, it emits and absorbs radiation at exactly the same rate. If it emits radiation poorly it must absorb radiation poorly. A good absorber is a good emitter. The ability of a body to emit radiation is directly proportional to its ability to absorb radiation. When the temperature of a body exceeds that of its surroundings, it emits more radiation than it absorbs. People can be seen with an infrared camera because they are hotter than their surroundings.

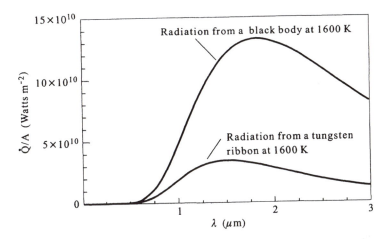

Fig. 8.1 The radiation emitted per unit area from a tungsten ribbon at 1600 K compared to the energy emitted by a black body at the same temperature. Notice that the radiation is much less than for a black body and that the peak is shifted to slightly shorter wavelengths.

The predominant wavelength in the radiation decreases as the temperature of the object is raised. An approximation to black-body radiation is seen when the tungsten filament in a light-bulb is heated by passing a current through it. As the current increases, the temperature of the filament rises, the filament glows more brightly and its colour changes. As the temperature rises the colour starts to glow dull red, then goes through bright red, orange, yellow, before the filament burns out.

There is much less radiation from the tungsten filament in a light-bulb than from a black body at the same temperature. Figure 8.1 shows the black-body radiation spectrum expected at 1600 K compared to that from a tungsten filament at the same temperature. The amount of radiation from the tungsten filament for all wavelengths is less than from the black body. The spectrum for the tungsten filament is also shifted slightly to shorter wavelengths. These differences arise because the tungsten filament does not absorb all radiation incident on it, so it emits less than a black body. No material substance is truly a black body.

A black body absorbs all the radiation incident on it. It is convenient to discuss black bodies in terms of an idealized model which is independent of the nature of any material substance. The best model is a large cavity with a very small hole in it which absorbs all the incident radiation. The radiation hits the walls and rattles around inside and comes to thermal equilibrium with the walls before emerging from the hole. The smaller the hole, the longer it takes before the radiation emerges. Kirchhoff (1860) gave a good operational definition: Given a space enclosed by bodies of equal temperature, through which no radiation can penetrate, then every bundle of radiation within this space is constituted, with respect to quality and intensity, as if it came from a completely black body at the same temperature.

Suppose that the walls of an oven surrounding a cavity are heated up to a temperature T. It is found experimentally that the radiation from the hole in the cavity depends only on the temperature of the oven walls and on the area, A, of the hole. The radiation is independent of the material that makes up the walls. The experiment was first done in 1879 by Josef Stefan, who found that the rate of radiation, dQ/dt, from area A is

$$\frac{dQ}{dt} = A\sigma T^4 \qquad (8.1.1)$$

where σ is a constant, now called the *Stefan constant*. Its value is $\sigma = 5.67 \times 10^{-8}$ $\mathrm{W\,m^{-2}\,K^{-4}}$. In 1884 Ludwig Boltzmann found a thermodynamic argument to support this law of radiation; nowadays eqn (8.1.1) is called the *Stefan–Boltzmann law* in their honour.

The Stefan–Boltzmann law tells us how much energy is emitted per second, but tells us nothing about how the radiated energy is divided into the infinite number of wavelengths that are present in black-body radiation. Kirchhoff challenged theorists and experimentalists to find out how the energy is distributed amongst all the wavelengths present.

The radiation that pours out of a cavity per second is proportional to the *energy density*, u, the amount of energy stored as radiation in a unit volume in the cavity. This energy density can be decomposed into the energy density for radiation in a small range of wavelengths. Let $u(\lambda)\,d\lambda$ be the radiation energy per unit volume with wavelength in the range λ to $\lambda + d\lambda$. We refer to $u(\lambda)$ as the *spectral density of radiation*. The rate of radiation from the hole must equal the rate at which radiation is incident on the hole from within the cavity. All the radiation travels at the speed of light, c. The calculation is the same as that given in Chapter 7. The rate at which radiation in a small range from λ to $\lambda + d\lambda$ is incident on a hole of area A is

$$\frac{dQ(\lambda)}{dt} = \frac{Acu(\lambda)\,d\lambda}{4\pi} \int_0^{2\pi} d\phi \int_0^{\pi/2} \sin(\theta)\cos(\theta)\,d\theta. \qquad (8.1.2)$$

This is the amount of radiant energy incident on the area from a half space in which energy is moving towards the hole. We integrate over λ and over angles to get

$$\frac{dQ}{dt} = \frac{Acu}{4} \qquad (8.1.3)$$

with

$$u = \int_0^\infty u(\lambda)\,d\lambda \qquad (8.1.4)$$

the total energy density. The rate of radiation per unit area is equal to $cu/4 = \sigma T^4$. In other words the energy density varies as the fourth power of the temperature, $u = 4\sigma T^4/c$.

Wilhelm Wien (1893) showed by thermodynamic arguments that the spectral density of black-body radiation has a particular form:

$$u(\lambda)\,\mathrm{d}\lambda = \frac{f(\lambda T)\,\mathrm{d}\lambda}{\lambda^5}. \tag{8.1.5}$$

We will refer to eqn (8.1.5) as *Wien's scaling law*. Equation (8.1.5) can be rewritten as

$$\frac{u(\lambda)}{T^5} = \frac{f(\lambda T)}{\lambda^5 T^5}, \tag{8.1.6}$$

which implies that if we were to plot $u(\lambda)/T^5$ as a function of λT, all experimental data should scale onto a single curve. Experiment could determine the unknown function $f(\lambda T)$.

The energy density is given by the expression

$$u = \int_0^\infty u(\lambda)\,\mathrm{d}\lambda = \int_0^\infty \frac{f(\lambda T)\,\mathrm{d}\lambda}{\lambda^5}. \tag{8.1.7}$$

There is a maximum in $u(\lambda)$ as a function of λ which occurs where

$$\frac{\mathrm{d}u(\lambda)}{\mathrm{d}\lambda} = \frac{T f'(\lambda T)}{\lambda^5} - \frac{5 f(\lambda T)}{\lambda^6} = 0. \tag{8.1.8}$$

Here f' is the derivative $\mathrm{d}f(\lambda T)/\mathrm{d}\lambda T$. This gives

$$\lambda T = \frac{5 f(\lambda T)}{f'(\lambda T)}. \tag{8.1.9}$$

The maximum in $u(\lambda)$ corresponds to a wavelength denoted as λ_m. Equation (8.1.9) must have a solution for a particular value of λT but the value depends on the scaling function $f(\lambda T)$. Modern experiments give

$$\lambda_m T = 0.002899 \,\mathrm{m\,K}. \tag{8.1.10}$$

(The units are metres kelvin, not millikelvin.) This relation, called *Wien's displacement law*, tells us that the maximum in $u(\lambda)$ shifts to shorter wavelengths as the temperature rises. The hotter the temperature, the smaller the wavelength at the maximum, λ_m.

Equation (8.1.2) indicates that the energy in the range to $\lambda + \mathrm{d}\lambda$ which is radiated from an oven with a tiny hole is proportional to $u(\lambda)$. The first experiment to measure $u(\lambda)$ was done by Paschen who could detect radiation in the near infrared in the range 1 to 8 μm for temperatures up 1600 K. More important was the experiment of Lummer and Pringsheim published in 1900 for

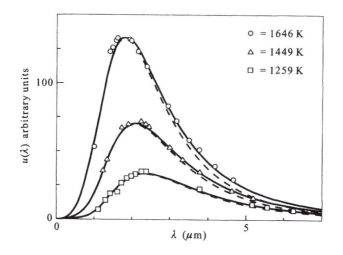

Fig. 8.2 Some of the original data of Lummer and Pringsheim (1900) taken at three different temperatures. The solid curve is the best fit of the data to Planck's formula, eqn (8.1.12), the dashed line the fit to Wien's formula, eqn (8.1.11).

wavelengths in the range 12 to $18\,\mu$m. We show in Fig. 8.2 some of the original set of data of Lummer and Pringsheim for $u(\lambda)$ as a function of λ. The data shown are for temperatures of 1646 K, 1449 K, and 1259 K and for wavelengths from 1 to $7\,\mu$m. Two things should be noted: at high temperatures, the total area under the curves is larger than at lower temperatures; the position of the maxima in the curves, which represents the predominant colour, shifts to shorter wavelength.

Is Wien's scaling law correct? The data of Lummer and Pringsheim when plotted as $u(\lambda)/T^5$ against λT should scale onto a single curve. Figure 8.3 shows that the data really do fall on a single curve. (It has to be admitted that when making this plot we have omitted the data which Lummer and Pringsheim indicated were poor as a result of absorption of water vapour and carbon dioxide in the air.)

There are two sets of theoretical curves shown in Fig. 8.2. The curves shown by dashed lines represent the model of Wilhelm Wien (1896) who proposed that $u(\lambda)$ was of the form

$$u(\lambda) = \frac{c_1 e^{-c_2/\lambda T}}{\lambda^5}, \qquad (8.1.11)$$

an equation which fits the data fairly well for the short wavelengths shown in Fig. 8.2, but which goes wrong for long wavelengths. The curves shown by solid lines in Fig. 8.2 are generated from Planck's formula

$$u(\lambda) = \frac{8\pi h c}{\lambda^5 \left(e^{hc/\lambda k_B T} - 1\right)}, \qquad (8.1.12)$$

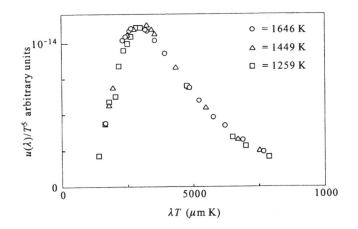

Fig. 8.3 The data of Lummer and Pringsheim shown in Fig. 8.2 plotted as $u(\lambda)/T^5$ as a function of λT. The data all fall on a single curve, verifying Wien's scaling law, eqn (8.1.5).

which fits over the entire range. Planck's formula will be discussed in detail in section 8.3.

When Planck created his theory he had just been told by Rubens of some new experiments which gave details about the energy density for long wavelengths where eqn (8.1.11) breaks down. Experiments of Rubens and Kurlbaum (1900) indicated that $u(\lambda)$ was proportional to the temperature for long wavelengths. Earlier there was an attempt by Lord Rayleigh to create a theory of black-body radiation for long wavelengths, or graver modes as Rayleigh called them. Rayleigh's first calculation of 1900 was based on 'the Boltzmann–Maxwell doctrine of the [equi]partition of energy'. Rayleigh considered the waves as if they were waves on a stretched string vibrating transversely. Rayleigh's calculation of 1900 was extremely brief, but he proposed that $u(\lambda)$ varies as $c_1 T\lambda^{-c_2/\lambda T}$. This formula works for long wavelengths but not for short ones. He expanded on the calculation in 1905 in a paper published in *Nature*; unfortunately he overestimated the number of modes by a factor of eight, a mistake that was pointed out by Sir James Jeans (1905). Einstein (1905), in the paper where he predicted the photoelectric effect, also derived the correct formula for long wavelengths. The theory is now usually referred to as the *Rayleigh–Jeans theory* even though Einstein deserves credit as well.

8.2   The Rayleigh–Jeans theory

The black body consists of electromagnetic radiation in thermal equilibrium with the walls of the cavity. When they are in thermal equilibrium, the average rate of emission of radiation by the walls equals their average rate of absorption of radiation. The condition for thermal equilibrium is that the temperature of the walls is equal to the temperature of the radiation. But what do we mean by the idea that *radiation has a temperature?*

We could imagine that the material of the walls contains charged particles which oscillate about their equilibrium positions. (It does not matter what the walls are made of, so we can mentally imagine them to consist of oscillating charges.) As was known from Maxwell's work on electromagnetism, a moving charge radiates an electromagnetic wave. We imagine the walls to be full of charged particles jiggling about, coupled to electromagnetic standing waves. Suppose each standing-wave mode of the electromagnetic field is coupled to an oscillator in the wall which oscillates with the same frequency as the standing wave. Each oscillator has two degrees of freedom, one for the kinetic energy, one for the potential energy, and so it has an average energy of k_BT according to the equipartition theorem. In thermal equilibrium the average energy of the oscillator and the average energy of the standing-wave mode of the electromagnetic field must be the same for the two to be in thermal equilibrium. Hence each mode of oscillation of the electromagnetic field has an energy k_BT and can be thought of as having a temperature T. This is the basis of the Rayleigh–Jeans theory.

It can be shown by experiment that the energy distribution of black-body radiation is independent of the shape of the cavity. For ease of calculation let us take the cavity to be a cube, and let us assume that the waves vanish at the cavity walls. The number of standing electromagnetic waves in a cube of length L with wave vector in the range k to $k + dk$ is

$$D(k)\, dk = \frac{L^3}{\pi^2} k^2 \, dk. \qquad (8.2.1)$$

There is an extra factor of two compared to eqn (7.2.4) because there are two transverse electromagnetic waves for each wave vector. We can express the spectrum of radiation in terms of wavelength. The density of states in wavelength for electromagnetic waves is

$$D(\lambda)\, d\lambda = \frac{8\pi L^3}{\lambda^4} \, d\lambda. \qquad (8.2.2)$$

Each mode of oscillation has an energy k_BT, so the energy in the range λ to $\lambda + d\lambda$ is $k_BTD(\lambda)\, d\lambda$. The energy density in this range is

$$u(\lambda)\, d\lambda = \frac{D(\lambda)k_BT}{L^3} \, d\lambda = \frac{8\pi k_BT}{\lambda^4} \, d\lambda, \qquad (8.2.3)$$

the Rayleigh–Jeans expression for the spectral density in the range λ to $\lambda + d\lambda$.

If we believe that energy is a continuous variable then the average energy per oscillator is k_BT, and the Rayleigh–Jeans formula for $u(\lambda)$ would follow. We are forced by experiment to concede that the theory is wrong for short wavelengths. But the theory is correct in two respects.

The first respect is that the Rayleigh–Jeans expression agrees with Wien's scaling formula:

$$u(\lambda) = \frac{8\pi k_B T}{\lambda^4} = \frac{f(\lambda T)}{\lambda^5} \qquad (8.2.4)$$

with $f(\lambda T) = 8k_B \lambda T$. Even a casual glance shows that this is not the correct scaling function. As λ decreases, eqn (8.2.4) predicts that $u(\lambda)$ increases as λ^{-4}. According to their theory, if you were to look in a glowing log fire there would be a huge amount of radiant energy of short wavelength; these microwaves would cook your eyes.

Worse than this is the observation that the total energy density is infinite, an observation which Paul Ehrenfest called the *ultraviolet catastrophe*. The total energy density is

$$u = \int_0^\infty u(\lambda)\,d\lambda = \int_0^\infty \frac{8\pi k_B T}{\lambda^4}\,d\lambda = \infty. \qquad (8.2.5)$$

The energy density is infinite, which is nonsense. If a cavity is full of radiation which is distributed according to the Rayleigh–Jeans formula, the amount of energy which is radiated should be infinite according to eqn (8.2.5). But Stefan found by experiment that the radiated energy varies as T^4. We cannot explain the Stefan–Boltzmann law using the Rayleigh–Jeans formula for the energy density.

However, there is a second respect in which the Rayleigh–Jeans formula is correct. It is now known to work perfectly for long wavelengths thanks to Jeans (and Einstein) getting the numbers correct. This partial agreement of the calculation with experiment showed that the theory is not totally wrong, but something is missing. The question is this: which classical law is wrong for small wavelengths?

8.3 Planck's distribution

Planck decided that he would not assume that the average energy of an oscillator in the wall was equal to $k_B T$. He thought he knew how $u(\lambda)$ varied for short wavelengths (Wien's formula) and he wanted $u(\lambda)$ to be proportional to T for long wavelengths, so he constructed a formula which would do both. That was how he obtained the equation

$$u(\lambda) = \frac{8\pi hc}{\lambda^5 \left(e^{hc/\lambda k_B T} - 1\right)}. \qquad (8.1.12)$$

Everything now works perfectly. When λ is small we get Wien's expression, eqn (8.1.11), but with the constants c_1 and c_2 defined:

$$u(\lambda) \approx \frac{8\pi hc\, e^{-hc/\lambda k_B T}}{\lambda^5}. \qquad (8.3.1)$$

The total energy density is

$$u = \int_0^\infty \frac{8\pi hc}{\lambda^5 \left(e^{hc/\lambda k_B T} - 1\right)} \, d\lambda. \tag{8.3.2}$$

To do the integral, change to a variable $z = hc/\lambda k_B T$, and use one of the integrals in Appendix B. The total energy density is

$$
\begin{aligned}
u &= 8\pi \left(\frac{k_B^4 T^4}{h^3 c^3}\right) \int_0^\infty \frac{z^3 dz}{(e^z - 1)} \\
&= \frac{8\pi^5}{15} \left(\frac{k_B^4 T^4}{h^3 c^3}\right).
\end{aligned} \tag{8.3.3}
$$

The energy density is finite and varies as T^4. We can explain the Stefan–Boltzmann law using eqns (8.1.3) and (8.3.3):

$$
\begin{aligned}
\frac{dQ}{dt} &= \frac{Acu}{4} \\
&= \frac{2\pi^5 A}{15} \left(\frac{k_B^4 T^4}{h^3 c^2}\right) = A\sigma T^4
\end{aligned} \tag{8.3.4}
$$

with σ given by

$$\sigma = \frac{2\pi^5 k_B^4}{15 h^3 c^2}. \tag{8.3.5}$$

The *Stefan constant* involves the ratio of $k_B^4/h^3 c^2$.

Wien's displacement law requires us to work out the value of λ at the maximum in $u(\lambda)$. To get this we have to differentiate $u(\lambda)$, find the maximum, and solve for $\lambda_m T$. This gives the equation

$$\frac{hc}{\lambda_m k_B T} = 5 \left(1 - e^{-hc/\lambda_m k_B T}\right),$$

which can be solved by iteration. As a first step treat the exponential term as small so that $hc/\lambda_m k_B T$ is 5. Insert this on the right-hand side to get the next iteration of $5 \left(1 - e^{-5}\right)$ and so on. After a few iterations we get

$$\frac{hc}{\lambda_m k_B T} = 4.965. \tag{8.3.6}$$

But $\lambda_m T$ is known, experimentally, to be 0.002899 m K. We know the speed of light, c, from independent measurements. Hence we know h/k_B.

Since the ratios h/k_B and h^3/k_B^4 involve different powers of h and k_B we can solve for both. Planck worked them out from the data then available. With the wisdom of hindsight we know that he got a far more accurate value of k_B than any of his contemporaries. The evaluation of k_B was one of the outstanding problems of the period.

From k_B and a knowledge of the gas constant, he could get Avogadro's number, the number of molecules in a mole. He got 6.17×10^{23}, a value which is pretty good for 1901. From a knowledge of the Faraday constant he got for the unit of charge a value of $e = 4.69 \times 10^{-10}$ e.s.u (they worked in electrostatic units then; in modern units his value is 1.56×10^{-19} C compared to the present day value of 1.602×10^{-19} C). Nobody believed this value for the electronic charge until Rutherford and Geiger (1908) did their experiments with α-particles whose charge they estimated to be 9.3×10^{-10} e.s.u, roughly double Planck's value for e. After that people trusted Planck's numbers.

These were the immediate successes of Planck's calculation. Moreover he could now predict the distribution $u(\lambda)$ and show that it agreed with experiment. The solid curves in Fig. 8.2 are the best fit of his theory to the data of Lummer and Pringsheim. The dashed lines are the fit to the theory if we ignore the one in the denominator and use eqn (8.1.11), the formula proposed by Wien. The discrepancy between the two expressions shows up best for long wavelengths at the highest temperatures. Planck's formula fits the experimental data amazingly well. He wanted to understand why this was so.

Planck's formula can be rewritten in terms of the average energy per mode of oscillation (by the term 'mode' we mean one of the standing waves in the cavity). The spectral density can be written as

$$u(\lambda) = \frac{8\pi}{\lambda^4} U(\lambda, T) \tag{8.3.7}$$

where the average energy per mode is

$$U(\lambda, T) = \frac{hc}{\lambda \left(e^{hc/\lambda k_{\mathrm{B}} T} - 1 \right)}. \tag{8.3.8}$$

This energy is equated to the average energy of the oscillator in the wall which corresponds to this standing-wave mode. Planck concentrated his attention on the statistical mechanics of the oscillators in the wall. He treated the oscillators in the wall as if they were thermally isolated from the rest of the universe so he could use the techniques of the microcanonical ensemble.

Suppose the entropy of an oscillator of frequency ν $(= c/\lambda)$ in the microcanonical ensemble is of the form

$$S = k_{\mathrm{B}} \left\{ \left(1 + \frac{U}{h\nu} \right) \ln \left(1 + \frac{U}{h\nu} \right) - \frac{U}{h\nu} \ln \left(\frac{U}{h\nu} \right) \right\} \tag{8.3.9}$$

where U is the energy of the oscillator. By using the equation $T^{-1} = (\partial S/\partial U)_V$ we recover eqn (8.3.8) with the frequency given by $\nu = c/\lambda$. Planck wanted to find a way of obtaining eqn (8.3.9) for the entropy by calculating W directly and using the relation $S = k_{\mathrm{B}} \ln(W)$. This is how he did it.

Consider a large number N of oscillators all with frequency ν. The total energy is $U_N = NU$ and the total entropy $S_N = NS = k_{\mathrm{B}} \ln(W_N)$. W_N is the number of arrangements for distributing the energy U_N amongst the N oscillators. Planck imagined that the total energy is made up of finite energy elements of size ϵ, so that $U_N = M\epsilon$ where M is a large integer. This is *Planck's quantum hypothesis*. W_N is taken to be the number of ways in which M indistinguishable energy elements (quanta) can be arranged amongst N distinguishable oscillators. This number is

$$W_N = \frac{(N-1+M)!}{M!\,(N-1)!}. \tag{8.3.10}$$

By using Stirling's approximation, then replacing $N-1$ by N, and by putting $M/N = U/\epsilon$ we get the average entropy per oscillator:

$$S = k_B \left\{ \left(1 + \frac{U}{\epsilon}\right) \ln \left(1 + \frac{U}{\epsilon}\right) - \frac{U}{\epsilon} \ln \left(\frac{U}{\epsilon}\right) \right\}, \tag{8.3.11}$$

which is the same as eqn (8.3.9) if we put $\epsilon = h\nu$. The success of this calculation indicates that the energy of the oscillators comes in integer amounts of $h\nu$.

Planck introduces two new ideas: not only is energy quantized, but the counting involves indistinguishable energy elements or quanta. This was an entirely new way of counting the number of arrangements of the energy amongst the oscillators. Planck's inspired guess was a crucial step in the birth of quantum mechanics. By trying to fit the data, in effect Planck had shown that the energy comes, not as a continuous variable, but in discrete bundles.

8.4   Waves as particles

Planck quantized the oscillators which he imagined to exist in the wall of the cavity. He was unaware that his quantization proposal could be applied to oscillations of the classical radiation field itself. This was the point of view which was put forward by Einstein (1905) in the paper in which he proposed an explanation for the photoelectric effect. As Einstein put it: 'the energy of a light ray spreading from a light source is not continuously distributed over an increasing space but consists of a finite number of energy quanta ... which move without dividing, and which can only be produced and absorbed as complete units.' He pictured the radiation as made up of particles, with each particle having an energy $h\nu$ when it is in a standing wave state. If there are no particles in the state $\phi_i(x)$ then the energy is zero, if there is one particle the energy is $h\nu$, if there are two particles, the energy is $2h\nu$, and so on. This explanation accounts simply for the quantization of energy—it is just a manifestation of the discreteness of particles.

Somehow we have to marry the concept of a particle with a well-defined energy and momentum to that of a wave. After all, electromagnetic waves show interference and diffraction effects, and behave in all optical experiments as continuous waves. In quantum mechanics the state of a particle is described by a wavefunction $\phi_i(x)$ which allows it to show interference effects. If there are two particles in the single-particle state then the effective amplitude of the wave is increased. When there are millions of particles in the same state, then the amplitude of the wave has increased so much as to become observable. It is this large-amplitude wave we observe as the classical wave in diffraction and interference experiments.

Planck's work suggested to Einstein that electromagnetic waves behave as if they were made of particles. The classical conception of an electromagnetic wave has serious difficulties in explaining the photoelectric effect. But assuming the electromagnetic wave to be made of particles makes the explanation relatively simple. An electromagnetic wave incident on a solid can be pictured as the bombardment of the solid by a stream of particles of light. When a particle of light (nowadays called a photon) strikes the atoms in the solid, it delivers its entire energy to a single electron. Some of these electrons which have absorbed a particle of light are energetic enough to leave the solid. Those electrons ejected normally to the surface of the solid have the largest energy. One can measure their energy by charging the solid to a positive potential and measuring the smallest potential where no electrons are emitted. This was verified in a series of experiments by Millikan, much to his surprise. As Millikan (1924) wrote in his Nobel lecture: 'In view of all these methods and experiments the general validity of Einstein's [photoelectric] equation is, I think, now universally conceded, and *to that extent the reality of Einstein's light-quanta may be considered as experimentally established.* But the conception of *localized* light quanta out of which Einstein got his equation must still be regarded as far from being established.' Millikan refused to accept the idea of localized particles of light, even though the idea led to predictions which were in complete accord with his experiments. He could not accept the quantum hypothesis. Eventually Einstein was awarded the Nobel Prize, not for his theory of special relativity, but for this suggestion that light could be considered as if it were made of particles, a truly revolutionary suggestion bearing in mind the success of Maxwell's theory of electromagnetism.

During the slow acceptance of Einstein's quantum hypothesis, the question arose whether other waves have quantized energy levels. The first application of this idea concerned waves in a crystalline solid. Debye (1912) used this idea to propose a model for the heat capacity of a solid. More direct evidence for quantized lattice waves comes from experiments in which neutrons are scattered off solids. When the neutron scatters off the lattice of atoms there are changes in the energy and momentum of the neutron which can be interpreted as due to the creation (or absorption) of a single particle of well-defined energy and momentum. The particles associated with a lattice wave are called phonons. Planck's work led, eventually, to the acceptance that all wave motion can be quantized and treated as if it were composed of particles. Examples include quantized spin waves called magnons, and quantized ripples on the surface of liquid helium, called ripplons.

8.5   Derivation of the Planck distribution

From one point of view, we can analyse the electromagnetic field in a box or cavity in terms of a lot of harmonic oscillators, treating each mode of oscillation according to quantum mechanics as a harmonic oscillator. From a different point of view, we can analyse the same physics in terms of identical Bose particles. And the results of both ways of working **are always in exact agreement**. There is no way to make up your mind whether the electromagnetic field is really to be described as a quantized harmonic oscillator or by giving how many photons are in each condition. The two views turn out to be mathematically identical.

Richard Feynman. The Feynman Lectures vol III (1965)

The classical wave equation is

$$\frac{1}{s^2}\frac{\partial^2 y}{\partial t^2} = \nabla^2 y \tag{8.5.1}$$

where s is the velocity of the wave. For waves in a box of sides L_x, L_y, L_z we look for a standing wave solution to this equation of the form

$$y_{\mathbf{k}}(x, y, z) = A \sin\left(\frac{n_1 \pi x}{L_x}\right) \sin\left(\frac{n_2 \pi y}{L_y}\right) \sin\left(\frac{n_3 \pi z}{L_z}\right). \tag{8.5.2}$$

(Compare to eqn (7.1.1).) The label $\mathbf{k}$ denotes the wave vector given by

$$\mathbf{k} = \hat{\mathbf{i}} k_x + \hat{\mathbf{j}} k_y + \hat{\mathbf{k}} k_z \tag{7.2.1}$$

with

$$k_x = \frac{\pi n_1}{L_x}; \ k_y = \frac{\pi n_2}{L_y}; \ k_z = \frac{\pi n_3}{L_z}. \tag{7.2.2}$$

In this sense the wave vector $\mathbf{k}$ acts as a shorthand for the set of numbers n_1, n_2, and n_3. The coordinate $y_{\mathbf{k}}$ satisfies the equation

$$\nabla^2 y_{\mathbf{k}}(x, y, z) = -k^2 y_{\mathbf{k}}(x, y, z) \tag{8.5.3}$$

with

$$k^2 = \pi^2 \left(\frac{n_1^2}{L_x^2} + \frac{n_2^2}{L_y^2} + \frac{n_3^2}{L_z^2}\right). \tag{8.5.4}$$

The standing wave satisfies the differential equation

$$\frac{\partial^2 y_{\mathbf{k}}}{\partial t^2} = -s^2 k^2 y_{\mathbf{k}}, \tag{8.5.5}$$

the equation of a harmonic oscillator of angular frequency $\omega(k) = sk$. If we were to use travelling waves, as described in Appendix D, then we would get exactly the same equation as (8.5.5) but with the quantity k^2 altered to

$$k^2 = 4\pi^2 \left(\frac{n_1^2}{L_x^2} + \frac{n_2^2}{L_y^2} + \frac{n_3^2}{L_z^2} \right).$$

The equation would still be that of a simple harmonic oscillator. The following analysis applies to both standing and travelling waves. Moreover, the simple harmonic oscillator equation is obtained for other waves, such as waves on the surface of a liquid, in which the angular frequency, $\omega(k)$, does not vary linearly with k. In general, $\omega(k)$ is an arbitrary function of the wave vector $\mathbf{k}$, but we shall simplify matters by assuming that it only depends on the magnitude of the wave vector, that is k.

What are the quantum mechanical energy eigenvalues for a harmonic oscillator? If we treat the waves as if they were simple harmonic oscillators, the energy eigenvalues are

$$\varepsilon_n(k) = (n + 1/2)\, \hbar\omega(k). \tag{8.5.6}$$

The separation between neighbouring energy levels is $\hbar\omega(k)$, so the energy is quantized. (Remember that $\hbar = h/2\pi$ so that $h\nu = \hbar\omega(k)$.) The number n represents the quantum number of the harmonic oscillator; in this picture it is not associated with the number of particles.

The quantity $\hbar\omega(k)/2$ is called the *zero-point energy* of the harmonic oscillator. In a classical description the oscillator has its lowest energy when it is at rest with zero kinetic and potential energies. In quantum mechanics the lowest energy is not zero, but $\hbar\omega(k)/2$.

An alternative picture is to postulate the existence of particles in a standing-wave state of energy $\hbar\omega(k)$. When no particles are present the energy is zero, when one is present the energy is $\hbar\omega(k)$, when two are present the energy is $2\hbar\omega(k)$, and so on. The particles must be bosons for we are allowing more than one to be in the standing-wave state $\phi_{\mathbf{k}}$.

This picture gives the same difference in energy levels as the picture based on the harmonic oscillator. So do we treat n as the quantum number of a harmonic oscillator, or do we treat it as the number of Bose particles present? If we are interested in the thermal properties, then only differences between thermodynamic states are important. The effect of the zero-point energy on the internal energy is unimportant, for it is constant. Of course the zero-point energy will affect the absolute value of the internal energy. In what follows we will ignore the zero-point energy for convenience and treat n as if it were the number of particles present, even when we are discussing harmonic oscillators.

The average number of particles in the single-particle state $\phi_{\mathbf{k}}$ is

$$\bar{n}(\mathbf{k}) = \sum_{n=0}^{\infty} n p_n(\mathbf{k}) \tag{8.5.7}$$

where

$$p_n(\mathbf{k}) = \frac{e^{-n\hbar\omega(k)/k_{\mathrm{B}}T}}{Z(\mathbf{k})} \tag{8.5.8}$$

is the probability that there are n particles in this single-particle state, and $Z(\mathbf{k})$ is the partition function for this state,

$$
\begin{aligned}
Z(\mathbf{k}) &= 1 + e^{-\hbar\omega(k)/k_B T} + e^{-2\hbar\omega(k)/k_B T} + e^{-3\hbar\omega(k)/k_B T} + \cdots \\
&= \frac{1}{1 - e^{-\hbar\omega(k)/k_B T}}.
\end{aligned}
\tag{8.5.9}
$$

The average number of particles is

$$
\bar{n}(\mathbf{k}) = \sum_{n=0}^{\infty} n \, e^{-n\hbar\omega(k)/k_B T} \left(1 - e^{-\hbar\omega(k)/k_B T} \right).
\tag{8.5.10}
$$

This sum can be done by the following trick:

$$
\begin{aligned}
S &= \sum_{n=0}^{\infty} n \, e^{-nx} = -\frac{d}{dx} \left(\sum_{n=0}^{\infty} e^{-nx} \right) \\
&= -\frac{d}{dx} \left(\frac{1}{1 - e^{-x}} \right) = \frac{e^{-x}}{(1 - e^{-x})^2}.
\end{aligned}
$$

Hence we get

$$
\bar{n}(\mathbf{k}) = \frac{1}{e^{\hbar\omega(k)/k_B T} - 1}
\tag{8.5.11}
$$

which is called the *Planck distribution function*, derived here for any wave which is treated as if it were made up of particles, not just for quantized electromagnetic waves. Equation (8.5.11) gives the average number of particles in the state $\phi_{\mathbf{k}}(x, y, z)$.

The average thermal energy of the particles of wave vector $\mathbf{k}$ is

$$
\bar{U}(\mathbf{k}) = \hbar\omega(k)\bar{n}(\mathbf{k}).
$$

This thermal energy is directly proportional to the Planck distribution function. Notice that at high temperatures ($k_B T > \hbar\omega$) the average energy of each mode is $k_B T$ which is just what is expected for a harmonic oscillator according to the equipartition theorem.

The total average thermal energy involves the sum of the average energies of all the modes of the system. The sum should include all the types of oscillation, such as longitudinal and transverse waves. If there is only one type (say longitudinal waves), then the thermal energy is

$$
\bar{U} = \sum_{\mathbf{k}} \hbar\omega(k)\bar{n}(\mathbf{k}).
\tag{8.5.12}
$$

The sum over states can be turned into an integral in the usual way, the answer depending on the dimensionality of the system. For example, the transverse waves on the surface of liquid helium have a frequency given by $\omega(k) = \left(\gamma_s k^3/\rho\right)^{1/2}$, where γ_s is the surface tension and ρ is the mass density of liquid helium. The surface has two dimensions, so the thermal contribution to the surface internal energy is

$$\bar{U} = \frac{A}{2\pi} \int_0^\infty \hbar\omega(k)\bar{n}(k)k \,\mathrm{d}k.$$

8.6 The free energy

The most direct way of calculating the thermodynamics of any system is through the partition function. For a collection of modes of oscillation we have to calculate the partition function for each mode and then multiply them together to construct the overall partition function. All wave modes are *distinguishable* because they have different wave vectors, so we do not have to divide by $N!$ in this case.

Let the partition function for the mode of wave vector $\mathbf{k}$ be $Z(\mathbf{k})$. Then the total partition function is

$$Z = Z(\mathbf{k}_1) \times Z(\mathbf{k}_2) \times Z(\mathbf{k}_3) \times Z(\mathbf{k}_4) \times \cdots \tag{8.6.1}$$

where the partition function for the mode of wave vector $\mathbf{k}$ is

$$Z(\mathbf{k}) = \frac{1}{1 - \mathrm{e}^{-\hbar\omega(k)/k_\mathrm{B}T}} \tag{8.5.9}$$

provided we ignore the zero-point energy. The *free energy of the waves* is

$$F = -k_\mathrm{B}T \ln(Z) = -k_\mathrm{B}T \sum_\mathbf{k} \ln(Z(\mathbf{k})). \tag{8.6.2}$$

We turn the sum into an integral in the usual way and get

$$F = k_\mathrm{B}T \int_0^\infty D(k) \ln(1 - \mathrm{e}^{-\hbar\omega(k)/k_\mathrm{B}T}) \,\mathrm{d}k. \tag{8.6.3}$$

This expression for the thermal contribution to the free energy implies that for finite temperatures the free energy is negative. As the temperature tends to zero, so the free energy tends to zero, the energy of the ground state.

For black-body radiation there are the two transverse waves, so we get for the free energy

$$F = \frac{k_\mathrm{B}TV}{\pi^2} \int_0^\infty k^2 \ln(1 - \mathrm{e}^{-\hbar ck/k_\mathrm{B}T}) \,\mathrm{d}k. \tag{8.6.4}$$

Integration by parts gives

$$F = -\frac{\hbar cV}{3\pi^2} \int_0^\infty \frac{k^3}{\mathrm{e}^{\hbar ck/k_\mathrm{B}T} - 1} \,\mathrm{d}k. \tag{8.6.5}$$

By changing to the variable $z = \hbar c k / k_B T$ and using one of the integrals in Appendix B we get

$$F = -\frac{\pi^2 k_B^4 T^4 V}{45 \hbar^3 c^3}. \tag{8.6.6}$$

Once the free energy has been found we can calculate the entropy as

$$S = -\left(\frac{\partial F}{\partial T}\right)_V = \frac{4\pi^2 k_B^4 T^3 V}{45 \hbar^3 c^3}. \tag{8.6.7}$$

The internal energy, $\bar{U} = F + TS$, is given by

$$\bar{U} = \frac{\pi^2 k_B^4 T^4 V}{15 \hbar^3 c^3}. \tag{8.6.8}$$

The pressure due to black-body radiation is given by

$$P = -\left(\frac{\partial F}{\partial V}\right)_T = \frac{\pi^2 k_B^4 T^4}{45 \hbar^3 c^3} = \frac{\bar{U}}{3V}. \tag{8.6.9}$$

The pressure of black-body radiation is a third of the internal energy density.

8.7   Einstein's model of vibrations in a solid

Einstein (1907) was the first to use quantum ideas to describe the thermal properties of solids. The characteristic property of a solid is that its atoms execute small vibrations about their equilibrium positions. When in thermal equilibrium the atoms are arranged on a regular lattice; this is called the crystalline state. The solid can also be in an amorphous state in which the atoms vibrate about points randomly situated in space rather than about lattice sites. This is not an equilibrium state, but it can persist almost indefinitely. Examples of amorphous material include ice and glass.

In classical mechanics all atoms should be at rest at $T = 0$. The potential energy due to the interaction between atoms is a minimum when the system is in equilibrium. At low temperatures atoms should execute small vibrations about their equilibrium positions. Suppose there are N atoms making up the solid, so the number of degrees of freedom is $3N$. Each oscillator has an average energy of $k_B T$. We expect the molar heat capacity of the solid to be $3N_A k_B = 3R$. The value $3R$ is found for several monatomic solids, an observation made by Dulong and Petit. It is called the *law of Dulong and Petit*. In fact the value of $3R$ is found for most solids at room temperature, but not for diamond which has a much smaller heat capacity. Einstein showed that the smaller value for diamond is a quantum effect. Quantum theory, which started with the quantization of light and with atomic physics, has a much more general range of applicability. It can be applied to solids, and to all condensed matter. That is the main significance of Einstein's work on the heat capacity of solids.

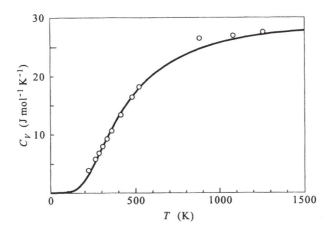

Fig. 8.4 The heat capacity of diamond compared to Einstein's theory (1907). The experimental values shown are those of Weber (1875) which were used by Einstein when he compared his theory to experiment.

Einstein treated the motion of the atoms in the solid as if they were oscillators. For N oscillators we expect a contribution to the free energy of the form

$$F = N\epsilon + k_\mathrm{B}T \sum_\alpha \ln\left(1 - \mathrm{e}^{-\hbar\omega_\alpha/k_\mathrm{B}T}\right) \tag{8.7.1}$$

where the sum is over the $3N$ vibrational modes. The term $N\epsilon$ is the energy of interaction between atoms in their equilibrium positions and includes the zero-point energy of the oscillators.

In Einstein's model all the frequencies are taken to be the same, and given the value ω_E, *the Einstein frequency*. The free energy becomes

$$F = N\epsilon + 3Nk_\mathrm{B}T \ln\left(1 - \mathrm{e}^{-\hbar\omega_E/k_\mathrm{B}T}\right). \tag{8.7.2}$$

The heat capacity is obtained by differentiating twice with respect to T

$$
\begin{aligned}
C_V &= -T\left(\frac{\partial^2 F}{\partial T^2}\right)_V \\
&= 3Nk_\mathrm{B}\left(\frac{\hbar\omega_E}{k_\mathrm{B}T}\right)^2 \frac{\mathrm{e}^{\hbar\omega_E/k_\mathrm{B}T}}{\left(\mathrm{e}^{\hbar\omega_E/k_\mathrm{B}T} - 1\right)^2}.
\end{aligned} \tag{8.7.3}
$$

For $k_\mathrm{B}T > \hbar\omega_E$ the heat capacity tends to the value of $3Nk_\mathrm{B}$; for $\hbar\omega_E > k_\mathrm{B}T$ the heat capacity decreases as $\mathrm{e}^{-\hbar\omega_E/k_\mathrm{B}T}$. The heat capacity vanishes as T tends to zero, as does the entropy which is consistent with the third law of thermodynamics.

Einstein could explain the temperature dependence of the heat capacity of solids, in particular diamond. We show in Fig. 8.4 the values that Einstein used for diamond, together with his best fit to the theory. Einstein's theory fits the experimental data fairly well as a first approximation; we have to go to very low temperatures to realize that it does not work for all temperatures.

8.8   Debye's model of vibrations in a solid

Einstein assumed in his model that there was only one frequency of vibration, even though he appreciated that atoms in a solid vibrate at a whole range of frequencies, not just at one. Each atom does not oscillate independently of its neighbours; they oscillate cooperatively. If one moves to the left it squashes its neighbour which also moves to the left and squashes the next neighbour, and so on. The displacement propagates along as a wave whose frequency varies with the wavelength. A range of frequencies is present.

The low-frequency longitudinal modes of the solid are sound waves of speed s. At a temperature T the important modes of vibration are those with frequency of order $k_B T/h$. For low temperatures we excite only the low-frequency sound waves; they have a wavelength $\lambda = 2\pi s/\omega$ which is much greater than the lattice spacing, a. The condition that $\lambda \gg a$ gives us the condition for low temperatures:

$$k_B T \sim \hbar\omega \ll \frac{\hbar s}{a}. \tag{8.8.1}$$

In this temperature region we expect Einstein's model to be in error.

Debye (1912) developed a theory of the statistical mechanics of the solid which was accurate in the low-temperature region. The number of modes of vibration with wave vector in the range k to $k+dk$ is $Vk^2 dk/2\pi^2$. In his theory there are longitudinal sound waves which we suppose travel at speed s_L, and two transverse sound waves which we take to travel at speed s_T. If we work in terms of frequency, the number of modes of vibration in the range ω to $\omega + d\omega$ is

$$D(\omega)\,d\omega = \frac{V}{2\pi^2}\omega^2\left(\frac{1}{s_L^3} + \frac{2}{s_T^3}\right)d\omega. \tag{8.8.2}$$

Let us define a quantity $\bar{s}$ via the equation

$$\frac{3}{\bar{s}^3} = \left(\frac{1}{s_L^3} + \frac{2}{s_T^3}\right). \tag{8.8.3}$$

The total number of modes with frequencies from ω to $\omega + d\omega$ is

$$D(\omega)\,d\omega = \frac{3V}{2\pi^2\bar{s}^3}\omega^2\,d\omega. \tag{8.8.4}$$

The free energy is

$$F = N\epsilon + \frac{3k_B TV}{2\pi^2\bar{s}^3}\int_0^\infty \omega^2 \ln\left(1 - e^{-\hbar\omega/k_B T}\right)d\omega. \tag{8.8.5}$$

We have taken the upper limit of integration to be infinity, an approximation which can be made only if the temperature is small. The integral over ω can be done by parts and we get

$$F = N\epsilon - \frac{\pi^2 k_{\mathrm{B}}^4 T^4 V}{30\hbar^3 \bar{s}^3},$$
(8.8.6)

from which it is easy to get the formula for the entropy:

$$S = \frac{2\pi^2 k_{\mathrm{B}}^4 T^3 V}{15\hbar^3 \bar{s}^3}.$$
(8.8.7)

Compare this expression to that obtained for the entropy of black-body radiation given by eqn (8.6.7). For black-body radiation there are two types of wave, both transverse, but here there are three types of wave: two transverse and one longitudinal. These waves are assumed to travel at the speed of sound, not at the speed of light. Apart from these differences, the calculation is the same.

Finally the contribution of the quantized lattice vibrations to the heat capacity of a solid is

$$C_V = \frac{2\pi^2 k_{\mathrm{B}}^4 T^3 V}{5\hbar^3 \bar{s}^3}.$$
(8.8.8)

For low temperatures the heat capacity is predicted to vary with temperature as T^3.

For high temperatures we can proceed in the following way. The free energy

$$
\begin{aligned}
F &= N\epsilon + k_{\mathrm{B}}T \sum_{\alpha} \ln\left(1 - e^{-\hbar\omega_\alpha/k_{\mathrm{B}}T}\right) \\
&\approx N\epsilon + k_{\mathrm{B}}T \sum_{\alpha} \ln\left(\hbar\omega_\alpha/k_{\mathrm{B}}T\right)
\end{aligned}
$$
(8.8.9)

involves a sum over α which has $3N$ terms. We can define an average frequency by

$$\ln(\bar{\omega}) = \frac{1}{3N} \sum_{\alpha} \ln\left(\omega_\alpha\right).$$
(8.8.10)

The free energy becomes

$$F = N\epsilon - 3Nk_{\mathrm{B}}T \ln\left(k_{\mathrm{B}}T/\hbar\bar{\omega}\right)$$
(8.8.11)

from which we can obtain the entropy of the system and its heat capacity. The entropy is just

$$S = 3Nk_{\mathrm{B}}\left\{\ln\left(k_{\mathrm{B}}T/\hbar\bar{\omega}\right) + 1\right\}.$$
(8.8.12)

It follows that at high temperatures the heat capacity is $3Nk_{\mathrm{B}}$ as expected.

In the Debye model an attempt is made to interpolate between these two limiting forms in a sensible way. Debye introduced a cutoff frequency, ω_D. He

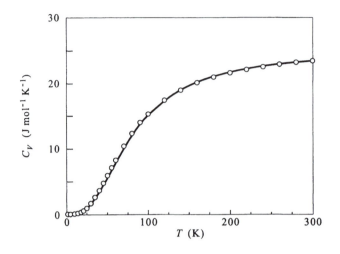

Fig. 8.5 The molar heat capacity of copper plotted as a function of temperature compared to the Debye theory.

assumed that the total number of modes with frequency between ω and $\omega + d\omega$ is given by

$$D(\omega)\,d\omega = \frac{3V}{2\pi^2 \bar{s}^3}\omega^2\,d\omega \qquad (8.8.4)$$

if $\omega < \omega_D$ and

$$D(\omega)\,d\omega = 0 \qquad (8.8.13)$$

if $\omega > \omega_D$. Here ω_D is a *cutoff* frequency chosen so that the total number of modes is $3N$. Thus

$$3N = \frac{3V}{2\pi^2 \bar{s}^3}\int_0^{\omega_D} \omega^2\,d\omega = \frac{V\omega_D^3}{2\pi^2 \bar{s}^3},$$

which can be rearranged to give

$$\omega_D = \bar{s}(6\pi^2 N/V)^{1/3}. \qquad (8.8.14)$$

The quantity ω_D is called the Debye frequency, and $\hbar\omega_D$ is called the Debye energy of the solid.

The Debye model makes a very crude approximation for the density of states of vibrational modes. Rather than calculate all the vibrational modes, Debye simplified the model of the system to the point where it represents a mere caricature of reality. A good theoretical model of a complex system should be like a caricature: it should emphasize those feature which are most important and should play down the inessential details. That is what Debye's theory does.

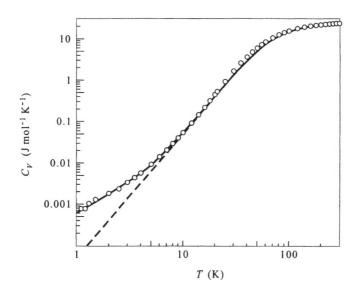

Fig. 8.6 The same data as that shown in Fig. 8.5 but here plotted on logarithmic instead of linear scales. The discrepancy at low temperatures between the data and Debye's theory is now apparent. The extra contribution to the heat capacity is caused by electrons. This contribution is analysed further in Chapter 10.

The free energy in the Debye model is

$$F = N\epsilon + \frac{3k_{\mathrm{B}}TV}{2\pi^2 \bar{s}^3} \int_0^{\omega_D} \omega^2 \ln\left(1 - e^{-\hbar\omega/k_{\mathrm{B}}T}\right) d\omega \qquad (8.8.15)$$

which can be rewritten in terms of the variable $x = \hbar\omega/k_{\mathrm{B}}T$ as

$$F = N\epsilon + \frac{3k_{\mathrm{B}}TV}{2\pi^2 \bar{s}^3} \left(\frac{k_{\mathrm{B}}T}{\hbar}\right)^3 \int_0^{\theta_D/T} x^2 \ln\left(1 - e^{-x}\right) dx, \qquad (8.8.16)$$

where $\theta_D = \hbar\omega_D/k_{\mathrm{B}}$ is the *Debye temperature*. From this equation it is possible to obtain an expression for the entropy and then the heat capacity of a solid.

We show in Fig. 8.5 some of the experimental results for copper of Franck *et al.* (1961) and of Martin (1960, 1967) and compare them with the predictions of the Debye theory. For copper the fit of the theory to experiment, as shown in Fig. 8.5, appears to be surprisingly good, given the crudeness of the approximations involved in Debye's theory. However, there are discrepancies which are revealed when the same data are plotted on a logarithmic scale. We show in Fig. 8.6 the same data plotted this way. It can be seen that there is an additional contribution to the heat capacity at very low temperatures which comes

Table 8.1 Values of the Debye temperature in kelvin in the limit that T tends to zero for several elements from Touloukian and Buyco (1970).

Ag	Bi	K	Ca	Cs	Ba	Na	Cu	Ga	C	Al	Si
228	119	89.4	234	40	110.5	157	342	317	234	423	647

from the thermal activation of conduction electrons. We will analyse this contribution in Chapter 10.

When we remove the electronic contribution, the Debye model gives a very good fit at low temperatures with one adjustable parameter, θ_D. In Table 8.1 we give values of θ_D in degrees kelvin for various elements, obtained by fitting the heat capacity to the Debye model at very low temperatures.

8.9 Solid and vapour in equilibrium

We can use our knowledge of the partition functions of solids and gases to create a theory for the equilibrium between these two phases using the idea that the free energy is a minimum. To make matters as simple as possible, consider a monatomic spinless gas like argon. The statistical mechanics of the solid will be treated using the Einstein model in the interest of simplicity. The interaction of the gas and the solid is taken to be weak so we can calculate the overall partition function by multiplying the partition functions of the solid and of the gas together.

The partition function of the solid when it has N_S atoms can be derived from eqn (8.7.2):

$$Z_S = e^{-N_S \epsilon / k_B T} \left\{ \frac{1}{1 - e^{-\hbar \omega_E / k_B T}} \right\}^{3N_S}.$$

Here $-\epsilon$ is the energy per atom needed to break the solid apart into separate gas atoms and ω_E is the Einstein frequency of vibration. The partition function of the gas is

$$Z_G = \frac{Z_1^{N_G}}{N_G!} \tag{8.9.1}$$

where Z_1 is the single particle partition function of the gas atom. The only contribution to the partition function Z_1 of the gas particle comes from its translational motion since the particle is monatomic and spinless.

The total partition function is given by the product of the gas and solid partition functions:

$$Z = Z_G Z_S. \tag{8.9.2}$$

The total number of particles, $N_G + N_S$, is kept constant for when an atom leaves the solid it goes into the vapour. It follows that $dN_G = -dN_S$. The free energy is

$$F = -k_B T \ln(Z_S) - k_B T \ln(Z_G), \tag{8.9.3}$$

and by minimizing this expression as a function of N_G keeping $N_G + N_S$ constant we get the equation

$$\frac{\partial F}{\partial N_G} = k_{\mathrm{B}}T\frac{\partial \ln(Z_S)}{\partial N_S} - k_{\mathrm{B}}T\frac{\partial \ln(Z_G)}{\partial N_G} = 0$$

which gives after simple algebra

$$\frac{Z_1}{N_G} = \frac{\mathrm{e}^{-\epsilon/k_{\mathrm{B}}T}}{\left(1 - \mathrm{e}^{-\hbar\omega_E/k_{\mathrm{B}}T}\right)^3}. \tag{8.9.4}$$

By using the formula for an ideal gas, $P = N_G k_{\mathrm{B}}T/V_G$, with V_G the volume of the gas, we get the following expression for the vapour pressure of the gas:

$$P = \frac{k_{\mathrm{B}}TZ_1}{V_G}\,\mathrm{e}^{\epsilon/k_{\mathrm{B}}T}\left(1 - \mathrm{e}^{-\hbar\omega_E/k_{\mathrm{B}}T}\right)^3. \tag{8.9.5}$$

Of course this a fairly crude model for the statistical mechanics of the solid. It can be improved perhaps by replacing the Einstein with the Debye model, and by including the effect of vacancies. If the atoms have a spin then we should include that contribution in the entropy of the solid and the gas. If the gas were made up of molecules we could include the effect of rotation and vibration in Z_1. In this way we can use the idea that the free energy is a minimum to develop a theory of the equilibrium vapour pressure of a substance.

8.10 Cosmic background radiation

There have been several demonstrations of the black-body spectrum, but few so stunning as that recently obtained which measures the *spectral density of background black-body radiation* which is left over from the big bang. The big bang was the start of the Universe, an event which occurred about 15 billion years ago, give or take a billion years. In the initial stages of the Universe, all matter was ionized. Radiation was constantly being absorbed and re-emitted by the ionized particles, until eventually the radiation and matter reached thermal equilibrium. Eventually the expansion of the Universe caused the radiation to cool below $3000\,\mathrm{K}$, and the ionized particles then combined to form atoms: the protons and electrons combined to form hydrogen atoms, the alpha particles combined with electrons to form helium atoms. From then on there was little interaction between radiation and matter. The matter in due course clumped together into galaxies and stars while the radiation just continued to cool with the expansion of the Universe.

The intensity and the spectrum of the background radiation (we ignore the radiation that has come from stars) only depend on its temperature. The radiation has expanded keeping its total energy constant. As a result the entropy of the radiation continues to increase as the Universe expands, even though the temperature continues to decrease. The expansion has continued for about 15 billion years and the radiation has now reached a temperature of $2.735\,\mathrm{K}$. The temperature of planets like the Earth would fall to $2.735\,\mathrm{K}$ if it were not for the Sun and for internal heating of the Earth due to radioactive decay processes.

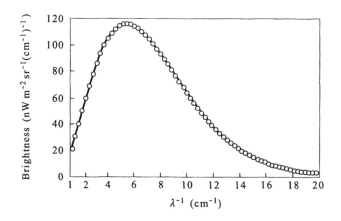

Fig. 8.7 The cosmic background radiation as measured by Mather *et al.* (1990) compared to Planck's formula for the spectrum of black-body radiation.

Until a few years ago experimental data on the background radiation were rather poor, but since 1989 the Cosmic Background Explorer (COBE) has measured the background microwave radiation with great accuracy. The data of Mather *et al.* (1990) are shown in Fig. 8.7; the solid curve is Planck's distribution which fits to within 1% for each data point over the range 1 to $20\,\text{cm}^{-1}$. Planck's formula, derived in 1900, fits experimental results from COBE exceptionally well.

Another property of black-body radiation is that it fills the space uniformly, with no preference for any particular direction. A medium which has no preferred direction is said to be *isotropic*. Some of the measurements have been designed to test whether the background radiation is isotropic or whether it varies according to the direction in space. Of course one has to take account of the small Doppler shift caused by the movement of the Earth. The rotation of the Earth around the Sun is important, but we also have to take account of the orbital motion of the Sun about the galaxy, along with the motion of the galaxy within the local group of galaxies. Apparently the local group of galaxies are falling towards the Virgo cluster of galaxies at a speed of $570\,\text{km}\,\text{s}^{-1}$.

When these effects are removed, what remains is a variation of the temperature of radiation with the direction in space. Smoot *et al.* (1992) have measured the black-body distribution in different directions in space to determine the fluctuations in temperature. These variations in the black-body distribution correspond to root mean square fluctuations in temperature of about $3 \times 10^{-5}\,\text{K}$ for different directions in space, although this number is slightly controversial. Such a small variation in temperature from the big bang in different directions in space is apparently due to the primordial density fluctuations of matter in the very early stages of the Universe, something which can be predicted using current inflationary theories of cosmology. The density fluctuations caused matter to clump together where the density was greatest forming gas clouds which evolved into stars. Planck's distribution, created at the birth of the twentieth century, is now used as a tool to test models of how the Universe expands.

8.11 Problems

1. Calculate the thermal contribution to the surface energy per unit area arising from ripplons whose dispersion relation is (see the end of section 8.5)

$$w(k) = \left(\frac{\gamma_s k^3}{\rho}\right)^{1/2}.$$

Express your answer in terms of a dimensionless integral.

2. If we were able to confine photons to a two-dimensional world they would be only one transverse wave. What would be the density of states, the spectral density as a function of energy, and the equivalent of the Stefan–Boltzmann law of radiation?

3. Suppose $u(\lambda)\, d\lambda$ is the energy density per unit volume emitted by a black body in the wavelength range λ to $\lambda + d\lambda$. At what wavelength does $u(\lambda)$ have its maximum value? Take the radiation from the Sun to be black-body radiation with a maximum in $u(\lambda)$ at 480 nm. What is the temperature of the Sun?

4. Show that the fluctuation in energy, ΔE, in the canonical ensemble is given by

$$\Delta E^2 = \overline{(E - \overline{E})^2} = k_B T^2 C_V$$

where T is the absolute temperature and C_V is the heat capacity at constant volume. Calculate ΔE^2 for a system of photons in an enclosure of volume V at temperature T.

5. Treat the Sun as a black body with temperature of about 5800 K, and use Stefan's law to show that the total radiant energy emitted by the Sun is 4×10^{26} W. Show that the rate at which radiant energy reaches the top of the Earth's atmosphere is about $1.4\,\mathrm{kW\,m^{-2}}$. [Radius of Sun $= 7 \times 10^8$ m; distance of Earth from Sun $= 1.5 \times 10^{11}$ m.]

6. Describe the Debye model for the internal energy of a solid, and obtain the result

$$U = \int_0^{\omega_D} \frac{V\omega^2}{2\pi^2 s^3} \frac{\hbar\omega}{e^{\hbar\omega/k_B T} - 1} d\omega$$

for each phonon polarization of speed s. Sketch the form of the integrand, and estimate the energy of the phonons which make the largest contribution to the internal energy of silicon at 30 K and at the Debye temperature (625 K). The average speed of sound in silicon is $7000\,\mathrm{m\,s^{-1}}$.

7. Imagine the Universe to be a spherical cavity with radius 10^{26} m. If the temperature in the cavity is 3 K, estimate the total number of thermally excited photons in the Universe, and the energy content of these photons. Some of the integrals which you need are given in Appendix B.

8. The Universe is full of black-body radiation at a temperature of 2.73 K. Suppose that the radius of the Universe is given by Ht where t is the age of the Universe and H is Hubble's constant. Assume that the total internal energy of black-body radiation in the Universe is constant. Give an expression for the internal energy of the radiation. Hence show that the entropy of black-body radiation in the Universe increases in time as $t^{3/4}$.

9. For spin waves the dispersion relation is $\omega(k) = \alpha k^2$. In a three-dimensional system how does the internal energy of the spin wave system vary with temperature?

10. Imagine a piano wire of mass M, under tension F, between fixed points a distance L apart. The wire at rest lies along the x axis. To simplify matters we assume that the displacement of the wire from equilibrium occurs only along the y direction. Thus $y(0) = y(L)$ since there is no displacement at the ends of the wire. The energy of the wire for an instantaneous displacement $y(x,t)$ is

$$E\left[y(x,t)\right] = \frac{1}{2}\int_0^L dx \left(\frac{M}{L}\left(\frac{\partial y}{\partial t}\right)^2 + F\left(\frac{\partial y}{\partial x}\right)^2\right).$$

We write $y(x,t)$ as the Fourier series

$$y(x,t) = \sum_{n=1}^N A_n(t)\sin\left(\frac{n\pi x}{L}\right).$$

Show that the energy can be written as

$$E\left[y(x,t)\right] = \sum_{n=1}^N \left(\frac{M}{4}\dot{A}_n^2 + \frac{n^2\pi^2 F}{4L}A_n^2\right).$$

Treat A_n and $\dot{A}_n$ as a set of $2N$ generalized coordinates which define the instantaneous position of the wire.

(a) Calculate the average thermal energy of the wire under tension, and hence obtain its heat capacity.
(b) Determine the average value of A_n, of A_n^2, and of $A_n A_{n'}$ with $n \neq n'$.
(c) Express the mean square $\overline{y(x,t)^2}$ in the form of a sum of some quantity over n. From this determine the mean square displacement of the wire at $x = L/2$. [You may use without proof the relation $\sum_{n=1}^\infty n^{-2} = \pi^2/8$.]

11. Elementary excitations in superfluid liquid helium can be treated as if they were particles with a complicated dispersion relation. For wave vectors less than $0.8 \times 10^{10}\,\mathrm{m}^{-1}$ the dispersion is nearly linear and can be described by the equation

$$\varepsilon(k) = \hbar sk.$$

For wave vectors around $2 \times 10^{10} \, \text{m}^{-1}$ there is a dip in the dispersion curve; here the excitations are called rotons. (Landau (1941) originally thought they were evidence for quantized rotational motion, hence the name.) The dispersion here is described by the equation

$$\varepsilon(k) = \Delta + \frac{\hbar^2 (k - k_0)^2}{2\mu}.$$

The roton energy, Δ, is $8.65 \, \text{K}$ and k_0 is about $2 \times 10^{10} \, \text{m}^{-1}$ at a pressure of $0 \, \text{bar}$.

The internal energy of a gas of excitations in helium has two parts: that due to phonons (longitudinal sound waves) and that due to rotons. Show that the phonons give, for low temperatures, a contribution to the free energy of the form

$$F_{\text{phonon}} = -\frac{\pi^2 k_{\text{B}}^4 T^4 V}{90 \hbar^3 s^3},$$

and that the rotons give a contribution to the free energy of the form

$$F_{\text{roton}} = -\frac{V k_{\text{B}} T k_0^2}{2\pi^2} \left(\frac{2\pi \mu k_{\text{B}} T}{\hbar^2} \right)^{1/2} e^{-\Delta / k_{\text{B}} T}.$$

12. Stephen Hawking (1974, 1976) calculated the entropy of a non-rotating, uncharged black hole and obtained the expression

$$S = k_{\text{B}} c^3 A / 4 G \hbar$$

where A is the area of the black hole equal to $4\pi R_S^2$ with R_S the Schwarzschild radius, $R_S = 2GM/c^2$. Here M is the mass of the star and G is Newton's gravitational constant. Show that this equation implies that the temperature T of a black hole is given by

$$T = \hbar c^3 / 8\pi G k_{\text{B}} M.$$

If the black holes radiate according to the Stefan–Boltzmann law they lose mass. Show that the rate of change of mass is given by the equation

$$c^2 \frac{\mathrm{d}M}{\mathrm{d}t} = -A\sigma T^4 = -\frac{16\pi G^2 M^2 \sigma}{c^4} \left(\frac{\hbar c^3}{8\pi G k_{\text{B}} M} \right)^4$$

Solve this equation to obtain the variation in the mass of a black hole as a function of time.

Given that the initial mass, $M(0)$, of the black hole is about $2 \times 10^{11} \, \text{kg}$, show that the black hole evaporates in a time of $t = 7 \times 10^{17} \, \text{s}$, which is roughly the age of the Universe.

9

Systems with variable numbers of particles

It seemed to me a superlative thing to know the explanation of everything, why it comes to be, why it perishes, why it is.   *Plato 429–346* BC

9.1   Systems with variable number of particles

A system can have a *variable number of particles* either because particles enter (or leave) or because a reaction takes place inside the system. We can differentiate between these two situations. *The first situation* occurs when two systems A and B are physically separated in space and particles are transferred into A from B or vice versa, as illustrated in Fig. 9.1.

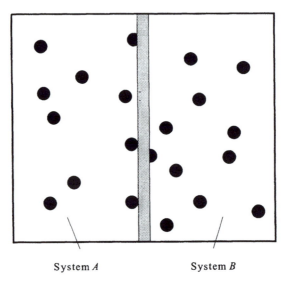

System *A*   System *B*

Fig. 9.1 Particles enclosed in two volumes, A and B, separated by a membrane which allows particles to pass from one system to the other.

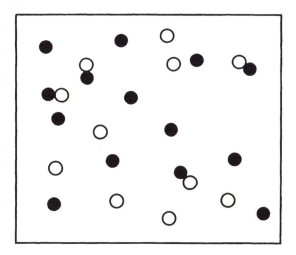

Fig. 9.2 Two gas atoms represented by open and solid circles. A molecule made of the atoms is represented by a joined pair of solid and open circles. The atoms and molecules are moving about the volume freely.

For example, there could be a large density of argon gas atoms in system A and a low concentration of argon atoms in system B, with a semipermeable membrane which allows some particles to transfer from A to B or from B to A. The density difference drives particles from A to B. Another example might be a sealed vessel containing a pure substance with both liquid and gas phases present. There is a meniscus separating the gas from the liquid which enables us to define which atoms are in the gas and which are in the liquid phase. The gas and the liquid are physically separate. The gas atoms can condense into the liquid or the liquid can evaporate into the gas. If there is a difference in chemical potentials between liquid and gas then there is a flow of particles from one phase to the other.

The total number of atoms is always conserved; this leads to a conservation law. The total number of atoms in A and B is constant in the experiment with the semipermeable membrane. Thus $N_A + N_B = $ constant. In the sealed container the total number of atoms, either in the liquid or the gas, is constant. When we alter the number of particles in each subsystem we do so in such a way that $dN_A + dN_B = 0$. Any increase in the number in A equals the decrease in the number in B. Thus

$$\frac{dN_A}{dN_B} = -1. \tag{9.1.1}$$

The second situation occurs when there are subsystems which occupy the same region of space as illustrated in Fig. 9.2. Consider a mixture of gases, composed of atoms and molecules, in which one subsystem is the gas of atoms and the other consists of the gas of molecules. The atoms react to form molecules, the molecules dissociate into atoms so that the numbers of atoms and molecules can change.

To make matters more specific consider the notional reaction

$$C + D \rightleftharpoons CD \tag{9.1.2}$$

where C and D represent atoms and CD represents a molecule. Notice that in this example, the molecule CD contains one atom of C and one atom of D. Let there be N_C of atoms C, N_D atoms of atoms D, and N_{CD} of molecule CD. The total number of each atom is constant. The number of C as atoms, plus the number of C forming part of the molecule is fixed, which means that $N_C + N_{CD} = $ constant. Similarly the number of D as atoms plus the number in molecules is fixed, which means that $N_D + N_{CD} = $ constant. There are two conservation laws for the reacting atoms. For any changes in the numbers we have $dN_C + dN_{CD} = 0$, which implies

$$\frac{dN_{CD}}{dN_C} = -1 \tag{9.1.3}$$

and $dN_D + dN_{CD} = 0$, which implies

$$\frac{dN_{CD}}{dN_D} = -1. \tag{9.1.4}$$

From these two expressions we get $dN_D = dN_C$, so that

$$\frac{dN_D}{dN_C} = 1. \tag{9.1.5}$$

The second situation is more complicated than the first one because there are two conservation laws, not just one.

We have said that C and D are gas atoms, but we need not be so restrictive. The same idea can be used to describe chemical reactions which occur in solutions. In water, the molecule of NaCl can dissociate into Na^+ and Cl^- ions, or the ions can recombine to form the molecule.

$$NaCl \rightleftharpoons Na^+ + Cl^-.$$

One subsystem consists of the molecule dissolved in water, the other subsystem consists of the different ions dissolved in water.

There are similar reactions in stars in which an electron combines with a proton to form a hydrogen atom:

$$e^- + p^+ \rightleftharpoons H.$$

Here the subsystems are the gases of hydrogen atoms, free electrons, and free protons.

Reactions of atoms and molecules are of central importance in chemistry, but the mathematical language and ideas we are using can also be applied to atomic processes in the Sun, a topic which is more the province of physics. The mathematical language can also be applied to reactions of elementary particles, reactions which occurred in the early stages of the Universe. The subject of reactions occurs frequently in science: solid-state physics, chemistry, stars, phase equilibrium. Reactions are not just the province of chemistry, although there they are particularly important.

9.2 The condition for chemical equilibrium

Clearly the number of quantum states, W, depends on the number of atoms in each subsystem. It follows that the entropy is also a function of the number of atoms in each subsystem. For the *first situation* with atoms A and B we have with $S = S_A + S_B$

$$
\begin{aligned}
\mathrm{d}S &= \left\{ \left(\frac{\partial S_A}{\partial N_A} \right) \mathrm{d}N_A + \left(\frac{\partial S_B}{\partial N_B} \right) \mathrm{d}N_B \right\} \\
&= \mathrm{d}N_A \left\{ \left(\frac{\partial S_A}{\partial N_A} \right) + \left(\frac{\partial S_B}{\partial N_B} \right) \frac{\mathrm{d}N_B}{\mathrm{d}N_A} \right\} \\
&= -\frac{\mathrm{d}N_A}{T} \left\{ \mu_A - \mu_B \right\}
\end{aligned}
\tag{9.2.1}
$$

where it has been assumed that the two systems have the same temperature and we have used the definition of the chemical potential

$$
\mu = -T \left(\frac{\partial S}{\partial N} \right)_{U,V}.
\tag{2.9.7}
$$

In chemical equilibrium the entropy is a maximum and any change in the entropy will be second order in $\mathrm{d}N_A$. Thus at the maximum there is no term linear in $\mathrm{d}N_A$ and so $\mu_A = \mu_B$, the condition for chemical equilibrium.

The *second situation* of reacting gases is a little more complex. We write the entropy as the sum of the entropies of the reacting gases

$$
S = S_C(N_C) + S_D(N_D) + S_{CD}(N_{CD}),
\tag{9.2.2}
$$

that is the sum of the entropy of atoms C, the entropy of atoms D, and the entropy of molecules CD. Let us alter the numbers of particles in each subsystem and find the condition that the entropy is a maximum. Suppose we take one atom of C and one of D and form one molecule of CD. In doing this we decrease both N_C and N_D by one and increase N_{CD} by one. Next consider $\mathrm{d}N_C$ atoms of C reacting with $\mathrm{d}N_D$ atoms of D to form $\mathrm{d}N_{CD}$ molecules of CD. Since the reaction always involves one atom of C reacting with one atom of D to form one molecule of CD we must have

$$
-\mathrm{d}N_C = -\mathrm{d}N_D = \mathrm{d}N_{CD}.
\tag{9.2.3}
$$

For chemical equilibrium the entropy is a maximum which means

$$
\begin{aligned}
\mathrm{d}S &= \mathrm{d}N_C \left\{ \left(\frac{\partial S_C}{\partial N_C} \right) + \left(\frac{\partial S_D}{\partial N_D} \right) \frac{\mathrm{d}N_D}{\mathrm{d}N_C} + \left(\frac{\partial S_{CD}}{\partial N_{CD}} \right) \frac{\mathrm{d}N_{CD}}{\mathrm{d}N_C} \right\} \\
&= -\frac{\mathrm{d}N_C}{T} \left(\mu_C + \mu_D - \mu_{CD} \right).
\end{aligned}
\tag{9.2.4}
$$

The condition for chemical equilibrium is

$$
\mu_C + \mu_D = \mu_{CD},
\tag{9.2.5}
$$

the sum of the chemical potentials of the reacting particles is equal to the sum of the chemical potentials of the products.

There are still more complex cases which can be treated the same way. For example, the reaction

$$
O_2 + 2H_2 \rightleftharpoons 2H_2O
$$

is slightly more complex because two hydrogen molecules react with one oxygen molecule. The condition for chemical equilibrium for this reaction is

$$
\mu_{O_2} + 2\mu_{H_2} = 2\mu_{H_2O}.
$$

The numbers in front of the chemical potential are called *stoichiometric constants*. They tell us how many particles of each sort combine together. In general, the condition for chemical equilibrium is an equation of the form

$$
\left(\sum \nu_i \mu_i \right)_{\text{reactants}} = \left(\sum \nu_j \mu_j \right)_{\text{products}}
\tag{9.2.6}
$$

where the ν_i are the stoichiometric constants.

What happens if there is no conservation law? Let us consider the first situation. If N_T is not conserved, then the numbers $\mathrm{d}N_A$ and $\mathrm{d}N_B$ are independent of each other which means that $\mathrm{d}N_B/\mathrm{d}N_A = 0$. When $\partial S/\partial N_A = 0$ the entropy is a maximum. But

$$
\frac{\partial S}{\partial N_A} = \frac{\partial S_A}{\partial N_A} + \frac{\partial S_B}{\partial N_B} \frac{\mathrm{d}N_B}{\mathrm{d}N_A} = 0.
$$

Since $\mathrm{d}N_B/\mathrm{d}N_A = 0$ it follows that $\partial S_A/\partial N_A = -\mu_A/T = 0$: the chemical potential of system A is zero. When the number of particles of one type is not conserved, then its chemical potential is zero.

An example is black-body radiation. Photons in equilibrium with the walls of a cavity can be absorbed by the walls, so the number of photons is not constant. Consequently photons have a chemical potential of zero. Similarly the number of phonons in a solid is not conserved: they can be absorbed by the walls containing the solid, or two phonons can collide and produce three outgoing phonons. As a result their chemical potential is zero. The same is true for all quantized 'classical' wave motions; all of them have a chemical potential which is zero.

9.3 The approach to chemical equilibrium

If system A is hotter than system B then energy flows from A to B; when system A is at a different chemical potential from system B, which way do the particles flow? To find the answer to this question we study the rate of entropy production which must be positive according to *Clausius' principle*.

Consider the *first situation*. We must have

$$
\begin{aligned}
\frac{dS}{dt} &= \frac{dN_A}{dt}\left\{\left(\frac{\partial S_A}{\partial N_A}\right) - \left(\frac{\partial S_B}{\partial N_B}\right)\right\} \\
&= -\frac{dN_A}{dt}\left\{\frac{\mu_A - \mu_B}{T}\right\} \geq 0.
\end{aligned}
\tag{9.3.1}
$$

If μ_A is larger than μ_B, we must have $dN_A/dt < 0$ so that particles leave system A. The side of the reaction with the larger chemical potential loses particles. If we have a gas on one side of the membrane (side A) with a chemical potential larger than the chemical potential of the gas on the other side, then the atoms on side A will be forced through the membrane until the chemical potentials become equal. Similarly if the atoms in the liquid phase in a sealed flask have a chemical potential larger than the chemical potential of the atoms in the gas phase then the liquid will evaporate until chemical equilibrium is reached.

The same ideas can be used for the *second situation*, that is for chemical reactions. For the notional reaction $C + D \rightleftharpoons CD$ the rate of entropy production is

$$
\begin{aligned}
\frac{dS}{dt} &= \frac{dN_C}{dt}\left\{\left(\frac{\partial S_C}{\partial N_C}\right) + \left(\frac{\partial S_D}{\partial N_D}\right) - \left(\frac{\partial S_{CD}}{\partial N_{CD}}\right)\right\} \\
&= -\frac{dN_C}{dt}\left\{\frac{\mu_C + \mu_D - \mu_{CD}}{T}\right\} \geq 0.
\end{aligned}
\tag{9.3.2}
$$

If $(\mu_C + \mu_D) > \mu_{CD}$ the reaction will proceed so as to decrease the number of C and D atoms and produce more molecules CD. If the chemical potential μ_{CD} is larger than $(\mu_C + \mu_D)$ then the reaction will go the other way.

9.4 The chemical potential

A knowledge of the chemical potential of reacting particles can be used to indicate which way a reaction proceeds. Since a large part of the wealth of most countries is based on the chemical industry, the chemical potential is something of more than academic interest.

9.4.1 Method of measuring μ

The chemical potential can be found obtained from the Gibbs free energy, something which can be measured experimentally. The Gibbs free energy, G, is given by

$$
G = H - TS = \bar{U} + PV - TS.
\tag{2.6.10}
$$

If we allow the number of particles to vary, the change in the average internal energy is

$$d\bar{U} = T\,dS - P\,dV + \mu\,dN. \tag{2.9.10}$$

The last term is the chemical work, that is the change in the internal energy of the system on increasing the number of particles. Each additional particle brings to the internal energy its chemical potential. A small change in G is

$$dG = -S\,dT + \mu\,dN + V\,dP. \tag{9.4.1}$$

It follows that for changes at constant temperature and pressure

$$\mu = \left(\frac{\partial G}{\partial N}\right)_{T,P}. \tag{9.4.2}$$

The Gibbs free energy is proportional to the number of particles in the system, so that

$$\left(\frac{\partial G}{\partial N}\right)_{T,P} = \frac{G}{N}$$

so that $\mu = G/N$. *The chemical potential is the Gibbs free energy per particle.*

We can measure the Gibbs free energy for any pure substance. Start as close to the absolute zero of temperature as possible and measure the enthalpy needed to heat the substance to any particular temperature; at the same time measure the heat capacity at constant pressure and all the latent heats; from these obtain the entropy and construct G as $(H-TS)$. This has been done for many chemicals so we can look up values of G in tables.

The condition for chemical equilibrium involves a relationship between the chemical potentials of the reactants and products. For the notional reaction

$$A + B \rightleftharpoons C + D$$

we have

$$\mu_A + \mu_B = \mu_C + \mu_D.$$

Now a mole of a substance contains Avogadro's number of particles, according to Avogadro's hypothesis. We can express the condition for chemical equilibrium in terms of molar Gibbs free energies as

$$G_A + G_B = G_C + G_D$$

and the Gibbs free energies are measurable quantities. This is a relation between measurable quantities so it can be tested experimentally. Notice that this relation is consistent with Mach's philosophy of science.

9.4.2   Methods of calculating μ

There are two main methods by which we can calculate the chemical potential. The first method can be used for systems which are thermally isolated; the second method can be used for systems which are in contact with a heat bath.

The first approach uses the definition of the chemical potential in the microcanonical ensemble. Suppose a system is thermally isolated from the rest of the universe. We count the number of accessible quantum states and use Boltzmann's hypothesis, $S = k_B \ln(W)$, to calculate the entropy. It follows from the definition of the chemical potential that

$$\mu = -T \left(\frac{\partial S}{\partial N} \right)_{U,V} = -k_B T \left(\frac{\partial \ln(W)}{\partial N} \right)_{U,V}. \tag{9.4.3}$$

This expression gives us a way of calculating μ: we take the partial derivative of $\ln(W)$ with respect to N at constant energy and volume and multiply by $-k_B T$. The main difficulty lies in calculating $\ln(W)$.

Let us start with a very simple example. Imagine that there are distinguishable atoms on lattice sites, with each atom having a degeneracy g. Then for N non-interacting atoms

$$W = g^N.$$

If there is a two-level system on each site, and the two levels have exactly the same energy, then g is 2. There are two states on the first site, two on the next, two on the next, all the way along. The degeneracy of the whole system is $2 \times 2 \times 2 \times \cdots = 2^N$ where N is the number of particles. Then

$$S = k_B \ln(g^N) = N k_B \ln(g)$$

and by differentiation we get

$$\mu = -k_B T \ln(g).$$

The larger the degeneracy, the lower the chemical potential. Remember systems evolve so that particles flow to the side of the reaction with the lower total chemical potential. Consequently the larger the degeneracy, the more quantum states available and the lower the chemical potential, with the result that when the system is in chemical equilibrium, more particles are created on that side of the reaction.

Another simple example concerns a gas of atoms, each atom having a spin of zero. Let us assume that the expression for the entropy of such a gas in the microcanonical ensemble is

$$S = N k_B \left\{ \ln \left(\frac{V}{N} \right) + \frac{3}{2} \ln \left(\frac{mU}{3\pi \hbar^2 N} \right) + \frac{5}{2} \right\}. \tag{9.4.4}$$

By differentiation with respect to N, for constant V and U, we get

$$\mu = -k_{\mathrm{B}}T\left\{\ln\left(\frac{V}{N}\right) + \frac{3}{2}\ln\left(\frac{mU}{3\pi\hbar^2 N}\right)\right\}. \tag{9.4.5}$$

The temperature is given by

$$\frac{1}{T} = \left(\frac{\partial S}{\partial U}\right)_{V,N} = \frac{3Nk_{\mathrm{B}}}{2U}. \tag{9.4.6}$$

Hence the chemical potential can be written as

$$\begin{aligned}
\mu &= -k_{\mathrm{B}}T\left\{\ln\left(\frac{V}{N}\right) + \frac{3}{2}\ln\left(\frac{mk_{\mathrm{B}}T}{2\pi\hbar^2}\right)\right\} \\
&= k_{\mathrm{B}}T\ln\left(\frac{n}{n_Q}\right), \tag{9.4.7}
\end{aligned}$$

where n is the number density, N/V, and n_Q is given by

$$n_Q = \frac{1}{\lambda_D^3} = \left(\frac{mk_{\mathrm{B}}T}{2\pi\hbar^2}\right)^{3/2}. \tag{9.4.8}$$

n_Q is the density which corresponds to one particle in a volume λ_D^3.

If there is a spin degeneracy of g for all the particles in the gas then the chemical potential changes to

$$\mu = k_{\mathrm{B}}T\ln\left(\frac{n}{gn_Q}\right) = k_{\mathrm{B}}T\ln\left(\frac{n}{n_{Q'}}\right) \tag{9.4.9}$$

with $n_{Q'} = gn_Q$. We call $n_{Q'}$ the *quantum density*. When there is no spin degeneracy the quantum density is n_Q. The spin degeneracy increases the quantum density and reduces the chemical potential.

The second method of calculating the chemical potential is from changes in the free energy. A small change in the free energy is

$$\mathrm{d}F = \mathrm{d}\bar{U} - T\,\mathrm{d}S - S\,\mathrm{d}T.$$

If we allow the number of particles to vary, the change in the average internal energy is

$$\mathrm{d}U = T\,\mathrm{d}S - P\,\mathrm{d}V + \mu\,\mathrm{d}N.$$

The last term is the chemical work, that is the change in the internal energy of the system on increasing the number of particles. Each additional particle brings to the internal energy its chemical potential. It follows that

$$\mathrm{d}F = -P\,\mathrm{d}V + \mu\,\mathrm{d}N - S\,\mathrm{d}T \tag{9.4.10}$$

so that for constant volume and temperature

$$\mu = \left(\frac{\partial F}{\partial N}\right)_{V,T} = -k_{\rm B}T\left(\frac{\partial \ln(Z)}{\partial N}\right)_{V,T}. \tag{9.4.11}$$

Provided we can calculate the N-particle partition function, we can calculate F and then get μ. This is a very direct and powerful way of calculating the chemical potential.

As an example, let us take the energy of a free particle to be

$$\varepsilon(k) = \Delta + \frac{\hbar^2 k^2}{2m}$$

where Δ is a constant and $\hbar^2 k^2/2m$ is the kinetic energy of the particle of mass m and momentum $\hbar \mathbf{k}$. The single-particle partition function is

$$Z_1 = \sum_k e^{-\varepsilon(k)/k_{\rm B}T}$$

where the sum over k represents the sum over the lattice of points in $\mathbf{k}$ space. The partition function for N particles is approximately

$$Z_N = \frac{Z_1^N}{N!}.$$

This approximation is valid for a low-density gas at high temperatures. The factor of $N!$ is needed because the particles are indistinguishable. All we have to do is work out F and differentiate with respect to N. We have

$$F = -k_{\rm B}TN\left\{\ln\left(Z_1\right) - \ln\left(N\right) + 1\right\}$$

and so

$$\mu = -k_{\rm B}T\left\{\ln\left(\frac{Z_1}{N}\right)\right\} = -k_{\rm B}T\left\{\ln\left(\frac{\sum_k e^{-\varepsilon(k)/k_{\rm B}T}}{N}\right)\right\}.$$

For a free particle in three dimensions we get

$$\mu = \Delta + k_{\rm B}T\ln\left(\frac{n}{n_{Q'}}\right)$$

the same result which we obtained before, except that the shift in the energy of the particle by Δ has resulted in a shift in the chemical potential by the same amount.

We can tackle more complicated problems. For example, we could include the vibrational and rotational motions when calculating the single-particle partition function. Then $Z_1 = Z_{\rm trans}Z_{\rm vib}Z_{\rm rot}$. Obviously the excitation of higher vibrational and rotational states reduces the chemical potential, leading to a change in the condition for chemical equilibrium. Thermal excitation of higher rotational states of gas molecules changes their chemical potential and affects the chemical equilibrium of reacting gases, a somewhat surprising result!

9.5 Reactions

The condition for chemical equilibrium is that the sum of the chemical potentials of the reacting particles is equal to the sum of the chemical potentials of the products. This statement can be used to obtain a relationship between the concentrations of the reacting particles, as will be illustrated by two examples. This relationship is known as the law of mass action.

Consider a simple reaction which occurs in the Sun, in which a hydrogen atom dissociates into an electron and a proton, or the proton and electron can recombine, losing a bit of energy, forming a hydrogen atom. This process of combination or dissociation is going on all the time.

The electron, proton, and hydrogen gases have a low density compared to their quantum densities so that the formula

$$\mu = k_B T \ln\left(\frac{n}{n_{Q'}}\right) \tag{9.4.9}$$

is a good approximation. To simplify matters let us suppose that the hydrogen atom always exists in the 1s state, its ground state. Then the chemical potential of the hydrogen atom is

$$\mu_H = k_B T \left\{ \ln\left(\frac{n_H}{4}\left[\frac{2\pi\hbar^2}{m_H k_B T}\right]^{3/2}\right) \right\} + \Delta$$

where n_H is the number density of hydrogen atoms and m_H is the mass of the hydrogen atom; the degeneracy of each atom is 4 since both the spin of the proton and the electron have a degeneracy of 2; the quantity $-\Delta$ is the binding energy of the hydrogen atom. It is the difference in energy between the 1s state and the lowest energy of the continuum of unbound states. For the hydrogen atom $-\Delta = 13.6\,\text{eV}$, which is equivalent to $1.6 \times 10^5 k_B$. Deep inside the Sun the temperature is about $10^7\,\text{K}$ which is much larger than $\Delta/k_B = 10^5\,\text{K}$, the binding energy expressed as a temperature. Very few hydrogen atoms are bound together.

The chemical potential of the proton is

$$\mu_p = k_B T \left\{ \ln\left(\frac{n_p}{2}\left[\frac{2\pi\hbar^2}{m_p k_B T}\right]^{3/2}\right) \right\}$$

where n_p is the number density of protons, and m_p is the mass of the proton ($\approx m_H$). Finally for the electron

$$\mu_e = k_B T \left\{ \ln\left(\frac{n_e}{2}\left[\frac{2\pi\hbar^2}{m_e k_B T}\right]^{3/2}\right) \right\}$$

where n_e is the number density of electrons, and m_e is the mass of the electron.

The condition for chemical equilibrium is

$$\mu_H = \mu_e + \mu_p.$$

The gases are all at the same temperature, so

$$\ln\left(\frac{n_H}{4}\left[\frac{2\pi\hbar^2}{m_H k_B T}\right]^{3/2}\right) + \frac{\Delta}{k_B T} = \ln\left(\frac{n_p}{2}\left[\frac{2\pi\hbar^2}{m_p k_B T}\right]^{3/2}\right)$$

$$+ \ln\left(\frac{n_e}{2}\left[\frac{2\pi\hbar^2}{m_e k_B T}\right]^{3/2}\right)$$

which can be written as

$$n_p n_e = n_H \left(\frac{m_e k_B T}{2\pi\hbar^2}\right)^{3/2} e^{\Delta/k_B T}.$$

There is overall charge neutrality, which means that the number of electrons equals the number of protons. It follows that

$$n_p = n_e = n_H^{1/2}\left(\frac{m_e k_B T}{2\pi\hbar^2}\right)^{3/4} e^{\Delta/2k_B T}.$$

The more negative Δ, the smaller n_p is.

In the centre of the Sun the temperature is about 10^7 K, so the ratio $\Delta/k_B T$ is about -0.016. The Sun has a mass density of hydrogen atoms of about 1.4×10^3 kg m^{-3}. Putting in these numbers gives the ratio n_p/n_H of about ten: only about 10% of the hydrogen atoms have not dissociated. With these numbers n_e is about 8×10^{30} m^{-3}. The corresponding quantum density, $n_{Q'}$, for electrons is about 1.6×10^{32} m^{-3}; the electron density is small compared to the quantum density, so the calculation is self-consistent.

As a second example, consider the reaction of electrons with holes in semiconductors. In the ground state, the valence band is completely filled with electrons and the conduction band is empty. The bottom of the conduction band is above the top of the valence band; there is an energy gap E_g. In an excited state an electron is excited into the conduction band leaving the absence of an electron—a hole in the valence band—as illustrated in Fig. 9.3. At finite temperatures there can be several electrons in the conduction band and several holes in the valence band. But how many?

An electron in the conduction band reacts with a hole in a valence band, the excess energy being taken away by a phonon, a particle whose chemical potential is zero since the number of phonons is not conserved. The condition for equilibrium between electrons and holes is

$$\mu_e + \mu_h = 0.$$

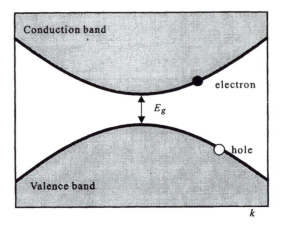

Fig. 9.3 A semiconductor in its ground state has a filled valence band of electrons and an empty conduction band. An excited state arises when an electron leaves the valence band and enters the conduction band. As the electron is excited it leaves a hole in the valence band. The hole behaves just as if it were a particle in its own right with a mass and charge. For finite temperatures the holes and electrons reach chemical equilibrium as discussed in the text.

Suppose the electrons in the conduction band have an energy

$$\varepsilon_e(k) = E_g + \frac{\hbar^2 k^2}{2m_e}$$

and the holes have an energy

$$\varepsilon_h(k) = \frac{\hbar^2 k^2}{2m_h}.$$

If both electrons and holes can be considered as low-density gases at high temperatures, the chemical potentials are

$$\mu_e = E_g + k_B T \ln\left(\frac{n_e}{n_{Q'_e}}\right)$$

where $n_{Q'_e} = 2n_{Q_e}$ is the quantum density of the electrons and

$$\mu_h = k_B T \ln\left(\frac{n_h}{n_{Q'_h}}\right)$$

where $n_{Q'_h} = 2n_{Q_h}$ is the quantum density of the holes. Chemical equilibrium requires that

$$n_e n_h = 4 n_{Q_e} n_{Q_h} e^{-E_g / k_B T}.$$

This equation is valid even in the presence of impurities which add or subtract electrons. The only conditions which must be fulfilled are that n_e is less than n_{Q_e}, and n_h is less than n_{Q_h}, conditions which are satisfied for semiconductors with sufficiently large band gaps. These conditions are needed so as to keep the electron and hole densities sufficiently small so that the chemical potential is given to a good approximation by the formula for a dilute gas.

In the absence of impurities, the hole and electron concentrations must be equal so as to preserve charge neutrality. Hence

$$n_e = 2 \left(n_{Q_e} n_{Q_h} \right)^{1/2} e^{-E_g / 2 k_B T}$$

and

$$\mu_e = \frac{E_g}{2} + \frac{k_B T}{2} \ln \left(\frac{n_{Q_h}}{n_{Q_e}} \right).$$

Provided that the electron and hole masses are not too different, the two quantum densities are about equal and the chemical potential sits midway in the band gap.

9.6 External chemical potential

So far all the quantities that affect the chemical potential relate to the energy levels of the particles. For example, in a hydrogen molecule there are rotational and vibrational energy levels, internal spin states for both the electrons and protons as well as the kinetic energy of translation. These internal quantum states of the molecule affect the chemical potential.

The molecule can also be coupled to an external field, such as gravity. When the particles interact with an external field, the energy levels can be written as

$$\varepsilon_i(\mathbf{r}) = \varepsilon_i^0 + \Delta(\mathbf{r})$$

where the shift in the energy level, $\Delta(\mathbf{r})$, is now a slowly varying function of position. By slowly varying we mean that the change in the value of $\Delta(\mathbf{r})$ over a typical distance travelled by the particle between collisions is much smaller than $k_B T$. As an example consider a gravitational field. If the gas molecule is at height z then $\Delta(\mathbf{r}) = mgz$ where m is the mass of the molecule and g is the acceleration due to gravity. Another example is an electric field, $\mathbf{E}$, applied to charged particles. In this case $\Delta(\mathbf{r}) = -e\mathbf{E}.\mathbf{r}$ where e is the charge on the particle and $\mathbf{r}$ is its position. Alternatively for a particle with magnetic moment μ in a magnetic induction field $\mathbf{B}(\mathbf{r})$ we have $\Delta(\mathbf{r}) = -\mu.\mathbf{B}(\mathbf{r})$.

External fields affect the chemical potential of the particles. The chemical potential can be written as

$$\mu = \Delta(\mathbf{r}) + \mu_{\text{int}}.$$

$\Delta(\mathbf{r})$ can be thought of as the *external chemical potential*, the part of the chemical potential that is affected by external macroscopic fields: it could arise from the coupling to an electric or magnetic field, or from the coupling to gravity. The rest, μ_{int}, is the *internal chemical potential* which is determined by the microscopic energy levels of the particle. It is the internal degrees of freedom like rotation, vibration, spin, and kinetic energy which affect μ_{int}. We have to be careful when talking about chemical potential to differentiate between the internal, the external, and the total chemical potential.

For example, a particle in a gravitational potential has a total chemical potential

$$\mu = \Delta(\mathbf{r}) + \mu_{\text{int}} = mgz + k_{\text{B}}T\ln\left(\frac{n}{n_{Q'}}\right)$$

made up of the external chemical potential, mgz, and the internal chemical potential. We can rewrite this equation as

$$n = n_{Q'}e^{(\mu-mgz)/k_{\text{B}}T}.$$

If the air in a room had the same temperature and chemical potential everywhere, from the floor to the ceiling, the air would be in equilibrium. There would be no flow of the air, no winds carrying the particles from the floor upwards; the air would be stationary, becalmed. The number density of oxygen molecules would then vary as

$$n \sim e^{-m_{\text{o}}gz/k_{\text{B}}T}$$

where m_{o} is the mass of the oxygen molecule. The different gas molecules in air—oxygen, nitrogen, water vapour, argon—each have a different mass, and so their densities will each decay exponentially with a length-scale z_L given by $z_L = k_{\text{B}}T/gm_i$ with m_i the mass of the molecule. The lightest particles have the largest z_L. If the gas is at a temperature of 290 K, then z_L is 8.5 km if the gas is nitrogen, and 32 km if the gas is hydrogen.

9.7 The grand canonical ensemble

We want to extend our treatment of statistical mechanics to systems with a variable number of particles. Suppose we place the system in contact with a particle reservoir so that particles can flow in and out of the system. (The reservoir also acts as a heat bath so that energy can flow in and out of the system as well.) The combined system is closed and isolated from the rest of the universe. The ensemble of quantum states in this case is called the *grand canonical ensemble*.

The system could be a gas of atoms which combine together to form molecules. The number of atoms in the system can change as the atoms can combine or dissociate. The system could even be a particular quantum state containing either one or no particles. The number of particles in the state is not fixed; it can fluctuate.

The definition of the temperature of the reservoir according to statistical mechanics is

$$\frac{1}{T} = k_{\text{B}}\left(\frac{\partial\ln(W_R)}{\partial U_R}\right)_{N_R}. \tag{9.7.1}$$

Similarly the chemical potential of the reservoir is

$$\mu = -k_B T \left(\frac{\partial \ln(W_R)}{\partial N_R} \right)_{U_R}. \tag{9.7.2}$$

The characteristic feature of the heat reservoir is that its temperature does not change when heat is added to it. Of course for any finite reservoir there will be a small temperature change, but for a large reservoir the change is so small it can be ignored. The characteristic feature of a particle reservoir is that when a few particles are added to it the chemical potential does not change. Therefore μ and T are constant for a large reservoir.

The value of $\ln(W_R)$ which is the solution of eqns (9.7.1) and (9.7.2) is

$$\ln(W_R) = K + \frac{(U_R - \mu N_R)}{k_B T} \tag{9.7.3}$$

where K is a constant. To verify this, just differentiate with respect to U_R keeping N_R constant, and vice versa. Taking the exponential of eqn (9.7.3) gives

$$W_R = \alpha \, e^{(U_R - \mu N_R)/k_B T} \tag{9.7.4}$$

where α is a constant. Equation (9.7.4) is a direct consequence of the definition of chemical potential and temperature in statistical mechanics.

Suppose system A is in contact with the reservoir, and the combined system is isolated from the rest of the universe. When energy moves from system A to the reservoir, none of this energy is lost to the rest of the universe. Similarly when particles cross from system A into the reservoir, none of them are lost to the rest of the universe. Because the combined system is isolated, it has a fixed total number of particles, N_T, and a fixed total energy, U_T. The energy of the reservoir is $U_R = U_T - U_A$ where U_A is the energy of system A, and the number of particles in the reservoir is $N_R = N_T - N_A$, where N_A is the number of particles in system A. Thus eqn (9.7.4) becomes

$$W_R = \alpha \, e^{\{(U_T - U_A) - \mu(N_T - N_A)\}/k_B T}.$$

Since U_T and N_T are constant, this expression can be written as

$$W_R = \alpha' \, e^{-(U_A - \mu N_A)/k_B T} \tag{9.7.5}$$

with $\alpha' = \alpha \, e^{(U_T - \mu N_T)/k_B T}$. The temperature is fixed, the chemical potential is fixed, U_T and N_T are constant and so α' is a constant. The only terms which vary in eqn (9.7.5) are those involving U_A and N_A. Previously we assumed that N_A could not change, but now we allow it to vary.

Consider a single quantum state, ψ_i, for system A. The total number of quantum states for the joint system when system A is in ψ_i is

$$W_i = \alpha' \, e^{-(E_i - \mu N_i)/k_B T} \tag{9.7.6}$$

where E_i is the energy of the quantum state, and N_i is the number of particles for that quantum state. More precisely the quantum state is an eigenstate of the number operator with eigenvalue N_i such that

$$\hat{N}\psi_i = N_i\psi_i. \tag{9.7.7}$$

(An implicit assumption behind this step is that the quantum state of system A is a simultaneous eigenstate of both the energy and the number of particles, an assumption which is not always valid.)

W_i is the total number of accessible states of the joint system which is thermally isolated from the rest of the universe when system A is in the quantum state ψ_i. The total number of quantum states for the joint system equals the sum of the W_i for all the different quantum states that system A can be in, allowing for fluctuations in both energy and number of particles. This number is

$$W = \sum_i W_i. \tag{9.7.8}$$

Since all quantum states are equally likely, the probability of being a particular quantum state, ψ_i, is

$$p_i = \frac{e^{-(E_i - \mu N_i)/k_B T}}{\sum_j e^{-(E_j - \mu N_j)/k_B T}}. \tag{9.7.9}$$

The probability is the number of ways of getting the particular quantum state divided by the total number of states with all possible energies and all possible numbers. Notice that the factor α' cancels out.

We can build a mental picture of an ensemble of replica systems, with both energy and particles flowing from each system to its neighbours. This ensemble of systems is called the grand canonical ensemble. When we want to calculate the average of any physical quantity, we average equally over the quantum states of this ensemble.

When we were concerned with the canonical ensemble, we called the denominator in Boltzmann's expression for the probability, the partition function, Z. All the thermodynamics could be found from Z. In the grand canonical ensemble the factor in the denominator is more complex:

$$\Xi = \sum_j e^{-(E_j - \mu N_j)/k_B T}. \tag{9.7.10}$$

Ξ is called the *grand partition function*; it is the analogue of Z for situations where the number of particles is allowed to vary. It can be thought of as a normalization constant: if we sum the probabilities over all quantum states ψ_i we must have $\sum p_i = 1$. It follows that Ξ is the factor which is needed to get the sum of the probabilities equal to one.

The number of conserved quantities has increased from one (the energy) to two (the energy and the number of particles). If we want to conserve extra quantities, such as the momentum or the angular momentum, or baryon number, then we can go through with the same mathematics, just slightly modified, and come out with a new probability distribution. However, the probability distribution which we have just derived, eqn (9.7.9), is very important and is given a solemn-sounding name: it is called the *grand canonical distribution function*. It looks rather daunting because we have to sum over the energies of the quantum states for all numbers of particles.

9.8 Absorption of atoms on surface sites

Suppose there is a surface on which there are sites where atoms can be absorbed; each site can be empty or it can have at most one absorbed atom. Let us take one site to be the system. For site i either no atom is present on the site in which case the energy is zero or there is one atom present on the site with energy ϵ_i. The grand partition function is the sum over two states. Either there are no particles so $(n_i = 0, E_i = 0)$ or there is one particle in the state in which case the number and energy are $(n_i = 1, E_i = \epsilon_i)$. The grand partition function for this state is

$$\Xi = 1 + e^{-(\epsilon_i - \mu)/k_B T}, \tag{9.8.1}$$

a very simple expression.

The probability that there is one particle present is

$$p_i = \frac{e^{-(\epsilon_i - \mu)/k_B T}}{\Xi} = \frac{1}{e^{(\epsilon_i - \mu)/k_B T} + 1}. \tag{9.8.2}$$

If $(\epsilon_i - \mu)$ is negative, the probability that the site is occupied is between 0.5 and 1. As the temperature rises, μ becomes more negative causing $(\epsilon_i - \mu)$ to become larger; the probability that the site is occupied decreases.

The average number, n_i, of particles on site i can easily be found now that we know the probabilities. The probability that no particle is present on site i is $(1 - p_i)$, the probability that one particle is present is p_i. Hence

$$n_i = 0 \times (1 - p_i) + 1 \times p_i = \frac{1}{e^{(\epsilon_i - \mu)/k_B T} + 1}. \tag{9.8.3}$$

In fact this expression for the average number of particles can be applied whenever there can be either zero or one particle in a particular state. We will use this expression in the next chapter to describe the average number of Fermi particles in a single-particle state.

9.9 The grand potential

Once we have calculated Ξ, it is relatively easy to calculate thermodynamic quantities. In the grand canonical ensemble there is a relation between the grand potential (a new thermodynamic energy) and the grand partition function. The procedure for obtaining this relationship is exactly the same as for the canonical ensemble.

The entropy is given by

$$S = -k_B \sum_i p_i \ln(p_i). \tag{5.3.4}$$

This formula, which expresses the entropy in terms of the probabilities, is always valid for a system in contact with a reservoir. From eqns (9.7.10) and (9.7.11) we can derive the formula for the probability in the grand canonical ensemble that the system is in quantum state ψ_i as

$$p_i = \frac{e^{-(E_i - \mu N_i)/k_B T}}{\Xi}. \tag{9.9.1}$$

Putting this expression in the term $\ln(p_i)$ in the equation for S, and using the relation that the sum of the probabilities is equal to one gives

$$
\begin{aligned}
S &= -k_B \sum_i p_i \left\{ -\frac{(E_i - \mu N_i)}{k_B T} - \ln(\Xi) \right\} \\
&= \frac{\bar{U}}{T} - \frac{\mu \bar{N}}{T} + k_B \ln(\Xi)
\end{aligned}
\tag{9.9.2}
$$

where $\bar{U}$ is the average energy and $\bar{N}$ is the average number of particles. Thus

$$\bar{U} - \mu \bar{N} - TS = -k_B T \ln(\Xi) = \Phi_G. \tag{9.9.3}$$

Φ_G is called the *grand potential*. We can derive from the grand potential all the thermodynamics for systems with variable numbers of particles, just as we could derive the thermodynamics from the free energy for systems with a constant number of particles.

An expression for dS can be obtained by writing the entropy $S(V, \bar{U}, \bar{N})$ as a function of the extensive quantities $V, U,$ and N:

$$
\begin{aligned}
dS &= \left(\frac{\partial S}{\partial U} \right)_{V,N} d\bar{U} + \left(\frac{\partial S}{\partial V} \right)_{U,N} dV + \left(\frac{\partial S}{\partial N} \right)_{U,V} d\bar{N} \\
&= \frac{d\bar{U} + PdV - \mu d\bar{N}}{T}.
\end{aligned}
\tag{2.9.9}
$$

Hence

$$d\bar{U} = -P \, dV + \mu \, d\bar{N} + T \, dS. \tag{9.9.4}$$

A small change in the grand potential can be written as

$$
\begin{aligned}
d\Phi_G &= d\bar{U} - d(\mu \bar{N}) - d(TS) \\
&= -P \, dV - \bar{N} \, d\mu - S \, dT.
\end{aligned}
\tag{9.9.5}
$$

This equation can be used to obtain expressions for the entropy, the pressure, and the average number of particles. Thus we have

$$S = - \left(\frac{\partial \Phi_G}{\partial T} \right)_{V,\mu} = \left(\frac{\partial \left(k_B T \ln(\Xi) \right)}{\partial T} \right)_{V,\mu}, \tag{9.9.6}$$

$$P = - \left(\frac{\partial \Phi_G}{\partial V} \right)_{T,\mu} = \left(\frac{\partial \left(k_B T \ln(\Xi) \right)}{\partial V} \right)_{T,\mu}, \tag{9.9.7}$$

and

$$\bar{N} = - \left(\frac{\partial \Phi_G}{\partial \mu} \right)_{T,V} = \left(\frac{\partial \left(k_B T \ln(\Xi) \right)}{\partial \mu} \right)_{T,V}. \tag{9.9.8}$$

Consequently, once Ξ has been calculated we can calculate thermodynamic quantities. In particular there is a way of calculating the average number of particles once we have worked out Ξ as a function of μ.

As an example let us calculate the average number of atoms absorbed on a surface site, the problem analysed in section 9.8. By differentiating $k_B T \ln(\Xi)$ with respect to μ with Ξ given by eqn (9.8.1) we get

$$
\begin{aligned}
n_i &= k_B T \left(\frac{\partial \ln \left(1 + e^{-(\varepsilon_i - \mu)/k_B T} \right)}{\partial \mu} \right)_{T,V} \\
&= \frac{e^{-(\varepsilon_i - \mu)/k_B T}}{1 + e^{-(\varepsilon_i - \mu)/k_B T}} \\
&= \frac{1}{e^{(\varepsilon_i - \mu)/k_B T} + 1}
\end{aligned}
\tag{9.9.9}
$$

the same result as we got in section 9.8.

9.10 Problems

1. Consider the reaction
$$3O_2 \rightleftharpoons 2O_3.$$
Identify the stoichiometric coefficients. What is the condition for chemical equilibrium between oxygen and ozone?

2. A system which is in equilibrium with a heat bath is rotating about the z axis in such a way that energy and angular momentum may be exchanged between the system and the bath. The total energy and angular momentum of the system and the bath are conserved. Let the system have a simultaneous set of eigenstates, $|\Psi_i\rangle$, of energy and angular momentum

$$\hat{H} |\Psi_i\rangle = E_i |\Psi_i\rangle,$$
$$\hat{L}_z |\Psi_i\rangle = L_{zi} |\Psi_i\rangle.$$

Develop the theory for this case by analogy with the treatment of the grand canonical ensemble. Define a parameter $\omega_z = -T \left(\partial S_R / \partial L_{zR} \right)_{U_R}$ for the reservoir. Show that the probability, p_i, of finding the system in this state can be written in the form

$$p_i = \frac{e^{-(E_i - \omega_z L_{zi})/k_B T}}{\Pi}$$

and give an expression for Π. Use the expression for the entropy of the system in terms of p_i to derive a formula for the equivalent to the grand potential.

3. Suppose there is a density n_D of electron donors in a semiconductor. Determine the electron density $n_e = n_D + n_h$ by using the condition for chemical equilibrium.

4. Imagine a solid with sites on which particles can be localized. Each site can have 0, 1, or 2 particles; each particle on the site can be in one of two states, ψ_1 or ψ_2 with energy ε_1 or ε_2. We neglect the interaction between particles even if they are localized on the same site.

   (a) Write down the grand partition function for one site if the particles are bosons, and deduce the grand partition function for a solid with N sites.
   (b) Repeat the above for particles which are fermions.
   (c) In both cases work out the average number of particles per site.

5. Assume that the reaction of proton plus electron to form hydrogen occurs in thermal equilibrium at 4000 K in a very low density gas, with overall charge neutrality. Find the chemical potential for each gas in terms of its number density. For simplicity you can ignore the spectrum of bound states of a hydrogen atom and consider only the ground state.
   Give the condition for chemical equilibrium and calculate the equilibrium concentration of electrons as a function of temperature and hydrogen concentration. Estimate the proton number density when the gas is half-ionized at 4000 K.

6. In the presence of an external potential, $V(r)$, the number density of particles sometimes can be written as

$$n(r) = n_0 \exp(-V(r)/k_B T).$$

Under what circumstances is this the case? A cylindrical vessel of gas is rotated with angular velocity ω about an axis perpendicular to that of the cylinder, and passing through its centre. The gas particles of mass m experience a centrifugal force $m\omega^2 r$, where r is the perpendicular distance to the axis of rotation. Show that the pressure of the gas varies as

$$p(r) = p_0 \exp(m\omega^2 r^2 / 2k_B T).$$

7. The partition function for an ideal gas of N atoms is approximately

$$Z_N = Z_1^N / N!.$$

Show, for example by considering the power series expansion of $e^{\lambda Z_1}$ where λ is equal to $e^{\mu/k_B T}$, that the grand partition function for this gas is

$$\Xi = e^{\lambda Z_1}.$$

8. The grand potential, Φ_G, is proportional to the volume of the system since both Φ_G and V are extensive quantities. A system is attached to a reservoir which fixes both the temperature and chemical potential. Prove that

$$\left(\frac{\partial \Phi_G}{\partial V} \right)_{\mu,T} = -P = \frac{\Phi_G}{V}.$$

Use this result to prove the relation

$$V \, dP - S \, dT - N \, d\mu = 0.$$

9. Suppose there is a flat surface of a solid in contact with a vapour which acts as the thermal and particle reservoir. The atoms stick to sites on the solid surface with at most one atom per site. Each site on the surface of the solid has an energy ϵ if an atom is stuck to it, and an energy of zero if no atom is stuck to it. If there are M sites with N atoms trapped on them the energy of the state is $N\epsilon$ and the degeneracy is $M!/N!(M - N)!$. Show that the grand partition function is

$$\Xi = \left(1 + e^{-(\epsilon - \mu)/k_B T} \right)^{M}.$$

10. $W(N)$ is a Gaussian function of N with a maximum at N^*. For any system show that

$$\Delta N^2 = k_B T \left(\frac{\partial \bar{N}}{\partial \mu} \right).$$

Derive this relation another way: by calculating $\partial^2 \ln(\Xi)/\partial\mu^2$, show that the mean square in the particle number is

$$\Delta N^2 = \sum_i p_i \left(N_i - \bar{N} \right)^2 = (k_B T)^2 \left(\frac{\partial^2 \ln(\Xi)}{\partial\mu^2} \right).$$

10

Fermi and Bose particles

Before I came here I was confused about this subject. Having listened to your lecture I am still confused. But on a higher level.   *Enrico Fermi*

10.1 Introduction

In Chapter 6 we gave a simplified treatment of a non-interacting gas of identical particles and claimed that the treatment was correct for a dilute gas at sufficiently high temperatures. In this chapter we use the techniques of the grand canonical ensemble to give an exact treatment for both Bose and Fermi particles. For sufficiently high temperatures we recover the results of Maxwell–Boltzmann statistics for the average number of particles in a single-particle state for both types of particles, fermions and bosons; for low temperatures there are substantial deviations which arise from the different statistics that need to be applied. For example, for Fermi particles the exclusion principle means that there cannot be two particles in the same single-particle state. There is no such restriction for Bose particles. The resulting formulae for the average number of particles in a single-particle state are quite different at low temperatures.

The behaviour we deduced in Chapter 6 for an ideal gas is expected to be valid for a dilute gas at sufficiently high temperatures. Even under such conditions our treatment was an approximation for we neglected the interactions between the particles. We expect this to be a good approximation if the gas molecules are far apart so that the kinetic energy of the molecules is much larger than the average potential energy of the interaction between them. These conditions are met for high temperatures and low densities. We can expect real monatomic gases at high temperatures to be described accurately by the theory. But can we use the model of non-interacting particles to describe real gases at low temperatures and high densities?

All gases that exist in nature involve particles which interact with each other. Provided the kinetic energy of the particles is much larger than the interaction between particles then formulae based on the non-interacting gas model may give a fair approximation to reality. As we will see, Fermi particles at low temperatures have a large kinetic energy, an energy which can be much larger than the interaction between particles. For example, when the particles are charged the interaction is of order $e^2/4\pi\epsilon_0 d$, where d is the spacing of the particles; we will show that the kinetic energy due to quantum effects is of order $\pi^2\hbar^2/2md^2$.

If the gas has a high density, as occurs for the electron gas in white dwarf stars, then the kinetic energy is much larger than the potential energy and it is a reasonable first approximation to ignore the interaction. For electrons in metals this is not the case, and the density of the electrons is such that the potential and kinetic energies are comparable. Yet the free-electron model works reasonably well for metals, especially for transport properties. The *Pauli exclusion principle* restricts the number of free states into which the electrons can be scattered, so reducing the scattering cross-section at low temperatures. In this sense the particles hardly interact. The free-electron model of the gas gives a reasonable description of transport properties such as the conductivity.

However, at very low temperatures the model of non-interacting Fermi particles goes seriously wrong when there are attractive interactions between the particles. Attractive interactions lead to pairing of the particles and to superfluidity for uncharged particles and superconductivity for charged ones.

For a gas of Bose particles, the kinetic energy decreases without limit as the temperature drops. Eventually the kinetic energy becomes smaller than the interactions which cannot be ignored. If there were no interactions, all the Bose particles would condense into the lowest energy single-particle state at the absolute zero. This phenomenon is called *Bose–Einstein condensation*. It is widely believed that Bose condensation is responsible for the superfluidity of liquid ^{4}He.

Liquid ^{4}He does not solidify unless the pressure is raised to 25 bar. The reason is that the light mass of the ^{4}He particle creates a large zero-point energy of each particle in the potential created by its neighbours, an energy which is minimized if the particles are kept far apart. The effect of the van der Waals potential between atoms is therefore weaker and so the liquid phase is favoured over the solid. In the liquid phase particles can exchange places so that effects arising from the indistinguishability of the particles can be seen. However, the interaction between the particles is very important and the non-interacting theory of the Bose gas is not realistic.

Nonetheless the theory of the non-interacting Bose gas is of interest for three reasons: it shows that the entropy of a gas can be zero at the absolute zero of temperature, something which had not been demonstrated before the work of Einstein (1924); it predicts that there is a phase transition; it is widely believed that Bose condensation is relevant to the superfluid state of liquid ^{4}He.

In this chapter we develop the statistical mechanics of indistinguishable particles from the grand partition function. Then we concentrate on a Fermi gas and calculate the entropy, pressure, and number of particles as a function of temperature and chemical potential. In the limit of high temperatures we recover the expressions for entropy and pressure that we obtained in Chapter 6. For low temperatures there are substantial differences in the pressure and the entropy. We then apply these ideas to various Fermi gases. Finally we present the theory of the non-interacting Bose gas and briefly discuss its relevance to superfluidity.

10.2   The statistical mechanics of identical particles

We use the techniques of the grand canonical ensemble to determine the distribution of particles in a single-particle quantum state. The treatment for Fermi particles differs from that for Bose particles.

10.2.1   Fermi particle

The system is taken to be a single spin-up Fermi particle in the standing-wave state $\phi_\mathbf{k}(x)$. Either there is no particle in this state in which case the energy is zero, or there is one particle present with energy $\varepsilon(k)$. (In this subsection dealing with a Fermi particle the symbol 'k' represents the single-particle state with wave vector $\mathbf{k}$ and spin s.) The contribution to the grand partition function from this state is

$$\Xi_k = 1 + e^{-(\varepsilon(k)-\mu)/k_\mathrm{B}T} \tag{10.2.1}$$

which gives a contribution to the *grand potential* of

$$\Phi_G(k) = -k_\mathrm{B}T \ln\left(1 + e^{-(\varepsilon(k)-\mu)/k_\mathrm{B}T}\right). \tag{10.2.2}$$

The total grand potential is the sum of the contributions from each wave vector multiplied by a factor of two for the two spin states.

From eqn (10.2.2) we can derive expressions for the number of particles in this state and their contribution to the entropy of the gas by using the formulae given in section 9.9. Let us start by calculating the number of particles in the state of wave vector $\mathbf{k}$ with spin s. By using eqn (9.9.8) we find that the average number of particles is given by

$$n(k) = -\left(\frac{\partial \Phi_G(k)}{\partial \mu}\right)_{T,V} = \frac{1}{e^{(\varepsilon(k)-\mu)/k_\mathrm{B}T} + 1}. \tag{10.2.3}$$

This is called *the Fermi–Dirac distribution function*. It was first derived by Fermi (1926a,b). As we shall show later on, at high temperatures $-\mu$ is positive and much larger than $k_\mathrm{B}T$. In the limit that $(\varepsilon(k)-\mu)/k_\mathrm{B}T$ is very large the Fermi–Dirac distribution function tends to

$$n(k) \approx e^{-(\varepsilon(k)-\mu)/k_\mathrm{B}T}$$

which is the value obtained using Maxwell–Boltzmann statistics, since

$$e^{\mu/k_\mathrm{B}T} = \frac{N}{Z_1}.$$

For low temperatures μ is positive. For energies such that $(\varepsilon(k)-\mu)/k_\mathrm{B}T$ is negative the Fermi–Dirac distribution function is nearly one; for energies such that $(\varepsilon(k)-\mu)/k_\mathrm{B}T$ is positive the Fermi–Dirac distribution function is nearly

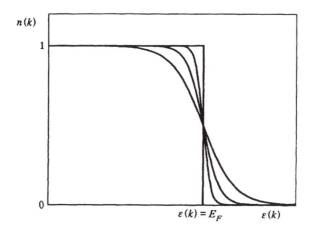

Fig. 10.1 The Fermi–Dirac distribution function, $n(k)$, for low temperatures. For $T = 0$ there is a sharp step in $n(k)$ at the Fermi energy, $\varepsilon(k) = E_F$. For finite temperatures the step is smoothed out. The curves shown here are for $\mu/k_BT = 50$, 20, and 10 with the curve for 50 lying closest to the step function.

zero. The variation of $n(k)$ with $\varepsilon(k)$ is shown in Fig. 10.1 for different values of μ/k_BT. At the absolute zero of temperature $n(k)$ changes from one to zero discontinuously at the energy $\varepsilon(k) = \mu$. This energy is called the *Fermi energy* and is denoted as E_F; the corresponding wave vector is called the Fermi wave vector and is denoted by k_F. In the ground state of the system all the low-energy single-particle states are occupied up to the Fermi energy and all higher-energy states are empty. The ground state can be arranged in one way only so it is non-degenerate.

The contribution to the entropy can be found by using eqn (9.9.6); it is given by

$$S(k) = -\left(\frac{\partial \Phi_G(k)}{\partial T}\right)_{V,\mu}$$

$$= k_B \ln\left(1 + e^{-(\varepsilon(k)-\mu)/k_BT}\right) + \frac{1}{T}\frac{(\varepsilon(k)-\mu)}{e^{(\varepsilon(k)-\mu)/k_BT} + 1}. \quad (10.2.4)$$

We can rewrite this using (10.2.3) as

$$S(k) = -k_B \left\{n(k)\ln[n(k)] + [1 - n(k)]\ln[1 - n(k)]\right\}. \quad (10.2.5)$$

The contribution to the entropy vanishes when $n(k) = 1$ or $n(k) = 0$. At the absolute zero of temperature $n(k)$ is one if $\varepsilon(k)$ is less than μ and zero if $\varepsilon(k)$ is greater than μ. The *Fermi gas at the absolute zero of temperature has no entropy* which makes perfect sense: the ground state is non-degenerate so W is one which means that $S = k_B \ln(W)$ is zero. A perfect Fermi gas has no entropy at the absolute zero. This statement is consistent with the third law of thermodynamics which says that the entropy tends to a constant at the absolute zero.

The contribution to the average energy of the gas for this wave vector is given by the average number of particles in the single-particle state multiplied by the energy:

$$U(k) = n(k)\varepsilon(k). \tag{10.2.6}$$

10.2.2   Bose particle

Consider a gas of spinless Bose particles, and concentrate on the single-particle standing-wave state $\phi_\mathbf{k}(x)$. *We treat the single-particle state as the system.* There is no restriction on the number of Bose particles in this single-particle state. When there are n particles in this single-particle state the energy is $n\varepsilon(\mathbf{k})$. The grand partition function for this single-particle state is a geometric series:

$$
\begin{aligned}
\Xi_\mathbf{k} &= 1 + e^{-(\varepsilon(\mathbf{k})-\mu)/k_\mathrm{B}T} + e^{-2(\varepsilon(\mathbf{k})-\mu)/k_\mathrm{B}T} + e^{-3(\varepsilon(\mathbf{k})-\mu)/k_\mathrm{B}T} + \cdots \\
&= \frac{1}{1 - e^{-(\varepsilon(\mathbf{k})-\mu)/k_\mathrm{B}T}}.
\end{aligned}
\tag{10.2.7}
$$

However, the geometric series can only be summed if successive terms get smaller, which means that $(\varepsilon(\mathbf{k}) - \mu)$ must be positive. The geometric series could not be summed if $(\varepsilon(\mathbf{k}) - \mu)$ were negative. Something drastic happens to prevent this occurring: Bose condensation.

The contribution to the *grand potential* for the wave vector $\mathbf{k}$ is

$$\Phi_G(\mathbf{k}) = k_\mathrm{B}T \ln\left(1 - e^{-(\varepsilon(\mathbf{k})-\mu)/k_\mathrm{B}T}\right). \tag{10.2.8}$$

From this equation we can derive expressions for the number of particles in this state and their contribution to the entropy. The number of particles, $f(\mathbf{k})$, in the state $\phi_\mathbf{k}(x)$ is given by

$$f(\mathbf{k}) = -\left(\frac{\partial \Phi_G(\mathbf{k})}{\partial \mu}\right)_{T,V} = \frac{1}{e^{(\varepsilon(\mathbf{k})-\mu)/k_\mathrm{B}T} - 1} \tag{10.2.9}$$

which is called the *Bose–Einstein distribution function.* We have given it the symbol $f(\mathbf{k})$ so as to differentiate it from the Fermi–Dirac distribution function.

In the limit that $(\varepsilon(\mathbf{k}) - \mu)/k_\mathrm{B}T$ is very large, as happens for sufficiently high temperatures, the Bose–Einstein distribution function tends to

$$f(\mathbf{k}) \approx e^{-(\varepsilon(\mathbf{k})-\mu)/k_\mathrm{B}T}$$

which corresponds to the value obtained using Maxwell–Boltzmann statistics. For low temperatures the Bose–Einstein distribution differs significantly from that of a Maxwell–Boltzmann gas.

The contribution to the entropy when $\mathbf{k}$ is non-zero is given by

$$S(\mathbf{k}) = -\left(\frac{\partial \Phi_G(\mathbf{k})}{\partial T}\right)_{V,\mu}$$

$$= -k_B \ln\left(1 - e^{-(\varepsilon(\mathbf{k}) - \mu)/k_B T}\right) + \frac{1}{T}\frac{(\varepsilon(\mathbf{k}) - \mu)}{e^{(\varepsilon(\mathbf{k}) - \mu)/k_B T} - 1}.$$

$$(10.2.10)$$

We can rewrite this using (10.2.9) as

$$S(\mathbf{k}) = k_B \left\{[1 + f(\mathbf{k})] \ln[1 + f(\mathbf{k})] - f(\mathbf{k}) \ln[f(\mathbf{k})]\right\}. \qquad (10.2.11)$$

The contribution to the entropy vanishes when $f(\mathbf{k}) = 0$. At the absolute zero of temperature all single-particle states are empty except the ground state, so the entropy of all other states is zero. All the particles are in the ground state, and this can be made in one way only so $W = 1$. The entropy of the ground state is zero. This discovery, due to Einstein, was the first demonstration that an ideal gas has zero entropy at the absolute zero of temperature, a result which is in accordance with the third law of thermodynamics.

Finally the contribution to the internal energy for Bose particles is

$$U(\mathbf{k}) = f(\mathbf{k})\varepsilon(\mathbf{k}). \qquad (10.2.12)$$

10.3 The thermodynamic properties of a Fermi gas

Now that we have calculated the distribution of particles amongst the different single-particle states we are in a position to determine the thermodynamic properties of these gases. For Fermi particles with spin 1/2, all we have to do is to sum up over wave vectors and spin.

The average number of particles can be calculated by using the equation

$$\bar{N} = 2 \sum_{\mathbf{k}} n(\mathbf{k}). \qquad (10.3.1)$$

where $n(\mathbf{k})$ is the Fermi distribution function for a particle of, say, spin-up in the single-particle state $\phi_{\mathbf{k}}$. We can determine the chemical potential by using eqn (10.3.1). Remember that the way to evaluate the sum over $\mathbf{k}$ space is to turn it into an integral over k weighted by the density of states, $D(k)$. For a three-dimensional system

$$D(k) = \frac{Vk^2}{2\pi^2},$$

and the number density, n, is for spin 1/2 Fermi particles

$$n = \frac{\bar{N}}{V} = \frac{1}{\pi^2} \int_0^\infty \frac{1}{e^{(\varepsilon(k) - \mu)/k_B T} + 1} k^2 \, dk. \qquad (10.3.2)$$

Suppose the energy of the particle is

$$\varepsilon(k) = \frac{\hbar^2 k^2}{2m}.$$

Let us rewrite this equation in terms of the variable $x = (\hbar^2 k^2 / 2mk_BT)^{1/2}$ as

$$n = \frac{1}{\pi^2} \left(\frac{2mk_BT}{\hbar^2} \right)^{3/2} \int_0^\infty \frac{x^2}{e^{x^2 - \eta} + 1} dx \qquad (10.3.3)$$

where $\eta = \mu/k_BT$. Equation (10.3.3) gives a relation between the number density of particles, the chemical potential, and the temperature. The quantity η depends on the ratio $n/T^{3/2}$.

We can also obtain an expression for the entropy of a gas of Fermi particles. By summing eqn (10.2.4) over different wave vectors and the two spin states we get

$$S = 2k_B \sum_k \left\{ \ln \left(1 + e^{-(\varepsilon(k) - \mu)/k_BT} \right) + \frac{1}{k_BT} \frac{(\varepsilon(k) - \mu)}{e^{(\varepsilon(k) - \mu)/k_BT} + 1} \right\}. \qquad (10.3.4)$$

Let us concentrate on a three-dimensional system with energy $\varepsilon(k) = \hbar^2 k^2 / 2m$. After integration by parts we get

$$
\begin{aligned}
S &= \frac{V}{\pi^2 T} \int_0^\infty k^2 \left(\frac{5\hbar^2 k^2}{6m} - \mu \right) \frac{1}{e^{(\varepsilon(k) - \mu)/k_BT} + 1} dk \\
&= \frac{Vk_B}{\pi^2} \left(\frac{2mk_BT}{\hbar^2} \right)^{3/2} \int_0^\infty x^2 \left(\frac{5}{3} x^2 - \eta \right) \frac{1}{e^{x^2 - \eta} + 1} dx. \qquad (10.3.5)
\end{aligned}
$$

By combining eqns (10.3.3) and (10.3.5) we can get an expression for the entropy per particle which is a function of η only:

$$\frac{S}{N} = k_B \frac{\int_0^\infty x^2 \left(\frac{5}{3} x^2 - \eta \right) \left(e^{(x^2 - \eta)} + 1 \right)^{-1} dx}{\int_0^\infty x^2 \left(e^{(x^2 - \eta)} + 1 \right)^{-1} dx}. \qquad (10.3.6)$$

Another quantity which we can calculate is the pressure of the gas. To get the pressure it is easiest to start with the grand potential which is

$$
\begin{aligned}
\Phi_G &= -2 \sum_k k_BT \ln \left(1 + e^{-(\varepsilon(k) - \mu)/k_BT} \right) \\
&= -\frac{Vk_BT}{\pi^2} \int_0^\infty k^2 \ln \left(1 + e^{-(\varepsilon(k) - \mu)/k_BT} \right) dk. \qquad (10.3.7)
\end{aligned}
$$

The thermodynamic potential is proportional to the volume of the system. We can obtain the pressure from eqn (9.9.7):

$$P = -\left(\frac{\partial \Phi_G}{\partial V} \right)_{T,\mu} = -\frac{\Phi_G}{V}$$

$$= \frac{k_B T}{\pi^2} \int_0^\infty k^2 \ln\left(1 + e^{-(\epsilon(k) - \mu)/k_B T}\right) dk. \tag{10.3.8}$$

This expression can be simplified first by integration by parts and then by converting to the variable $x = (\hbar^2 k^2 / 2mk_B T)^{1/2}$. In this way we get

$$P = \frac{2k_B T}{3\pi^2} \left(\frac{2mk_B T}{\hbar^2}\right)^{3/2} \int_0^\infty \frac{x^4}{e^{x^2 - \eta} + 1} dx. \tag{10.3.9}$$

By dividing eqn (10.3.9) by eqn (10.3.3)

$$\frac{P}{nk_B T} = \frac{2 \int_0^\infty x^4 \left(e^{(x^2 - \eta)} + 1\right)^{-1} dx}{3 \int_0^\infty x^2 \left(e^{(x^2 - \eta)} + 1\right)^{-1} dx} \tag{10.3.10}$$

we see that the ratio $P/nk_B T$ is a function of the variable η only.

These expressions for η, S, and P can be evaluated numerically. However, analytic expressions can be obtained for both the high- and the low-temperature limits.

10.3.1 High-temperature region

For high temperatures the chemical potential is negative and $-\eta = -\mu/k_B T$ is much larger than one. It is then a good approximation to ignore the one in the denominator of eqn (10.3.3) to get

$$n = \frac{1}{\pi^2} \left(\frac{2mk_B T}{\hbar^2}\right)^{3/2} \int_0^\infty x^2 e^{-(x^2 - \eta)} dx \tag{10.3.11}$$

which can easily be integrated to give

$$\eta = \frac{\mu}{k_B T} = \ln\left(\frac{N}{2V}\right) + \frac{3}{2} \ln\left(\frac{2\pi\hbar^2}{mk_B T}\right). \tag{10.3.12}$$

This is the equation for the chemical potential that we obtained in Chapter 9 (see eqn (9.4.9)).

The entropy per particle can be found from eqn (10.3.6) by taking the limit that $-\eta$ tends to infinity. We then get

$$\begin{aligned} \frac{S}{N} &= k_B \frac{\int_0^\infty x^2 \left(\frac{5}{3} x^2 - \eta\right) e^{-(x^2 - \eta)} dx}{\int_0^\infty x^2 e^{-(x^2 - \eta)} dx} \\ &= k_B \left\{\frac{5}{2} - \eta\right\} \\ &= k_B \left\{\ln\left(\frac{2V}{N}\right) + \frac{3}{2} \ln\left(\frac{mk_B T}{2\pi\hbar^2}\right) + \frac{5}{2}\right\} \end{aligned} \tag{10.3.13}$$

which is the *Sackur–Tetrode* expression for the entropy of a gas with spin 1/2 (compare it to eqn (6.5.5)).

The pressure of the gas in this limit can be calculated by using eqn (10.3.10) with $-\eta$ tending to infinity. We get

$$\frac{P}{nk_\mathrm{B}T} = \frac{2\int_0^\infty x^4 \, e^{-(x^2-\eta)} \, dx}{3\int_0^\infty x^2 \, e^{-(x^2-\eta)} \, dx} = 1, \tag{10.3.14}$$

as expected.

10.3.2   Properties at the absolute zero

At the absolute zero of temperature the Fermi–Dirac distribution function is zero for energies $\varepsilon(k) > \mu$ and one for $\varepsilon(k) < \mu$. The energy for which $\varepsilon(k) = \mu$ is called the Fermi energy and the corresponding wave vector is called the *Fermi wave vector*.

How do we determine the Fermi energy? Imagine a cube of volume V, say a cube of a metal such as copper, in which are placed N Fermi particles such as conduction electrons. We place the fermions in standing-wave states inside the cube so as to minimize the total energy of the system. These standing-wave states can be represented by a lattice of points in $\mathbf{k}$ space. The energy of each state is given by $\varepsilon(k) = \hbar^2 k^2 / 2m$, so the single-particle states of lowest energy have the smallest values of k^2. The first particle goes into the standing-wave state of lowest energy. When the first state is filled the second particle must go into the next unoccupied state of lowest energy. After several billion fermions have been added there is a region of $\mathbf{k}$ space with all states inside the region occupied and all the states outside, empty. For standing waves the region is an eighth of a sphere of radius k in $\mathbf{k}$ space: the states must lie in the positive octant since k_x, k_y, and k_z must be positive. The next batch of fermions must go into the lowest energy, unoccupied states which lie in a thin shell with wave vector ranging from k to $k + dk$. As more fermions are added the occupied region becomes larger and larger. When the last particle is added it goes into a state of wave vector k_F, the Fermi wave vector. The surface of this sphere (really it is an eighth of a sphere for standing waves) is called the *Fermi surface*. It is illustrated in Fig. 10.2.

How do we calculate k_F? The total number of single-particle states with wave vector less than k_F for both spin-up and spin-down is equal to the total number of particles. For a three-dimensional system the density of states is $D(k) = Vk^2/2\pi^2$ so

$$N = 2\frac{V}{2\pi^2}\int_0^{k_F} k^2 \, dk = \frac{V}{3\pi^2}k_F^3. \tag{10.3.15}$$

It follows that

$$k_F = \left(3\pi^2 n\right)^{1/3} \tag{10.3.16}$$

where n is the number density, N/V. The Fermi wave vector k_F depends on the number density of Fermi particles in the system.

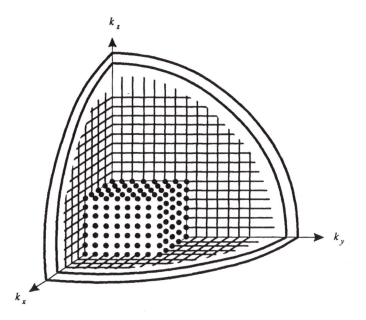

Fig. 10.2 The single-particle states in **k** space are all occupied for wave vector less that k. The single-particle states in the range k to $k+\mathrm{d}k$ are occupied by the next batch of Fermi particles. Eventually, when the last fermions enter, all states with wave vector up to k_F are occupied, all states with larger wave vector are empty.

We can define several quantities from k_F. The *Fermi momentum* is defined as $p_F = \hbar k_F$. More important is the *Fermi energy*, the energy of the last particle to be added to the system. The relativistic formula for the Fermi energy for a particle of mass m is $E_F = (\hbar^2 k_F^2 c^2 + m^2 c^4)^{1/2} - mc^2$; in the non-relativistic limit $(\hbar k_F \ll mc)$ the Fermi energy is $E_F = \hbar^2 k_F^2 / 2m$. The velocity of the particles in this limit is called the *Fermi velocity*, $v_F = \hbar k_F / m$.

The Fermi energy can be represented as a *Fermi temperature* through the equation $T_F = E_F / k_B$. This temperature defines the high- and low-temperature regions for a Fermi system. If the temperature of the system is large compared to T_F then the Fermi–Dirac distribution tends to the Maxwell–Boltzmann distribution. If the temperature is very much lower than T_F then to a first approximation we can pretend that the system is in its ground state at $T = 0$. In the ground state the single-particle states with k less than k_F are full, those with k larger than k_F are empty.

The formula for the internal energy is

$$\bar{U} = 2 \sum_{\mathbf{k}} n(\mathbf{k}) \varepsilon(\mathbf{k}).$$

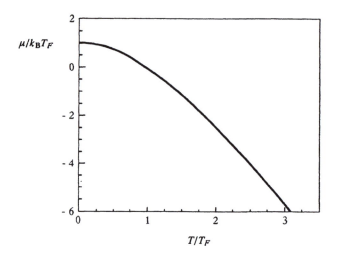

Fig. 10.3 The variation of the chemical potential (in units of $k_B T_F$) as a function of T/T_F. For temperature much less than T_F the chemical potential is very nearly the Fermi energy. For higher temperatures the chemical potential decreases and becomes negative.

Since $n(\mathbf{k})$ is one for k less than k_F and zero for k greater than k_F the internal energy for a free particle in three dimensions at the absolute zero is

$$\bar{U} = \frac{V}{\pi^2} \int_0^{k_F} k^2 \frac{\hbar^2 k^2}{2m} \, dk$$

$$= \frac{V\hbar^2 k_F^5}{10\pi^2 m} = \frac{3NE_F}{5}. \qquad (10.3.17)$$

The pressure can be found by integrating eqn (10.3.8) by parts. When the system is at the absolute zero of temperature the pressure is

$$P = \frac{1}{3\pi^2} \int_0^{k_F} \frac{\hbar^2 k^4}{m} \, dk = \frac{\hbar^2 k_F^5}{15\pi^2 m} = \frac{2nE_F}{5}. \qquad (10.3.18)$$

There is a finite pressure at the absolute zero. In contrast there is no entropy at the absolute zero of temperature.

10.3.3   Thermal properties of a Fermi gas at low temperatures

A Fermi system is said to be *degenerate* when the temperature is much less than T_F. For temperatures lower than T_F the system would like to have several particles in each low-energy state, but the exclusion principle forbids this, and forces the particles to fill states of higher energy. For high temperatures the system is non-degenerate because the average number of particles in each state is very

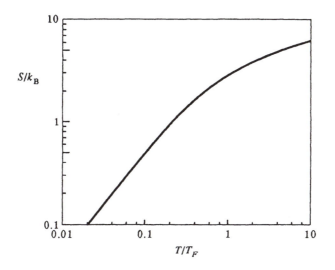

Fig. 10.4 The variation of the entropy per particle (in units of k_B) as a function of T/T_F. For low temperatures the entropy increases linearly with temperature.

much less than one. The exclusion principle has little effect in this temperature range. For temperatures of order T_F there is a smooth transition between the two limits. The mathematical description in the intermediate regime is a bit more complicated.

For the general case we can solve for η numerically. Let us write eqn (10.3.3) as

$$\left(\frac{T_F}{T}\right)^{3/2} = 3 \int_0^\infty \frac{\mathrm{d}x\, x^2}{e^{x^2 - \eta} + 1}. \tag{10.3.19}$$

To use this equation we choose a value of η and evaluate the integral on the right-hand side so as to calculate T_F/T; then we multiply by T/T_F so as to get $\mu/k_B T_F$ as a function of T/T_F. The results are shown in Fig. 10.3. Notice that the chemical potential decreases as the temperature rises so that at sufficiently high temperatures $-\mu/k_B T$ is much larger than one.

We can calculate the entropy per particle numerically using the same technique. First we divide eqn (10.3.5) by eqn (10.3.3) to get the entropy per particle as a function of η. But the quantity η is known as a function of T/T_F, so we can plot the entropy per particle as a function of T/T_F. The results are shown in Fig. 10.4. As the temperature tends to zero so the entropy varies as T, which implies that the heat capacity is proportional to the temperature. A detailed calculation gives for the heat capacity

$$C_V = \frac{N k_B \pi^2}{2} \left(\frac{k_B T}{E_F}\right) \tag{10.3.20}$$

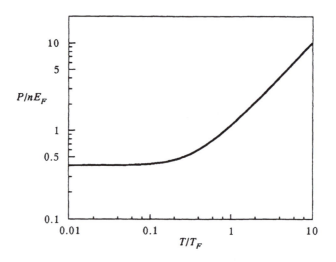

Fig. 10.5 The ratio of the pressure of a non-interacting Fermi gas to nE_F as a function of T/T_F. For low temperatures the pressure of the Fermi gas is constant, but as the temperature rises, the pressure increases. At very high temperatures the pressure increases linearly with temperature.

for $T \ll T_F$.

We can interpret this formula in the following way. When the temperature rises the internal energy of the system increases. The extra energy excites the particles into higher energy states. Particles with wave vectors just inside the Fermi surface are excited into unoccupied states with wave vectors just outside. The increase in the energy of each of these thermally excited particles is of order $k_B T$. Only those particles whose energy is less than the Fermi energy by about $k_B T$ are capable of finding an unoccupied state in a neighbouring region of k space, so that only a fraction of order $k_B T/E_F$ of the particles are thermally excited. Consequently the increase in internal energy is of order $N k_B^2 T^2/E_F$, leading to the heat capacity given by eqn (10.3.20).

The pressure of the Fermi gas can be calculated in the same way. It is shown in Fig. 10.5. The pressure changes smoothly from the low-temperature value to that of an ideal gas.

10.4 Examples of Fermi systems

10.4.1 Dilute ^{3}He solutions in superfluid ^{4}He

Dilute solutions of the isotope ^{3}He (a fermion) in superfluid ^{4}He form a weakly interacting Fermi gas. The superfluid ^{4}He acts as an inert background medium, except that it changes the bare mass of the ^{3}He atom from m_3, the ^{3}He atomic mass, to $m^* = 2.26m_3$ at 0 bar pressure. The effective mass varies with the pressure. Moreover the backflow of superfluid helium around the ^{3}He atom changes

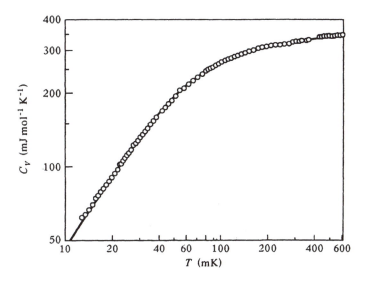

Fig. 10.6 The molar heat capacity of a 2.6% solution of ^{3}He in superfluid ^{4}He. Data from Owers-Bradley *et al.* (1988). At low temperature the heat capacity varies linearly with temperature. The heat capacity is not constant for high temperatures, as one might expect, because the dispersion relation of ^{3}He particles is not simply $\hbar^2 k^2/2m^*$.

the dispersion relation which is of the form

$$\varepsilon(k) = \varepsilon_0 + \frac{\hbar^2 k^2}{2m^*}\left(1 - \frac{k^2}{k_0^2}\right) \tag{10.4.1}$$

where k_0 is a constant.

There are two great advantages of studying dilute mixtures of ^{3}He in superfluid ^{4}He: the first is that ^{3}He particles interact very weakly with each other; the second is that the concentration can be varied from zero to about 6.5% of ^{3}He. It is an ideal system to test the theory of a non-interacting Fermi gas.

Measurements of the heat capacity of dilute ^{3}He–^{4}He mixtures have been made for different concentrations by Owers-Bradley *et al.* (1988). We show some of their results for a fractional concentration of ^{3}He 2.6% in Fig. 10.6. The solid line is the theoretical curve for a non-interacting Fermi gas with Fermi temperature of 231 mK. The experimental results are in perfect agreement with the theory.

Measurements of the osmotic pressure have been made by Landau *et al.* (1970). In Fig. 10.7 we show their results together with the best fit to the theory. The fit uses exactly the same choice of k_0 and m^* as for Fig. 10.6, but we have also included a small constant interaction between the particles. If such a term were not included the fit would not be quite so good.

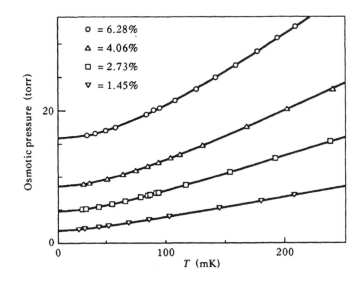

Fig. 10.7 The osmotic pressure of ^{3}He in ^{3}He–^{4}He mixtures for various concentrations as a function of temperature. Notice the deviation from the pressure of an ideal classical gas for low temperatures. Experimental data taken from Landau *et al.* (1970).

A good review of the properties of ^{3}He–^{4}He is by Ebner and Edwards (1970); a more recent one is by Edwards and Pettersen (1992).

10.4.2   Electrons in metals

The alkali metals such as lithium, sodium, and potassium have one valence electron which becomes detached from the atom and forms a band of conduction electrons. The concentration of electrons is equal to the concentration of atoms in the solid which can be determined from a knowledge of the mass density of the metal and its atomic weight.

Let us calculate the Fermi energies for electrons in a typical metal such as lithium. The number density of conduction electrons in lithium is $n = 4.6 \times 10^{28}$ m^{-3} so the Fermi wave vector is

$$k_F = \left(3\pi^2 \times 4.6 \times 10^{28}\right)^{1/3} = 1.11 \times 10^{10} \, \text{m}^{-1}.$$

The Fermi energy is $E_F = 7.5 \times 10^{-18}$ J $\equiv 4.7$ eV provided we ignore corrections to the electron's mass from band structure effects in the metal. Typical Fermi energies of electrons in a metal are of the order of a few electronvolts. We give some typical Fermi velocities, energies, and temperatures for several metals in Table 10.1. The parameter n_0 tells how many electrons are released from each atom: one for monovalent elements, two for divalent, and so on. The *Fermi temperature*, E_F/k_B, is much higher than the typical melting temperature of metals.

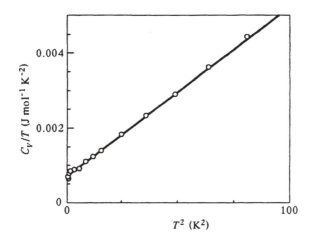

Fig. 10.8 The ratio of the molar heat capacity of copper to the temperature plotted as a function of the temperature squared. The intercept on the vertical axis gives the contribution to the heat capacity from the electrons in the copper, a contribution which is proportional to the temperature; the slope of the graph gives the contribution to the heat capacity from the phonons of the solid, a contribution which varies as T^3.

For example, the melting temperature is 55 000 K for lithium, much less than the Fermi temperature of 500 000 K.

The Fermi velocity tells us the speed of these particles. If they are travelling at a speed close to the speed of light we must use relativistic expressions for energy and momentum. Let us assume the electrons in lithium are not relativistic, so that their speed is $v_F = \hbar k_F / m_e = 1.3 \times 10^6 \, \mathrm{m\,s^{-1}}$. This is about a kilometre a second, less than the speed of light by a factor of 200. The electrons in lithium are not going at relativistic speeds.

The *heat capacity of metals* has a contribution from the conduction electrons as we saw in Fig. 8.6. We have plotted the same data in Fig. 10.8 but this time we have drawn C/T as a function of T^2. According to Debye's theory the

Table 10.1 The Fermi wave vector, velocity, and energy for electrons in different metals

Element	n_0	$n \, (\mathrm{m^{-3}})$	$k_F \, (\mathrm{m^{-1}})$	$v_F \, (\mathrm{m\,s^{-1}})$	$E_F \, (\mathrm{eV})$
Li	1	4.6×10^{28}	1.1×10^{10}	1.3×10^6	4.7
Na	1	2.5×10^{28}	0.9×10^{10}	1.1×10^6	3.1
K	1	1.3×10^{28}	0.7×10^{10}	0.9×10^6	2.0
Rb	1	1.1×10^{28}	0.7×10^{10}	0.8×10^6	1.8
Cs	1	0.9×10^{28}	0.6×10^{10}	0.7×10^6	1.5
Be	2	24.6×10^{28}	1.9×10^{10}	2.2×10^6	14.0
Mg	2	8.6×10^{28}	1.4×10^{10}	1.6×10^6	7.1

contribution from lattice vibrations should vary as T^3; the contribution from conduction electrons should vary as T. If we plot the data as C/T against T^2 then the data should fall on a straight line. Figure 10.8 shows that the data for copper do fall on a straight line. The intercept on the C/T axis gives an estimate of the contribution of conduction electrons to the heat capacity; the slope gives an estimate of the contribution of phonons to the heat capacity.

10.4.3 Electrons in stars

Consider a reasonably sized star with a mass of 3×10^{30} kg at a temperature of 10^7 K. The star is composed mainly of dissociated hydrogen atoms, that is protons and electrons. To calculate the Fermi energy of the electrons we need to know their number density. Suppose nearly all the hydrogen has dissociated. The total number of electrons would then be

$$N = \frac{3 \times 10^{30}}{1.67 \times 10^{-27}} \approx 1.8 \times 10^{57}.$$

Suppose the radius of the star is $R = 3 \times 10^7$ m, a typical radius for a star. The volume of the star is $4\pi R^3/3$, the average number density of electrons in the star is

$$\frac{N}{V} = \frac{3 \times 1.8 \times 10^{57}}{4\pi(3 \times 10^7)^3} \approx 1.6 \times 10^{34}\,\mathrm{m}^{-3}.$$

If we assume the number density is uniform over the star, the Fermi wave vector of the electrons in the star is

$$k_F = \left(3\pi^2 n\right)^{1/3} = \left(3\pi^2 \times 1.6 \times 10^{34}\right)^{1/3} = 8 \times 10^{11}\,\mathrm{m}^{-1},$$

and the Fermi momentum is $\hbar k_F = 8 \times 10^{-23}\,\mathrm{kg\,m\,s}^{-1}$.

The condition that the particle behaves non-relativistically is that the speed is less than the speed of light, c. Now the speed $\hbar k_F/m_e$ is about $0.9 \times 10^8\,\mathrm{m\,s}^{-1}$, smaller than c by a factor of about three, so the electron behaves more or less non-relativistically. The Fermi energy is roughly $\hbar^2 k_F^2/2m_e = 3.8 \times 10^{-15}$ J, and the Fermi temperature is $T_F = 2.8 \times 10^8$ K. Notice that T_F is much larger than 10^7 K. Although you might think that a temperature of 10^7 K is very hot, it is cold compared to the Fermi temperature of the star. Such a star at a temperature of 10^7 K is a degenerate Fermi system.

10.4.4 Electrons in white dwarf stars

In a normal star hydrogen nuclei fuse together to form helium, the reaction giving out huge amounts of nuclear energy. It is the release of all this energy that keeps the star hot and inflated. When the star has used up all its hydrogen, it loses this source of energy, cools down, and collapses. When this happens the atoms of helium are strongly compressed together under gravitational forces, and become very compressed. An electron orbiting a helium atom would become more tightly localized, which leads, via the uncertainty principle, to an increase in its kinetic energy.

If the density is increased sufficiently, the kinetic energy of the electrons overcomes the potential energy which binds them to the atomic nuclei and the electrons become free particles. The helium atom turns into an alpha particle plus two free electrons. There is a huge density of electrons which are relatively cold since the hydrogen fuel has been used. The large star has shrunk to what is called a *white dwarf*.

The kinetic energy per electron, when they are free particles, is of order $\hbar^2 (3\pi^2 N/V)^{2/3}/2m_e$. The electron density N/V is equal to Z/d^3 where d is the spacing of the nuclei and Z is the nuclear charge. Thus a typical kinetic energy of a particle is of order $\hbar^2 Z^{2/3}\pi^2/m_e d^2$. The potential energy per particle is of order $-Ze^2/4\pi\epsilon_0 d$. The magnitude of the potential energy is smaller than the kinetic energy if

$$\frac{\hbar^2 (Z)^{2/3} \pi^2}{m_e d^2} > \frac{Ze^2}{4\pi\epsilon_0 d}. \tag{10.4.2}$$

When this inequality is satisfied (within a numerical factor) the electrons overcome their attraction to their parent nucleus and become free.

We can estimate from this inequality the density above which the electrons become delocalized. Let m' be the mass of the star per electron. Since each atom when completely ionized yields Z electrons, the quantity Zm' is the mass per nuclei. The mass density obeys the inequality

$$\rho = \frac{Zm'}{d^3} > Z^2 m' \left(\frac{m_e e^2}{4\pi\epsilon_0 \hbar^2}\right)^3 = \frac{Z^2 m'}{a_0^3} \tag{10.4.3}$$

where a_0 is the Bohr radius, $4\pi\epsilon_0 \hbar^2/m_e e^2$.

For white dwarf stars which are made up mainly of helium, $Z = 2$. It follows that $m' = 2m_H$ where m_H is the mass of a hydrogen atom. Hence ρ must be larger than $10^5 \, \text{kg m}^{-3}$.

This treatment assumes that the electrons are non-relativistic, and for this to be a good approximation the Fermi momentum $p_F = \hbar k_F$ must be less than $m_e c$, a restriction which places an upper bound on the mass density of about $2 \times 10^9 \, \text{kg m}^{-3}$. For stars of densities between these limits the electrons behave as a non-relativistic degenerate Fermi gas in which the Coulomb attraction to the protons can be ignored.

In a typical white dwarf the density is $10^9 \, \text{kg m}^{-3}$, the mass is $10^{30} \, \text{kg}$, and the temperature is $10^7 \, \text{K}$. For these values the Fermi temperature is $2 \times 10^9 \, \text{K}$. A typical white dwarf is a star which has cooled sufficiently so that the electrons form a degenerate Fermi gas. The best known white dwarf star is in Sirius B. Its mass, roughly equal to that of the Sun, is $2 \times 10^{30} \, \text{kg}$; its radius of $5600 \, \text{km}$ is smaller than that of the Earth.

What determines the radius of the white dwarf star? First let us consider the case where the electrons are non-relativistic. The pressure in the star is of order $-\alpha G M^2/R^4$, where G is the gravitational constant $= 6.67 \times 10^{-11} \, \text{N m}^2 \, \text{kg}^{-1}$, α is a numerical factor of order unity, and R is the star's radius. For mechanical stability this pressure must balance the pressure of the Fermi gas, $2nE_F/5$. Now the mass of the star, M, is Nm' where N is the number of electrons so that $N = M/m'$. The volume of the star is $4\pi R^3/3$. Thus the kinetic energy is of order $\hbar^2 M^{5/3}/m_e(m')^{5/3}R^2$ and the pressure is of order

$$P_F \sim \frac{\hbar^2 M^{5/3}}{m_e(m')^{5/3}R^5} \sim \frac{GM^2}{R^4} \sim P_{\text{gravity}}. \tag{10.4.4}$$

From the condition that these two pressures are equal we find

$$M^{1/3}R = \frac{\gamma\hbar^2}{m_e(m')^{5/3}G} \tag{10.4.5}$$

where γ is a numerical factor which a more fastidious calculation (Weinberg 1972) gives as 4.5.

As the mass of the white dwarf star increases, its radius shrinks; it is much smaller than the radius of normal stars of the same mass, hence the name 'dwarf'. The shrinking continues until the density of the white dwarf star has reached the point where the electron density is so high that the electrons become relativistic, with kinetic energy

$$\varepsilon(k) = ((\hbar ck)^2 + m_e^2 c^4)^{1/2} - m_e c^2$$

which is approximately $\hbar ck$. The total kinetic energy is then

$$\frac{3}{4}E_F N = \frac{3}{4}\hbar c \left(3\pi^2 N/V\right)^{1/3} N = \frac{\gamma'\hbar c}{R}\left(\frac{M}{m'}\right)^{4/3} \tag{10.4.6}$$

where γ' is a numerical factor of order one. The crucial point is that the energy now varies as R^{-1}, and as a result the pressure varies as R^{-4}. In equilibrium this pressure must equal the gravitational pressure, which also varies as R^{-4}. From the condition that the pressures are equal we get

$$M^{2/3} = \frac{\delta\hbar c}{(m')^{4/3}G} \tag{10.4.7}$$

with a numerical factor, δ, which turns out to be 2.13. The mass then saturates to a critical value which is independent of radius, as shown in Fig. 10.9. The critical mass is called the *Chandrasekhar mass*. The value of the critical mass is 1.45 times the mass of the Sun, that is 2.9×10^{30} kg. Provided the star has a mass less than the Chandrasekhar mass it will form a stable white dwarf when it has burnt all its hydrogen and formed helium. For example, the Sun, when it has burnt all its energy, will become a white dwarf of radius 7×10^6 m.

If the mass of the star is greater than the critical mass, the star collapses when all of the hydrogen has burnt to form helium. As it collapses the electron density

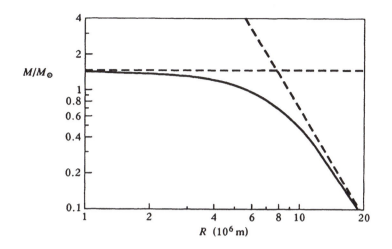

Fig. 10.9 The mass of a white dwarf star (in units of the Sun's mass) as a function of the radius of the star. The two asymptotic limits of extreme relativistic and of classical behaviour are shown as dashed lines. The solid line interpolates between these limits.

increases and the electron's chemical potential rises. Eventually the chemical potential is sufficiently large that the electron can react with a proton to form a neutron plus a neutrino in the reverse of the usual beta decay process. The star, as is collapses, loses electrons and protons and becomes enriched with neutrons—it evolves into a *neutron star*.

The analysis of the neutron star proceeds in the same way. For low densities the mass varies as R^{-3}, just as for the mass of a white dwarf. For larger densities the neutrons become relativistic and there is an upper critical mass. The theory is further complicated because we need to use a relativistic theory of gravity for such high masses. However, such an upper critical mass is found in a proper theory; for neutron star masses larger than the critical mass the neutron star itself collapses and may form a black hole.

10.5 A non-interacting Bose gas

Consider a non-interacting, degenerate Bose system with the particles of zero spin having an energy $\varepsilon(k) = \hbar^2 k^2/2m$. At high temperatures the chemical potential of the particles is approximately

$$\mu = k_{\mathrm{B}} T \ln \left(n \lambda_D^3 \right)$$

where λ_D is the thermal de Broglie wavelength. This result is valid only when $V \gg N\lambda_D^3$, that is for a low-density gas at high temperatures. The distribution of particles is described by the Maxwell–Boltzmann distribution

$$f(k) = \mathrm{e}^{-(\varepsilon(k) - \mu)/k_{\mathrm{B}} T}. \tag{10.5.1}$$

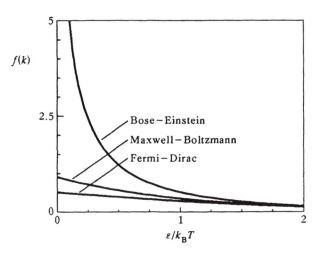

Fig. 10.10 A comparison of the Bose–Einstein, Maxwell–Boltzmann, and Fermi–Dirac distribution functions when μ/k_BT is -0.1. Notice that there are many more Bose particles in states of low energy than there are Fermi particles.

The distribution becomes narrower than the Maxwell–Boltzmann distribution as the temperature decreases, as illustrated in Fig. 10.10. To keep the total number of particles constant we must increase μ. Quantum corrections appear when $-\mu$ is of order k_BT. At a critical temperature, T_c, the chemical potential approaches zero, which it cannot pass. If it became positive the states of low energy would have $f(k) < 0$, which makes no sense. The critical temperature is given by the condition

$$N = \sum_k \frac{1}{e^{\varepsilon(k)/k_BT_c} - 1} = \frac{V}{2\pi^2} \int_0^\infty \frac{k^2}{e^{\varepsilon(k)/k_BT_c} - 1} dk. \tag{10.5.2}$$

The integral over k can be put in dimensionless form as

$$\frac{2\pi^2 N}{V} = \left(\frac{2mk_BT_c}{\hbar^2}\right)^{3/2} \int_0^\infty \frac{x^2}{e^{x^2} - 1} dx \tag{10.5.3}$$

which can be evaluated to give the equation

$$k_BT_c = 3.31 \frac{\hbar^2}{m} \left(\frac{N}{V}\right)^{2/3}. \tag{10.5.4}$$

The quantity k_BT_c sets a scale of energy.

Let us give an order of magnitude estimate of T_c for liquid ^{4}He at zero pressure. The mass density of liquid helium is $145 \, \text{kg m}^{-3}$ and so the number density is $2.2 \times 10^{28} \, \text{m}^{-3}$. Hence T_c is about $3.1 \, \text{K}$ if we consider liquid helium to be a non-interacting Bose gas. The superfluid transition appears at 2.18 K, which is the right order of magnitude.

We are considering a system with a fixed volume held next to a heat bath with a constant temperature ($= T_c$) into which we place N Bose particles. There is no physical reason why we cannot keep adding particles to the system. But according to eqn (10.5.2) the total number of particles cannot change. If we add particles where do they go?

The answer is that they condense into the single-particle state of lowest energy, $\varepsilon(k = 0)$, which we take to have an energy of zero. This is called *Bose condensation*. The number of particles in this state is

$$f(k = 0) = \frac{1}{e^{(\varepsilon(k=0)-\mu)/k_BT} - 1} \approx -\frac{k_BT}{\mu}. \qquad (10.5.5)$$

The number of particles in this single-particle state becomes macroscopically large as μ tends to zero from negative values. The particles which we add condense into the ground state just as particles in an ordinary gas can condense out into drops of liquid. The difference is that droplets of liquid fall as rain, whereas the particles which condense are spread out uniformly throughout the volume. The appearance of such condensed particles indicates the appearance of a new phase of matter.

The thermodynamics of the partly condensed system is easy to calculate once we know where the extra particles have gone. The integral in eqn (10.5.2) counts the number of particles which are not in the ground state. Let us call this number $(N - N_0)$, where N_0 is the number of particles in the ground state. The population of the excited states is just

$$(N - N_0) = \frac{V}{2\pi^2} \int_0^\infty \frac{k^2}{e^{\varepsilon(k)/k_BT} - 1} dk = N \left(\frac{T}{T_c} \right)^{3/2}. \qquad (10.5.6)$$

The first few states, for which we should replace the integral by a sum, have a negligible weight in the thermodynamic limit (that is as N tends to infinity for constant N/V). From this equation we can deduce the number of Bose condensed particles

$$N_0 = N \left\{ 1 - \left(\frac{T}{T_c} \right)^{3/2} \right\}. \qquad (10.5.7)$$

For temperatures below T_c the chemical potential is effectively zero. The internal energy of the gas only depends on the particles in states with finite k. The internal energy is

$$
\begin{aligned}
\bar{U} &= \sum_k n(\mathbf{k})\varepsilon(\mathbf{k}) = \frac{V}{2\pi^2} \int_0^\infty \frac{k^2\varepsilon(k)}{e^{\varepsilon(k)/k_BT} - 1} dk \\
&= \frac{Vk_BT}{2\pi^2} \left(\frac{2mk_BT}{\hbar^2} \right)^{3/2} \int_0^\infty \frac{x^4}{e^{x^2} - 1} dx. \qquad (10.5.8)
\end{aligned}
$$

The internal energy is proportional to $T^{5/2}$, so that the heat capacity varies as $T^{3/2}$.

For temperatures above T_c the chemical potential is negative and this leads to a change in the heat capacity. For very high temperatures the heat capacity at constant volume is $1.5Nk_B$. There is a peak in the heat capacity at T_c.

We can also calculate the pressure of an ideal Bose gas from the grand potential:

$$P = -\frac{\Phi_G}{V} = -\frac{k_B T}{2\pi^2} \int_0^\infty k^2 \ln\left(1 - e^{-(\varepsilon(k)-\mu)/k_B T}\right) dk. \tag{10.5.9}$$

This gives the pressure of an ideal gas for very high temperatures. For temperatures below T_c the chemical potential is zero. The pressure is then independent of the volume of the system and so the system has an infinite compressibility.

A system which is *stable* must be at the maximum value of the entropy; an equivalent statement is that, to be stable, the system must be at the minimum of the Helmholtz free energy. For the free energy to be a minimum the quantity

$$\left(\frac{\partial^2 F}{\partial V^2}\right)_T = -\left(\frac{\partial P}{\partial V}\right)_T$$

must be finite and positive—otherwise it is not a minimum. For a non-interacting Bose gas $-(\partial P/\partial V)_T$ is zero and so the system is marginally unstable. There must be interactions between the particles which are repulsive for short distances in order to ensure a finite value of $-(\partial P/\partial V)_T$ and the stability of the system.

Another indication that the theory is imperfect comes from the observation that $\partial\mu/\partial N$ is zero in the Bose condensed phase below T_c. In problem 9.10 you are asked to prove that the standard deviation in N obeys the equation

$$\Delta N^2 = k_B T \left(\frac{\partial \bar{N}}{\partial \mu}\right). \tag{10.5.10}$$

For temperatures below T_c the quantity $\partial\mu/\partial N$ is zero and so the standard deviation ΔN is infinite. Particles can enter and leave the system in enormous numbers, all of the particles going into the lowest energy single-particle state, without anything stopping them. The density of particles in the system can fluctuate to an infinite extent, thanks to Bose condensation in the non-interacting gas. However, in a gas of interacting Bose particles the fluctuations are of finite extent because the compressibility of such a gas is finite.

For a long time there was no direct evidence for the existence of Bose condensation, even though it is believed to be present in liquid helium. The situation changed with the experiment of Anderson *et al.* (1995). They cooled a gas of rubidium-87 atoms that was confined by magnetic fields. They saw direct evidence for Bose condensation below $0.170\,\mu\text{K}$ for a number density of gas atoms of $2.5 \times 10^{18}\,\text{m}^3$.

To detect the presence of Bose–Einstein condensation they suddenly switched off the magnetic field which was used to confine the particles. At a slightly later time they photographed the spatial density of particles. Those with high speeds move rapidly away from the site of the confining potential; those in the zero-momentum state remain stationary. If there are a large number of particles in this state they will show up as a peak in the spatial density in the region of the confining potential. That is what is observed, but only if the temperature is lower than a critical temperature, T_c. As the temperature is lowered below T_c, so this peak in the number density in the region of the confining potential becomes larger, just as expected.

10.6 Problems

1. Calculate the Fermi wave vector, k_F, for both one- and two-dimensional systems.

2. Show that the average energy per particle in a non-relativistic Fermi gas at the absolute zero of temperature in three dimensions is

$$\bar{U} = \frac{3E_F}{5}$$

where E_F is the Fermi energy. Hence deduce the Fermi pressure

$$P_F = \frac{2E_F}{5}\frac{N}{V}.$$

Estimate the Fermi pressure for a neutron star in which the neutrons pack to a mass density of 10^{18} kg m^{-3}.

3. What is the Fermi temperature of the protons in the star described in section 10.4.3? The mass of the star is 3×10^{30} kg and its radius is 3×10^7 m. Are the protons degenerate at a temperature of 10^7 K?

4. Calculate the average energy per particle in a non-relativistic Fermi gas at the absolute zero of temperature in two dimensions. What is the equivalent of the Fermi pressure in this case?

5. Obtain an expression for μ for a Fermi gas at high temperatures when $n(k)$ is given by

$$n(k) = e^{-(\epsilon(k)-\mu)/k_B T} - e^{-2(\epsilon(k)-\mu)/k_B T} + \cdots .$$

6. Calculate the average energy per particle for a Fermi gas in three dimensions at $T = 0$ in terms of the Fermi energy in the extreme relativistic case where $\hbar k_F \gg mc$.

7. In sodium there are about 2.6×10^{28} conduction electrons per cubic metre which behave as a free electron gas. From these facts estimate the Fermi energy of the gas and an approximate value of the molar electronic heat capacity at 300 K.

8. An ideal gas of N spin $1/2$ fermions is contained in a volume V. A small, constant magnetic induction field $B = \mu_0 H$ is applied along the z direction so that the energies are

$$\varepsilon = \frac{\hbar^2 k^2}{2m} \pm \mu_B B$$

with the minus sign if the spins are parallel to the field. Show that the Pauli susceptibility is

$$\chi = M/H = \mu_0 \mu_B^2 D(E_F)/V$$

where $D(E_F)$ is the density of states in energy at the Fermi level. Use the fact that $D(E) \sim V E^{1/2}$ to estimate χ for copper, given that $E_F = 7.0$ eV and the density of free electrons is 8.5×10^{28} m^{-3}.

9. Consider an ideal Fermi gas whose atoms have a mass of 5×10^{-27} kg, nuclear spin $1/2$, and nuclear magnetic moment $\mu_N = 10^{-26}$ J T^{-1} so that the spin energy levels are $\pm \mu_N B$. At $T = 0$, what is the largest density for which the gas can be completely polarized by a magnetic induction field of 10 T?

10. Consider a non-interacting Bose gas of N atoms of mass m and zero spin which are restricted to motion on a flat surface of area A. The system is two-dimensional.

    (a) Write an equation which fixes the chemical potential, μ, of the atoms as a function of N, A, and the temperature T when the system is 'normal'.
    (b) Show that the integral which arises in the expression for μ diverges as $\mu \to 0^-$. From this observation deduce that there always exists a negative value of μ which satisfies your equation.
    (c) Does Bose condensation exist for a non-interacting Bose gas in two dimensions?

11. A white dwarf star is made up of degenerate electrons at a temperature much lower than the Fermi temperature. Calculate the electron density for which the Fermi momentum is $0.1 m_e c$, where m_e is the electronic mass and c is the velocity of light. Determine the pressure of the degenerate electron gas under these circumstances.

12. Give numerical estimates for the Fermi energy for:

    (a) electrons in a typical metal with $n = 2 \times 10^{28}$ m^{-3},
    (b) protons in a heavy nucleus with $n = 0.55 \times 10^{44}$ m^{-3},
    (c) ^{3}He atoms in liquid ^{3}He (atomic volume $= 46 \times 10^{-30}$ m^3).

    In each case treat the Fermi particles as if they were non-interacting.

13. Determine an approximate expression for the critical mass of a neutron star, corresponding to the Chandrasekhar mass of a white dwarf.

14. The adiabatic bulk modulus of a system is defined as

$$B = -V(\partial P/\partial V)_s.$$

Show that the adiabatic bulk modulus of a free electron gas at $T = 0$ is given by

$$B = 2nE_F/3.$$

Calculate the bulk modulus of the materials listed in the table and discuss discrepancies between theoretical and experimental values.

Metal	Number density (m^{-3})	E_F (eV)	Bulk modulus $(N\,m^{-2})$
K	1.3×10^{28}	0.9	2.81×10^9
Cs	0.9×10^{28}	0.7	1.43×10^9
Cu	8.5×10^{28}	7.0	134.3×10^9

15. In the very early stages of the Universe, it is usually a good approximation to neglect particle masses and chemical potentials compared to $k_B T$. Write down formulae for the number densities and energy densities of non-interacting fermions under these conditions.

16. In a white dwarf star of mass 2×10^{30} kg all the atoms of hydrogen are ionized. The star has a radius of 2×10^7m. Find the Fermi energy of the electrons in eV. If the star is at a temperature of 10^7 K are the electrons degenerate? If the radius of the star were reduced to 10^4m would the electrons be relativistic? What would their Fermi energy be?

17. Calculate the Fermi wave vector for an electron density of $4.2 \times 10^{27}\,m^{-3}$. Compute the Fermi energy in eV. Given that the electrons are replaced by neutrons, compute the magnitude of the Fermi energy.

18. Estimate the temperature at which the lattice and electronic contributions to the molar heat capacity are equal for potassium. Consult Tables 8.1 and 10.1 for values of θ_D and E_F.

19. Given that the mass of the Sun is 2×10^{30} kg, estimate the number of electrons in the Sun given that it is mainly composed of atomic hydrogen. In a white dwarf star of one solar mass the atoms are all ionized and contained in a sphere of radius 2×10^7 m. What is the Fermi energy of the electrons in eV? Are they relativistic? If the temperature of the white dwarf star is 10^7 K are the electrons and nucleons in the star degenerate?

11
Phase transitions

First we guess it. Then we compute the consequences of the guess to see what would be implied if the law we guess is right. Then we compare the result of the computation to nature, with experiment or experience, compare it directly with observation, to see if it works. If it disagrees with experiment it is wrong. In that simple statement is the key to science. It does not make any difference how beautiful your guess is. It does not make any difference how smart you are, who made the guess, or what your name is—if it disagrees with experiment it is wrong. That is all there is to it.   *Richard Feynman*

11.1 Phases

When a system is cooled, new states of matter can appear. For example, a gas when cooled enough forms a liquid; a liquid when cooled can form a solid, or it can become a superfluid, a fluid which flows without resistance. Sometimes the states which appear are quite unexpected: the electrons in some ceramics were found to flow without resistance at relatively high temperatures. We describe the states of matter—gas, liquid, solid, and superfluid—as different phases of the system; the transitions between them are called *phase transitions*.

What is meant by the term 'phase'? The answer given by Gibbs is a *state of matter that is 'uniform throughout, not only in chemical composition but also in physical state.'* Uniformity is crucial: the physical properties, such as density or conductivity, must be the same throughout the system, as must be the chemical composition. Consider, as an example, a mixture of solid A in solid B which forms a single solid phase. The notion of uniformity means that any representative sample of the solid of macroscopic size appears to be the same as any other representative sample. The size of the sample must be macroscopic to minimize the effect of fluctuations: if there are N_A of particle A in the sample the size of the fluctuation is proportional to $N_A^{1/2}$. If the sample consists of about 20 atoms, the size of the fluctuation is about 5 atoms; by chance one sample might have 15 atoms, another such sample might have only 5. In contrast when there are 10^{20} particles, the fluctuation is of order 10^{10} which is negligible compared to 10^{20}. The notion of uniformity only makes sense for a sample which is sufficiently large that fluctuations can be ignored. If the fluctuations become infinite, of course, the notion of uniformity no longer makes sense, and we cannot identify a phase.

Usually we can tell that there is a phase transition because there is an abrupt change in one or more properties of the system. Solids are rigid materials whereas liquids can flow. The electrical resistivity of a material goes from zero to a finite

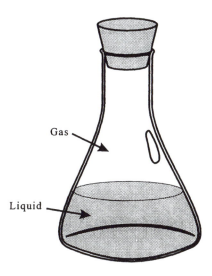

Fig. 11.1 Liquid and gas coexisting in a container with constant volume.

value in a superconducting to normal phase transition. In a ferromagnetic phase transition the magnetic properties of a system change abruptly from those of a paramagnet to those of a ferromagnet. An abrupt change in properties is one sign of a phase transition.

Another sign of a phase transition is the appearance of two phases coexisting side by side. For example, water, in a sealed container, can exist in a liquid phase in equilibrium with its vapour phase as shown in Fig. 11.1. There is a thin phase boundary (meniscus) separating the two phases. The water forms a uniform phase below the meniscus, the gas forms a uniform phase above. If the temperature is raised sufficiently the water boils and the container is filled with the vapour only. When the vapour phase is cooled sufficiently, liquid forms and separates out at the bottom of the vessel. We say that there is *phase separation*. Phase separation is another sign of a phase transition.

In this chapter we shall explore the nature of phase transitions, using as our guide the simple idea that the most stable phase of matter is the one with the lowest *thermodynamic potential*, Φ. Whichever phase has the lowest thermodynamic potential is stable; phases which have a higher thermodynamic potential are unstable and decay into the stable phase.

11.2 Thermodynamic potential

What is meant by the expression 'the thermodynamic potential'? To answer this question let us consider a collection of N_A identical particles (system A) placed in contact with a heat bath with temperature T_R and at constant pressure P_R. In Chapter 5 we showed that the equilibrium state of such a system can be found by minimizing the appropriate thermodynamic potential. For a system with a fixed number of particles and constant pressure the maximum in the entropy of the joint system occurs where

$$G(P_R, T_R, N_A; U_A, V_A) = U_A + P_R V_A - T_R S_A \qquad (11.2.1)$$

is a minimum. The condition for this to be a minimum is that the partial derivatives with respect to U_A and V_A are zero. Differentiating with respect to U_A gives

$$\left(\frac{\partial G(P_R, T_R, N_A; U_A, V_A)}{\partial U_A} \right)_{V_A} = 1 - \frac{T_R}{T_A}$$

which is zero when $T_A = T_R$ so that the system is in thermal equilibrium. Differentiating with respect to V_A gives

$$\left(\frac{\partial G(P_R, T_R, N_A; U_A, V_A)}{\partial V_A} \right)_{N_A} = P_R - \frac{T_R}{T_A} P_A.$$

When the system is in thermal equilibrium we get $P_A = P_R$, the condition for mechanical equilibrium.

The thermodynamic potential is defined as the function which is minimized subject to all the constraints that are imposed on the system. Consequently, the function $G(P_R, T_R, N_A; U_A, V_A)$ is the thermodynamic potential for the system. $G(P_R, T_R, N_A; U_A, V_A)$ varies as a function of V_A and U_A; however, we will assume that the system is always in thermal equilibrium so that the variation with U_A can be ignored. When the system is in equilibrium (so that $P_A = P_R$, $T_A = T_R$) the thermodynamic potential is the Gibbs free energy. We can think of the thermodynamic potential as a generalized Gibbs free energy which we can use when the system is displaced from equilibrium.

The same analysis can be done for other systems with different constraints. The thermodynamic potential is then a function of those parameters which are allowed to vary. In Chapter 9 we showed that a system of constant volume placed in contact with a reservoir which allows exchange of energy and particles is in equilibrium if the generalized grand potential, Φ_G, is a minimum. The generalized grand potential, $\Phi_G(V, T, \mu; N)$, is a function of N, the number of particles in system A; the reservoir fixes the temperature and the chemical potential. The number of particles varies because particles can be exchanged with the reservoir. The system evolves freely by changing the number of particles until the generalized grand potential is a minimum.

Notice that we have dropped the suffices R and A for convenience; they can readily be restored since we know the constraints imposed on the system. For example, if we write $\Phi_G(V, T, \mu; N)$ we know that we are working in the grand canonical ensemble with fixed volume, that the reservoir fixes T and μ, and that the number of particles in system A can vary. Thus $\Phi_G(V, T, \mu; N)$ is equivalent to $\Phi_G(V_A, T_R, \mu_R; N_A)$. If we want to we can drop more symbols by writing the generalized grand potential as $\Phi_G(N_A)$ and assume the reader knows which constraints are implied when using the grand potential.

If the system is magnetic and held in a constant external magnetic field, the thermodynamic potential is $F_H(H, T, N; M)$, the generalized magnetic free energy. The magnetization of the system is not a conserved quantity: M can change, so the system can minimize the thermodynamic potential as a function of magnetization. The minimum value of the thermodynamic potential is the magnetic free energy $F_H(H, T, N)$.

In all of the cases outlined above there is a quantity which can be varied freely to find the minimum in the thermodynamic potential. For some systems there is a conservation law which complicates matters. For example, the thermodynamic potential for a system at constant temperature in a constant volume which cannot exchange particles with a reservoir is the generalized Helmholtz free energy, $F(V, T; N)$. The total number of particles is conserved: particles cannot be created or destroyed in the system. The system cannot evolve freely by changing the number of particles so as to reach the minimum in the generalized Helmholtz free energy. If we were to plot F as a function of N for fixed T and V there could be one or more minima present. The system may not sit in the lowest minimum; such a system could lower the thermodynamic potential if it could alter the number of particles, but it cannot do this because of the conservation law. Instead the system separates into two phases in such a way that the total number of particles is conserved. We defer discussion of phase separation until section 11.6.

11.3   Approximation

In order to simplify the notation we describe all thermodynamic potentials as a simple mathematical function $\Phi(\phi)$ where ϕ is the quantity which can vary. Thus $\Phi(\phi)$ could be the generalized Gibbs free energy as a function of volume, so that $\phi = V$; or $\Phi(\phi)$ could represent the magnetic free energy, F_H, with ϕ the magnetization. By so doing we are making a simplification. If ϕ represents the magnetization, the simplification is that the thermodynamic potential only depends on the average value of the magnetization and is independent of fluctuations in magnetization. Clearly we are making an approximation: the true thermodynamic potential can be evaluated by summing over all accessible states of the system; by making this approximation we are restricting the sum over states to those for which ϕ has a given value.

Having made this approximation (or 'guess' as Feynman's quotation implies), what consequences follow? If the thermodynamic potential $\Phi(\phi)$ has a single minimum (see Fig. 11.2), the system reaches equilibrium by varying ϕ until the system has reached the minimum value $\Phi(\phi)$. For example, if the magnetization of the system is displaced from the equilibrium value, the system will respond by relaxing back to equilibrium exponentially with a time constant T_1.

The thermodynamic potential is analogous to the potential energy, $V(x)$, of a particle in a one-dimensional well. The particle lowers its energy by sitting at the bottom of the potential well; the thermodynamic system lowers its 'free energy' by sitting at the bottom of the thermodynamic potential.

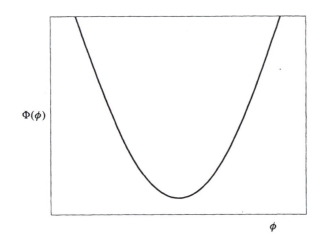

Fig. 11.2 The variation of $\Phi(\phi)$ with respect to ϕ showing a single minimum.

If $\Phi(\phi)$ has two minima the more stable state is the one with the lower energy. The other minimum is unstable and will eventually decay into the lower minimum. We say that the system is *metastable*. One minimum could correspond to the liquid phase of matter, the other minimum to the gas phase. Suppose the two minima evolve as the temperature (or pressure) increases so that one minimum is lower over one range of temperatures, the other is stable over another range. When the system evolves from one stable minimum to the other the phase changes, say from liquid to gas. It follows that if we accept the proposition that the thermodynamic potential can be written as a simple function $\Phi(\phi)$ we have created a mathematical formalism which is capable of describing phase transitions.

The underlying principle is this: the system, when in equilibrium, exists in those microstates which lie close to the lowest possible value of the thermodynamic potential. Loosely speaking, the system minimizes the thermodynamic potential.

Phase transitions can be of two types: in a *first-order phase transition* there is a discontinuity in one or more of the first derivatives of the thermodynamic potential; in a continuous transition there is no such discontinuity.

Suppose the thermodynamic potential $\Phi(x_1, x_2, \ldots)$ depends on a set of constrained parameters x_i. If any of the first derivatives $\partial\Phi/\partial x_i$ are discontinuous at the transition then it is a first-order phase transition. For example, suppose the thermodynamic potential is the Gibbs free energy G with constrained parameters $x_1 = T$ and $x_2 = P$. By slightly changing the constraints we generate a small change in G as

$$dG = -S\,dT + V\,dP \qquad (2.6.11)$$

so that

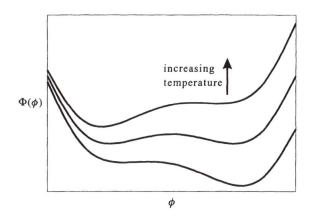

Fig. 11.3 The variation of $\Phi(\phi)$ showing two minima with a maximum in between. As the temperature varies the minima change position relative to each other. Where the two minima are equal, both are equally stable and a phase transition occurs.

$$S = -\left(\frac{\partial G}{\partial T}\right)_P; \qquad V = \left(\frac{\partial G}{\partial P}\right)_T.$$

If the first derivative of G with respect to temperature is discontinuous there is an abrupt change in the entropy as the phase changes; the change in the entropy is associated with a latent heat. If the first derivative with respect to pressure is discontinuous there is a change in molar volume. The liquid–solid transition is an example of a first-order transition, as usually there is a latent heat and a change in the molar volume as the system goes from solid to liquid.

If all the first derivatives of the thermodynamic potential are continuous at the transition we call it a *continuous transition* (formerly it was called a second-order phase transition). The entropy is continuous at a continuous transition. We defer discussion of continuous transitions until Chapter 12.

11.4   First-order phase transition

Suppose we alter the temperature and find that one minimum is lower over one range of temperatures, the other is lower over the remainder, as is illustrated in Fig. 11.3. The system evolves as the temperature changes from well 2 being stable to well 1 being stable. The evolution of the minima shown in Fig. 11.3 describes a *first-order phase transition*. We can think of well 1 describing one state of matter, well 2 describing another. For example, well 1 could correspond to the low density of the gas phase, well 2 to the high density of the liquid phase. If the system changes with temperature so that it evolves from well 2 being stable to well 1 being stable we can expect an abrupt change in the properties of the system. For example, the liquid could convert to a gas with a change in its density, viscosity, thermal conductivity, refractive index, and so on.

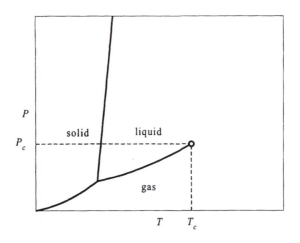

Fig. 11.4 The phase diagram for a system with just liquid, gas, and solid phases. The lines separate regions where different phases are stable. When a system crosses a line the phase changes and there is usually a latent heat.

In general, there is a latent heat in such a transition: the two minima Φ_1 and Φ_2 vary with temperature at different rates so that the entropy of 'well 1'

$$S_1 = -\left(\frac{\partial \Phi_1}{\partial T}\right) \qquad (11.4.1)$$

and of 'well 2'

$$S_2 = -\left(\frac{\partial \Phi_2}{\partial T}\right) \qquad (11.4.2)$$

are different. When the system changes from one phase to another the entropy changes which gives rise to a latent heat

$$L_{12} = T_p\,(S_1 - S_2) \qquad (11.4.3)$$

where T_p is the temperature of the phase transition.

Well-defined boundaries exist between the regions where the different phases are stable. For the liquid–gas transition we can draw a line on the pressure–temperature diagram in Fig. 11.4 such that the liquid phase is the more stable on one side of the line and the gas phase is the more stable on the other. They are called *phase boundary lines*. For pressure–temperature coordinates which lie on the phase boundary line, both phases are equally stable (the minima are equal). The two phases can coexist in a sealed container with a meniscus showing the division between the liquid and gas phases. Similarly there are phase boundary lines between the liquid and solid phases, and between the solid and gas phases. On these lines the two phases are equally stable, that is the minima in $\Phi(\phi)$ are equal. The phase boundary lines can intersect at a *triple point* where the three minima are all equally deep. The three phases can coexist at the triple point.

When the system crosses these lines there is a first-order phase transition with a discontinuity in one of the first derivatives of the thermodynamic potential, leading, for example, to a latent heat, one of the characteristic signs of a first-order phase transition. The discontinuity could be in another derivative of the thermodynamic potential. For example, in a ferromagnetic system there is a discontinuity in $(\partial F_H / \partial H)_T$ for temperatures below the critical temperature where $H = 0$.

Another characteristic property of a first-order phase transition is that the system displays *metastability*. Suppose well 2 is stable at a certain temperature and the system evolves as shown in Fig. 11.3. The temperature changes so that beyond a certain point well 2 becomes metastable: the minimum is higher than that of well 1. In order for the system to find its way to well 1 it must cross a free-energy barrier by thermal activation, a process which can take a long time if the barrier is high compared to $k_B T$. If the system is cooled still further the barrier shrinks to zero and the system finds its way rapidly to well 1.

Sometimes the barrier is sufficiently high that the metastable state persists indefinitely; it then seems that the metastable state is a thermodynamically stable state. A liquid can be cooled below the solidification temperature and kept there for a very long time before a solid is nucleated somewhere and spreads throughout the liquid. We say the liquid is supercooled. The metastable, supercooled liquid will eventually form a solid if it is cooled further or if nucleating sites are artificially introduced. Similarly water vapour in the air can be supercooled: the system would lower its free energy if it could form water droplets (rain) but there is a free-energy barrier to be overcome before the rain can form. Seeding the cloud with nucleating sites will sometimes help the rain to form.

The lowest temperature that a supercooled liquid can reach before the solid phase is formed is lower than the temperature at which the solid melts. If we were to take the system around a cycle in which the solid first is heated and then cooled we would obtain two different estimates of the transition temperature. In fact, for the same temperature we could either have the solid or the supercooled liquid, depending on the point in the cycle the system has reached. This behaviour, which is known as *hysteresis*, is a characteristic of a first-order phase transition.

11.5 Clapeyron equation

We can get additional information from the phase boundary lines on the phase diagram. Consider the phase diagram of a one-component system such as water. When the system is on the phase boundary line, the two phases can coexist in equilibrium. If the two phases 1 and 2 are in equilibrium it implies that their chemical potentials, μ_1 and μ_2, are equal, or that their Gibbs free energies, $G = N\mu$, are the same. Suppose that one point on the phase boundary line occurs at temperature T and pressure P; another at temperature $T + dT$ and pressure $P + dP$. We must have

$$G_1(T, P) = G_2(T, P) \tag{11.5.1}$$

and

$$G_1(T + dT, P + dP) = G_2(T + dT, P + dP). \tag{11.5.2}$$

By making a Taylor expansion of both sides of eqn (11.5.2) and then using eqn (11.5.1) we get

$$\left(\frac{\partial G_1}{\partial T}\right)_P dT + \left(\frac{\partial G_1}{\partial P}\right)_T dP = \left(\frac{\partial G_2}{\partial T}\right)_P dT + \left(\frac{\partial G_2}{\partial P}\right)_T dP$$

which can be rearranged to give

$$\left[\left(\frac{\partial G_1}{\partial P}\right)_T - \left(\frac{\partial G_2}{\partial P}\right)_T\right] dP = -\left[\left(\frac{\partial G_1}{\partial T}\right)_P - \left(\frac{\partial G_2}{\partial T}\right)_P\right] dT.$$

From the relation

$$dG = -S\, dT + V\, dP \tag{2.6.11}$$

we get $(\partial G/\partial P)_T = V$ and $(\partial G/\partial T)_V = -S$. It follows that along the coexistence line

$$\begin{aligned}
\frac{dP}{dT} &= -\frac{[(\partial G_1/\partial T)_P - (\partial G_2/\partial T)_P]}{[(\partial G_1/\partial P)_T - (\partial G_2/\partial P)_T]} \\
&= \frac{S_1 - S_2}{V_1 - V_2} = \frac{L_{12}}{T_p \Delta V}.
\end{aligned} \tag{11.5.3}$$

Here $L_{12} = T_p(S_1 - S_2)$ is the molar latent heat absorbed at the transition temperature, T_p, and $\Delta V = V_1 - V_2$ is the difference in the molar volumes of the two phases. Equation (11.5.3) is known as the *Clapeyron equation*. All the terms on both sides of this equation can be determined experimentally and satisfy the equation to good precision, a triumph for thermodynamics. This is a result that would have satisfied Mach.

The Clapeyron equation indicates that when the latent heat is zero the slope dP/dT is zero. In general, there is a latent heat at a first-order transition, but there are exceptions: the P–T coexistence curve between liquid and solid ³He shows a minimum at a temperature of 0.31 K. At the minimum in the melting curve the latent heat is zero, but it still is a first-order transition. The essential ingredient of a first-order transition is that one of the first derivatives of the thermodynamic potential is discontinuous at the transition. At the minimum in the melting curve between liquid and solid ³He the derivative of the Gibbs potential with respect to pressure is discontinuous.

11.6 · Phase separation

We must indeed all hang together, or, most assuredly, we shall all hang separately.
  Benjamin Franklin. Said on signing the Declaration of Independence.

A gas kept at constant pressure will condense into a liquid as the temperature is lowered. To keep the pressure constant we could confine the gas in a cylinder with a movable piston; as the liquid forms, the volume of the container shrinks and the piston moves down. The thermodynamic potential for a system held at constant pressure is the Gibbs free energy. As the temperature is lowered the number density of molecules, $n = N/V$, changes so that the system minimizes the Gibbs free energy as a function of n.

The situation is more complicated if we hold the volume of the system constant. As the gas is cooled the pressure decreases placing a strain on the containing vessel. The average density of particles is constant if the vessel maintains a constant volume without allowing any atoms to leak into it. For a system held at constant volume the thermodynamic potential is the Helmholtz free energy. The average density cannot vary to find the minimum in the Helmholtz free energy for it is fixed. How does the system minimize its Helmholtz free energy?

A similar question can be posed for solids made up of two different atoms, A and B. Let us call X the fractional concentration of A:

$$X = \frac{N_A}{N_A + N_B}.$$

At high temperatures the solid has atoms A and B uniformly distributed throughout the solid so that the fractional concentration is the same everywhere. The particles 'hang together'. At sufficiently low temperature the system can become unstable if it is energetically favourable for A atoms to sit next to other A atoms rather than B atoms. The particles want to 'hang separately'. If the fractional concentration is fixed how can the system respond to minimize its free energy?

In both cases outlined above there is a conserved quantity which we denote as ϕ: the average density of the gas is fixed for a system held at constant volume; the average fractional concentration of particle A in the mixture is fixed. How does a system minimize the thermodynamic potential by varying a parameter whose average value is fixed?

The answer is phase separation: the system breaks up into two phases, each phase having a different value of the parameter ϕ. It can happen that the phases become completely separated, with one phase uniform throughout one region of the space, the other throughout the rest. For example, in a liquid–gas system, the gas rises to the top of the container, the liquid falls to the bottom. Similarly, a crystal can form a phase of mainly A atoms in one part of the container and mainly B atoms in the remainder. The system will only break up into two phases if by so doing the thermodynamic potential is reduced.

Let us assume that the thermodynamic potential can be written as a simple function of the parameter ϕ which may represent the average density, n, or the

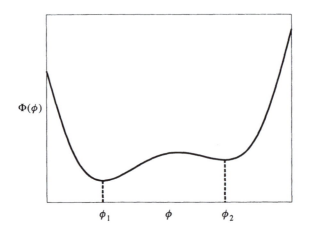

Fig. 11.5 $\Phi(\phi)$ showing two minima.

average fractional concentration, X. The average value of ϕ is fixed; we ignore the presence of local fluctuations in ϕ.

If $\Phi(\phi)$ has a single minimum the system cannot find a way of lowering the thermodynamic potential; but if $\Phi(\phi)$ has two minima the system sometimes can lower the thermodynamic potential by breaking into two phases with values ϕ_1 and ϕ_2 in such way that the average value of ϕ is constant.

Suppose ϕ is a value which lies close to the maximum in $\Phi(\phi)$ as shown in Fig. 11.5. The maximum sits between two minima which we call well 1 and well 2. Clearly the thermodynamic potential would be lower if the system were near the bottom of well 1 or near the bottom well 2. Because the parameter ϕ is conserved the system cannot evolve to the bottom of either well; but the system can evolve so that it experiences both wells: part of the system ends up sitting in well 1, the remainder in well 2.

Let the fraction of the system which sits in well 1 be x, the fraction in well 2 be $(1-x)$. To ensure that the value of ϕ integrated over the volume V is constant we must have an average parameter

$$\frac{1}{V}\int d^3\mathbf{r}\,\phi(\mathbf{r}) = \overline{\phi} = x\phi_1 + (1-x)\phi_2. \tag{11.6.1}$$

It is clear that the thermodynamic potential is lowered this way; what is not obvious is the optimum choice of ϕ_1 and ϕ_2.

We can rewrite eqn (11.6.1) as

$$\phi_2 = \frac{\overline{\phi} - x\phi_1}{1-x}.$$

The thermodynamic potential can be written as

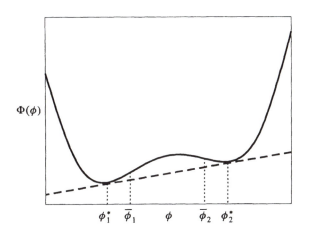

Fig. 11.6 The double tangent drawn to $\Phi(\phi)$. The values of ϕ where the double tangent touches the curve are denoted ϕ_1^* and ϕ_2^*. The spinodal limits, $\overline{\phi}_1$ and $\overline{\phi}_2$, are also shown.

$$\Phi(\phi_1, x) = x\Phi(\phi_1) + (1-x)\Phi(\phi_2)$$
$$= x\Phi(\phi_1) + (1-x)\Phi\left(\frac{\overline{\phi} - x\phi_1}{1-x}\right).$$

This expression for $\Phi(\phi_1, x)$ is a minimum with respect to both x and ϕ_1. If we minimize by varying ϕ_1 we get the relation

$$x\Phi'(\phi_1) - x\Phi'(\phi_2) = 0. \qquad (11.6.2)$$

This relation tells us that the slope of $\Phi(\phi)$ at ϕ_1 is the same as the slope at $\Phi(\phi_2)$. If we minimize by varying x we get the relation

$$\Phi(\phi_1) - \Phi(\phi_2) - (1-x)\Phi'(\phi_2)\left(\frac{\phi_1}{1-x} - \frac{\overline{\phi} - x\phi_1}{(1-x)^2}\right) = 0.$$

When we rearrange this we get

$$\Phi'(\phi_2) = \frac{\Phi(\phi_1) - \Phi(\phi_2)}{(\phi_1 - \phi_2)}. \qquad (11.6.3)$$

These two conditions can be satisfied by drawing a *double tangent* to the curve $\Phi(\phi)$, touching it at $\phi = \phi_1^*$ and $\phi = \phi_2^*$ (see Fig. 11.6). The slope of the double tangent equals the slope at both ϕ_1^* and ϕ_2^*; moreover, the slope of the double tangent equals the ratio of $(\Phi(\phi_1^*) - \Phi(\phi_2^*))$ to $(\phi_1^* - \phi_2^*)$.

If the value of ϕ lies in the range ϕ_1^* to ϕ_2^* the system is unstable and can lower its thermodynamic potential by phase separation. The change in thermodynamic potential is given by the difference between the curve $\Phi(\phi)$ and the value given by the double tangent.

The system is completely unstable if $\Phi''(\phi) > 0$ for then the thermodynamic potential is not a minimum. The values of ϕ where $\Phi''(\phi) = 0$, the points of inflection, are denoted as $\overline{\phi}_1$ and $\overline{\phi}_2$ and are called the *spinodal limits*. If ϕ lies in the range $\overline{\phi}_1$ to $\overline{\phi}_2$ the system must decay. These values of ϕ are not the same as those given by the double tangent construction, ϕ_1^* and ϕ_2^*. It follows that the system can have values of ϕ which appear at first sight to be stable ($\Phi''(\phi) < 0$) but which lower their energy by phase separation. These are metastable states of the system. Metastable states can persist for a long time without decaying, so that they appear to be true equilibrium states. The decay of metastable states is examined in more detail in Chapter 13.

11.7 Phase separation in mixtures

Consider a mixture of two particles, with N_A of the first and N_B of the second. Let us define the fraction of the particles in the mixture which are particle 1 as X:

$$X = \frac{N_A}{N}$$

where $N = N_A + N_B$ is the total number of particles. For constant volume the thermodynamic potential, $\Phi(\phi)$, is the Helmholtz free energy which can be written as $F(X)$. The quantity X is fixed for the numbers N_A and N_B are conserved. If the curve of $F(X)$ as a function of X allows us to make the double tangent construction the system is unstable against the formation of two phases.

Phase separation can occur in a mixtures of liquids or in a mixture of solids. If you mix three parts olive oil and one part vinegar together and shake the liquid violently for a while the mixture becomes uniform in appearance; you have made an emulsion, commonly known as French dressing, to put on your salad. Such an emulsion remains stable for only a short while unless you have got the ingredients in the optimum proportion and you have shaken the mixture violently for a long time. Even then it will eventually separate back into two phases, the oil-rich liquid sitting on the denser vinegar. The reason for the phase separation is that the water is attracted to other water molecules more strongly than it is to an oil molecule. Other liquids, such as water and alcohol, will mix perfectly. The reason is the very strong attraction between water and alcohol molecules.

To illustrate phase separation in mixtures let us consider a solid composed of atoms A and B with the atoms arranged randomly on the sites of the crystal. There are $N_A = NX$ of atom A and $N_B = N(1-X)$ of atom B. Let the bond energy between two neighbouring A atoms be ε_{AA}, the bond energy between neighbouring A and B atoms be ε_{AB}, and the bond energy between two neighbouring B atoms be ε_{BB}. If the number of AA bonds is N_{AA}, the number of BB bonds is N_{BB}, and the number of AB bonds is N_{AB}, then the internal energy is

$$U = N_{AA}\varepsilon_{AA} + N_{BB}\varepsilon_{BB} + N_{AB}\varepsilon_{AB}.$$

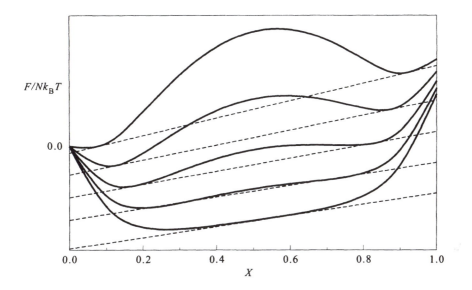

Fig. 11.7 The free energy for binary mixtures in the regular solution model. We have plotted the curve $F/Nk_BT = X\ln(X) + (1-X)\ln(1-X) + 0.1X + 2tX(1-X)$ where $t = T/T_c$. The curves shown from bottom to top are for $t \doteq 1.0$, 0.9, 0.8, 0.7, and 0.6. For $t < 1$ we can draw a double tangent; for $t \geq 1$ we cannot.

There are z nearest neighbours for each atom and a total of $Nz/2$ bonds in the crystal. Let us assume that the particles are randomly placed: the probability that an A atom is on a particular site in the crystal is independent of the arrangement of neighbour atoms. The probability that a site has an A atom is $p_A = X$, the probability that a site has a B atom is $p_B = (1-X)$. The chance of a neighbouring sites having two A atoms is $p_{AA} = X^2$; the probability of neighbouring sites having two B atoms is $p_{BB} = (1-X)^2$; the chance of a neighbouring pair of sites having an A and a B atom is $p_{AB} = 2X(1-X)$. Thus $N_{AA} = NzX^2/2$, $N_{BB} = Nz(1-X)^2/2$, and $N_{AB} = NzX(1-X)$. The internal energy is

$$U = \frac{Nz}{2}\left(X\varepsilon_{AA} + (1-X)\varepsilon_{BB}\right) - NzvX(1-X) \qquad (11.7.1)$$

where we have introduced the parameter $v = \frac{1}{2}\left(\varepsilon_{AA} + \varepsilon_{BB}\right) - \varepsilon_{AB}$ with v negative: the attractive interaction between A and B atoms is weaker than that between A atoms or that between B atoms. As a result we expect like atoms to cluster together.

The entropy per particle of the mixture in this model is (using eqn 5.3.4)

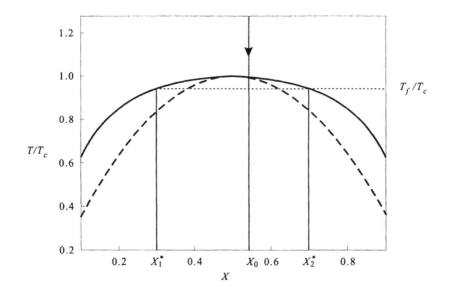

Fig. 11.8 The phase diagram for binary mixtures. Below the solid curve the system can separate into the two concentations X_1^* and X_2^* as shown. The dashed curve is the spinodal limit where $d^2F/dX^2 = 0$. If the system lies between the dashed and solid curves, it is in a metastable state.

$$
\begin{aligned}
S/N &= -k_{\mathrm{B}}p_A \ln(p_A) - k_{\mathrm{B}}p_B \ln(p_B) \\
&= -k_{\mathrm{B}}[X \ln(X) + (1-X)\ln(1-X)]. \quad\quad (11.7.2)
\end{aligned}
$$

The Helmholtz free energy, $U - TS$, in this, the *regular solution model*, is

$$
\begin{aligned}
F(X) &= \frac{Nz}{2}(X\varepsilon_{AA} + (1-X)\varepsilon_{BB}) - NzvX(1-X) \\
&\quad + Nk_{\mathrm{B}}T[X \ln(X) + (1-X)\ln(1-X)]. \quad\quad (11.7.3)
\end{aligned}
$$

The overall concentration, X, is fixed. However, the phases can separate if the shape of $F(X)$ is such that we can make the double tangent construction, as shown in Fig. 11.7, and if X lies in the range X_1^* to X_2^*.

For high temperatures there is no instability. As the temperature is lowered, the instability first appears when $X = 1/2$ giving a critical temperature $T_c = z|v|/2k_{\mathrm{B}}$. When $X = 1/2$ and $t = T/T_c > 1$ there is no phase separation. As the temperature is lowered so that $t < 1$ the mixture separates into two phases with concentrations X_1^* and X_2^* given by the double tangent construction as shown in Fig. 11.7. In this way we generate the phase diagram shown in Fig. 11.8.

Suppose we start with an initial concentration X_0. We now lower the temperature so that t is less than one. If X_0 is lies in the range X_1^* to X_2^* the system can separate into two phases: one of fractional concentration X_1^*, the other of fractional concentration X_2^*. If X_0 lies outside the range X_1^* to X_2^* the system remains as a single phase.

The dashed curve on Fig. 11.8 shows the spinodal limit. If the concentration lies in the region between the solid and dashed curves the system can stay in a metastable state. In this region the system does not have to separate into two phases, but may do so due to fluctuations. If the temperature is lowered sufficiently so that the system meets the spinodal limit, the system must phase separate for it is then completely unstable: the curve has a single minimum and a point of inflection at the spinodal limit. The spinodal limit is never reached in practice: the system finds a way of decaying into the most stable state before the spinodal limit is reached, either by thermal fluctuations or by nucleation of the stable phase on impurities or defects.

The regular solution model for the Gibbs free energy has been used to analyse isotopic phase separation in solid ^{3}He–^{4}He mixtures by Edwards and Balibar (1989) with considerable success. By using the regular solution model, they are able to explain the observed phase diagram over a wide range of concentrations, pressures, and temperatures. Edwards and Balibar go further and describe the two liquid phases: the ^{3}He-rich phase and the ^{4}He-rich phase. They are able to account accurately for the phase diagram of liquid and solid ^{3}He–^{4}He mixtures over a range of pressures with only a few parameters.

11.8   Liquid–gas system

Phase separation occurs for a system when held at constant volume. When a pure system has evolved to produce the two phases lying side by side, the two phases are in equilibrium, an expression which implies thermal, mechanical, and chemical equilibrium has been reached. To be in chemical equilibrium the chemical potential of the particle in one phase must be the same as in the other. For the liquid–gas system we must have

$$\mu_L = \mu_G.$$

The chemical potential of the particle is

$$\mu = \left(\frac{\partial F(n)}{\partial N} \right)_{V,T} = \frac{1}{V} \frac{\mathrm{d}F(n)}{\mathrm{d}n} = \frac{\mathrm{d}f(n)}{\mathrm{d}n}$$

where $f(n)$ is the free energy density. Consequently the slope of the free energy density as a function of the number density must be the same for both liquid and gas.

One way of finding the densities of liquid and gas is by using the *double tangent construction*. We draw a straight line which just touches the curve $f(n)$ at two points so that it is a double tangent. The densities at the two points

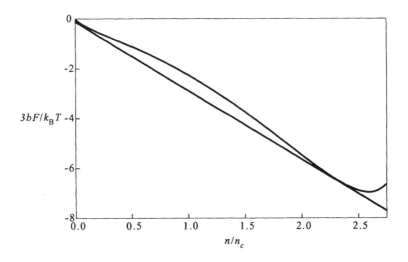

Fig. 11.9 Values of $3bF/Vk_BT$ for a van der Waals gas as a function of n/n_c. The solid line is a double tangent. We have used values for ^{4}He given in Appendix F to generate this curve. The value used for T/T_c is 0.6.

of contact define the densities n_1^* and n_2^*; the slopes of $f(n)$ at these densities are equal which means that the two chemical potentials, μ_1 and μ_2, are the same at the value μ^*. The double tangent ensures that the system is in chemical equilibrium.

One model of the liquid–gas phase transition is due to van der Waals. In Appendix F we show that the Helmholtz free energy in the van der Waals model can be written as

$$\frac{3bF}{Vk_BT} = -\frac{n}{n_c}\left(\ln\left[\left(\frac{2\pi mk_BT_cb^{2/3}}{h^2}\right)^{3/2}(T/T_c)^{3/2}\left(\frac{3n_c}{n}-1\right)\right]+1+\frac{9n/n_c}{8T/T_c}\right).$$

where $n_c = 1/3b$ and T_c are the critical density and temperature. It is a straightforward matter to plot this function as a function of n/n_c for different $t = T/T_c$, and to draw the double tangent. This is illustrated in Fig. 11.9 for $T/T_c = 0.6$ for ^{4}He using the tabulated values of critical temperature and b given in Appendix F. Similar curves can be drawn for different temperatures and for other gases.

Maxwell's construction is a graphical construction which is equivalent to the double tangent construction. If $f(n)$ has two minima and one maximum, the curve of $\mu(n) = f'(n)$ has one maximum and one minimum. We can intersect the curve $\mu(n)$ by a horizontal line in such way that the two shaded areas between the curve and the horizontal line are equal and opposite, as shown in Fig. 11.10 for the van der Waals gas. This is known as Maxwell's construction. The horizontal line specifies the equilibrium chemical potential, μ^*. To prove this assertion we calculate the sum of the two areas:

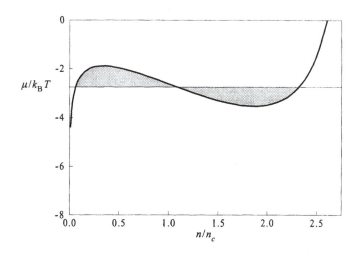

Fig. 11.10 The variation of $\mu(n)/k_{\mathrm{B}}T$ with n/n_c for $T/T_c = 0.6$ for ^{4}He in the van der Waals model (see Appendix F). The horizontal line is drawn so that the shaded areas are equal.

$$\int_{n_1^*}^{n_2^*} (\mu(n) - \mu^*)\, dn = \int_{n_1^*}^{n_2^*} \left(\left(\frac{\partial f(n)}{\partial n} \right) - \mu^* \right) dn = [f(n) - \mu^* n]_{n_1^*}^{n_2^*} = 0.$$

In other words the areas are equal as long as $f - \mu^* n$ is equal at both ends, as must be the case when μ^* is the slope given by the common tangent construction. Maxwell's construction is therefore equivalent to the double tangent construction.

11.9 Problems

1. For a liquid (L) in equilibrium with its vapour (G) the relation

$$\frac{\mathrm{d}P}{\mathrm{d}T} = -\frac{L_{GL}}{T_p \left(V_G - V_L \right)}$$

is valid. By assuming that the gas obeys an ideal gas law, obtain an expression for the vapour pressure as a function of temperature if the latent heat is independent of temperature.

2. A phase transition occurs at a temperature T_c for a system which is held in a constant magnetic induction, B. What is the appropriate thermodynamic potential? Derive a relation between $\mathrm{d}B/\mathrm{d}T_c$ and the latent heat of the transition.

3. Show that Maxwell's construction for a system whose thermodynamic potential is the generalized grand potential can be written as

$$\int_{V_1^*}^{V_2^*} P\,dV = 0.$$

4. Assume that the variation of the pressure of air with height, z, obeys the simple expression

$$P(z) = P(0)\exp(-mgz/k_BT)$$

where m is the molecular weight of air. The boiling point of water varies with height above sea-level due to the variation of the pressure of the air. Show that the rate of change of the boiling point, T_b, with height, z, is given by

$$\frac{dT_b}{dz} = -\frac{P(z)mg}{k_BT}\frac{dT_b}{dP} = -\frac{P(z)mg}{k_BT}\frac{T_b(\rho_L - \rho_G)}{\rho_L\rho_G L_{LG}}$$

where ρ_L is the mass density of water (10^3 kg m^{-3}), ρ_G is the vapour density (0.6 kg m^{-3} at 100°C) and L_{LG} is the latent heat per unit mass. If the latent heat of vaporization of water per unit mass is 2.4×10^6 J kg^{-1} find the rate of change of the boiling point of water with altitude in degrees kelvin per kilometre. The density of air at sea level is 1.29 kg m^{-3}.

5. The concentrations where $d^2 F/dX^2 = 0$ define the spinodal limit. Show that the spinodal limit for the regular solution model is

$$X(1 - X) = \frac{k_BT}{2z\,|v|}.$$

What is the maximum temperature which satisfies this relation?

6. Suppose that the thermodynamic potential for a system with a conserved quantity ϕ is of the form

$$\Phi(\phi) = \Phi(0) + c(\phi^2 - \phi_a^2)^2.$$

What is the slope of the double tangent to the curve? Obtain ϕ_1^* and ϕ_2^*, the values of ϕ where the double tangent touches the curve. What are the spinodal limits, $\bar{\phi}_1$ and $\bar{\phi}_2$, for this thermodynamic potential? Show that the spinodal limits lie between ϕ_1^* and ϕ_2^*. For what value of ϕ is the system in a metastable state?

12
Continuous phase transitions

But elements of symmetry are either present or absent; no intermediate case is possible.

Lev Landau

12.1 Introduction

A continuous phase transition occurs when the minimum in the thermodynamic potential evolves smoothly into two (or more) equal minima. An example is the liquid–gas phase transition held at constant pressure such that the system is on the liquid–gas coexistence line. The thermodynamic potential is the generalized Gibbs free energy, G. Because the system is on the coexistence line the two wells are equally deep. Suppose the system starts at the triple point and changes gradually towards the critical point; as it changes the depth of the two wells decreases and the two minima move closer together. At the critical point the two minima merge into a single minimum and the liquid–gas coexistence line terminates. We denote the end-point of the coexistence line as having temperature T_c, volume V_c, and pressure P_c.

For temperatures above the critical temperature there is no evidence of two phases; the system has a uniform density and is therefore a single phase which we will call the fluid phase: it makes no sense to describe it as a liquid or a gas. As the system approaches the critical point, either from above or from below, one finds bubbles of gas and drops of liquid intermingled on all length-scales from the macroscopic to the atomic. These bubbles represent fluctuations of the system away from the average density. In the simple picture we are using the system is described by the average density; fluctuations are ignored.

Another example of a continuous transition is provided by a uniaxial ferromagnet in which the spins point either in the $+z$ or $-z$ directions, leading to a magnetic moment for each spin of $+\mu$ or $-\mu$. In the absence of an external magnetic field the generalized free energy, F_H, is a symmetric function of the magnetization, $F_H(M) = F_H(-M)$. For temperatures above the transition temperature, T_c, there is a single minimum in $F_H(M)$ at the value $M = 0$; for temperatures below the transition temperature two minima appear in $F_H(M)$ with magnetizations of $+M_0$ or $-M_0$.

Suppose the system starts at low temperatures with a positive magnetization. As the temperature rises the depth of the two wells decreases and the minima move closer together. At the critical point the two minima merge and the magnetization is zero.

For temperatures above the critical temperature the magnetization is zero. As the system cools, and approaches the critical point, one finds domains of positive magnetization and negative magnetization intermingled on all length-scales. In the simple picture we are using the system is described by the average magnetization; all fluctuations are ignored.

12.2 Ising model

A magnetic material such as iron has a magnetization at low temperatures even when there is no applied magnetic field. The magnetization arises from magnetic dipoles in the iron which are partially aligned leading to a net magnetic moment. The alignment of the dipoles corresponds to partial ordering of the system. When the temperature is raised sufficiently the dipoles become progressively more randomly oriented and the magnetization decreases. Above a transition temperature, known as the Curie temperature, the magnetization disappears.

The simplest model which exhibits a magnetic phase transition is the *Ising model* (Ising 1925). The main assumption of the Ising model is that the magnetic dipoles are assumed to point in the $+z$ and $-z$ directions only, a gross simplification. The dipoles arise from unpaired electrons of spin $1/2$ localized on lattice sites, j; the dipoles have magnitude $g\mu_B s_{zj}/\hbar$ where μ_B is the Bohr magneton, g for electrons is 2 to a good approximation, and s_{zj} is the z component of spin for the electron on site j. Rather than use s_z we multiply by $g = 2$ and use $\sigma_j = 2s_{zj}/\hbar$ with σ_j having eigenvalues of $+1$ if the spin is up and -1 if the spin is down.

The energy of the system contains two terms: the particles interact with their nearest neighbours through an *exchange interaction* of the form $-J\sigma_j\sigma_{j+1}$; each spin interacts with the magnetic induction field, $B = \mu_0 H$, through a term $-\mu_B\mu_0 H\sigma_j$. The total Hamiltonian is

$$\hat{H} = -\mu_B\mu_0 H \sum_j \sigma_j - \frac{J}{2} \sum_{(i,j)_{nn}} \sigma_j\sigma_i. \tag{12.2.1}$$

The sum is over $\sigma_j\sigma_i$ is for nearest neighbour pairs only; the factor of $1/2$ is to avoid counting the same pair twice. The constant J was introduced by Ising on empirical grounds and justified by Heisenberg using quantum mechanics.

In this model neighbouring spins have an interaction (the exchange interaction) whose sign depends on whether the spins are parallel $\sigma_j = \sigma_i$ or antiparallel $\sigma_j = -\sigma_i$. If $J > 0$ the interaction is positive if the spins are antiparallel and negative if they are parallel. For $J > 0$ the lowest energy state of the system (the ground state) occurs when all the spins are parallel so that the system is ferromagnetic. If $J < 0$ the lowest energy state is antiferromagnetic with neighbouring spins pointing in opposite directions.

12.2.1 Mean field theory

Our objective is to minimize the free energy by varying a quantity such as the average magnetization. How do we do this? In *mean field theory* the local magnetization is treated as a constant; all fluctuation effects are neglected. One way to achieve this is to write the average value of the spin as $\bar{\sigma}$ and put

$$\sigma_j \sigma_i = (\sigma_j - \bar{\sigma} + \bar{\sigma})(\sigma_i - \bar{\sigma} + \bar{\sigma}) \simeq \bar{\sigma}^2 + \bar{\sigma}(\sigma_i - \bar{\sigma}) + \bar{\sigma}(\sigma_j - \bar{\sigma}) \qquad (12.2.2)$$

thereby ignoring the term $(\sigma_j - \bar{\sigma})(\sigma_i - \bar{\sigma})$ which leads to correlation between neighbouring spins. In mean field theory we ignore correlations. The mean field Hamiltonian is

$$
\begin{aligned}
\hat{H}_{\mathrm{mf}} &= -\mu_B \mu_0 H \sum_j \sigma_j - \frac{J}{2} \sum_{(i,j)_{nn}} \left(\bar{\sigma}^2 + \bar{\sigma}(\sigma_i - \bar{\sigma}) + \bar{\sigma}(\sigma_j - \bar{\sigma}) \right) \\
&= -(\mu_B \mu_0 H + zJ\bar{\sigma}) \sum_j \sigma_j + \tfrac{1}{2} zJ\bar{\sigma}^2 \qquad (12.2.3)
\end{aligned}
$$

where z is the number of nearest neighbours for site j. Equation (12.2.3) is the Hamiltonian of a system of *non-interacting* spins in an effective (or mean) field

$$B_{\mathrm{eff}} = \mu_0 H_{\mathrm{eff}} = \mu_0 H + \frac{zJ\bar{\sigma}}{\mu_B}.$$

To complete the theory we need to calculate the average spin, or equivalently the average magnetization per unit volume since

$$M = \mu_B n \bar{\sigma}.$$

How do we get $\bar{\sigma}$?

Method 1

One way of proceeding is through the use of the microcanonical ensemble, as described in Chapter 4. Non-interacting spins in an effective field B_{eff} give rise to a magnetization (see eqn 4.4.6)

$$M = \mu_B n \bar{\sigma} = \mu_B n \tanh \left(\frac{\mu B_{\mathrm{eff}}}{k_B T} \right). \qquad (12.2.4)$$

It follows that the mean field is

$$B_{\mathrm{eff}} = \mu_0 H + \frac{zJM}{\mu_B^2 \mu_0 n}. \qquad (12.2.5)$$

The effective induction field B_{eff} is the sum of the applied field and a field due to the exchange interaction with the neighbouring spins. Equations (12.2.4) and (12.2.5) can be combined to form a single, non-linear equation for the magnetization.

Method 2

Another way of proceeding is through the use of the canonical ensemble described in Chapter 5. Each spin has an energy $\pm\mu_B B_{\text{eff}}$ in the mean field; the partition function for N two-level systems is just

$$Z = e^{-NzJ\bar{\sigma}^2/2k_BT} \left(e^{\mu_B B_{\text{eff}}/k_BT} + e^{-\mu_B B_{\text{eff}}/k_BT}\right)^N$$

so the free energy is

$$F_H = \tfrac{1}{2}NzJ\bar{\sigma}^2 - Nk_BT\ln\left(e^{\mu_B B_{\text{eff}}/k_BT} + e^{-\mu_B B_{\text{eff}}/k_BT}\right). \tag{12.2.6}$$

This is the thermodynamic potential of the system. By minimizing this expression as a function of $\bar{\sigma}$ we find

$$\bar{\sigma} = \tanh\left(\frac{\mu_B B_{\text{eff}}}{k_BT}\right).$$

We get the same expression for $\bar{\sigma}$ as eqn (12.2.4).

Method 3

We can make an estimate of the thermodynamic potential by assuming that the probability of spin-up on any site is $p_\uparrow = \tfrac{1}{2}(1 + \bar{\sigma})$ and the probability of spin-down is $p_\downarrow = \tfrac{1}{2}(1 - \bar{\sigma})$. It is then relatively easy to calculate the entropy of the spins and the internal energy of the system. The entropy for one site is

$$
\begin{aligned}
S_1 &= -k_B \sum_i p_i \ln(p_i) \\
&= -k_B \left\{\tfrac{1}{2}(1 + \bar{\sigma})\ln((1 + \bar{\sigma})/2) + \tfrac{1}{2}(1 - \bar{\sigma})\ln((1 - \bar{\sigma})/2)\right\}.
\end{aligned}
$$

The entropy for N sites is NS_1. The internal energy is

$$U = -\frac{z}{2}JN\bar{\sigma}^2 - \mu_B\mu_0 HN\bar{\sigma}.$$

The generalized magnetic free energy, $F_H = U - TS$, is the thermodynamic potential.

$$
\begin{aligned}
F_H(\bar{\sigma}) &= \frac{Nk_BT}{2}\left\{(1 + \bar{\sigma})\ln\left((1 + \bar{\sigma})/2\right) + (1 - \bar{\sigma})\ln\left((1 - \bar{\sigma})/2\right)\right\} \\
&\quad -\frac{z}{2}JN\bar{\sigma}^2 - \mu_B\mu_0 HN\bar{\sigma}. \tag{12.2.7}
\end{aligned}
$$

$F_H(\bar{\sigma})$ can be minimized by varying $\bar{\sigma}$ to give the equation of state

$$\frac{k_BT}{2}\ln\left(\frac{1 + \bar{\sigma}}{1 - \bar{\sigma}}\right) = \mu_B\mu_0 H + Jz\bar{\sigma}$$

which can be rearranged to give

$$\bar{\sigma} = \tanh\left(\frac{\mu_B B_{\text{eff}}}{k_B T}\right).$$

All three methods give the same equation for the average spin, $\bar{\sigma}$.

Provided $\bar{\sigma}$ is small, the thermodynamic potential (eqn 12.2.7) can be expanded as a power series in $\bar{\sigma}$ (or M) as

$$F_H(M) = -Nk_B T \ln(2) - \mu_B \mu_0 H N \bar{\sigma} - \frac{z}{2} J N \bar{\sigma}^2 + Nk_B T \left(\tfrac{1}{2}\bar{\sigma}^2 + \tfrac{1}{12}\bar{\sigma}^4 + \cdots\right)$$

or as

$$F_H(M) = F_0 - V\mu_0 HM + \tfrac{1}{2}aM^2 + \tfrac{1}{4}bM^4 + \cdots \qquad (12.2.8)$$

with

$$
\begin{aligned}
a &= Vk_B(T - T_c)/\mu_B^2 n, \\
T_c &= zJ/k_B, \\
b &= Vk_B T/3\mu_B^4 n^3.
\end{aligned}
$$

The parameter a is positive for $T > T_c$ and negative for $T < T_c$. The transition temperature is proportional to the strength of the exchange coupling.

12.3   Order parameter

In the Ising model the magnitude of the magnetization of the system tells us how well the spins are aligned. Perfect alignment, with all spins pointing one way, corresponds to complete order in the system. When the temperature rises, the order decreases as the spins become more randomly arranged; nevertheless, provided there is still a net magnetization there are vestiges of order. For temperatures above T_c the order is lost, and spins point equally in the $+z$ and $-z$ directions.

The magnetization of the uniaxial ferromagnet acts as an *order parameter*: it tells us how well the spins are aligned with each other. The evolution of M with temperature in the absence of an external magnetic field is shown in Fig. 12.1. The upper curve denotes the equilibrium magnetization taken in the limit that the external magnetic field H is positive but tending to zero; the lower curve, in the limit that the external magnetic field is negative but tending to zero.

In the absence of an external magnetic field the Hamiltonian of the Ising model is

$$\hat{H} = -\frac{J}{2} \sum_{(i,j)_{nn}} \sigma_j \sigma_i. \qquad (12.3.1)$$

The Hamiltonian is unchanged if we reverse the sign of all the spins

$$\sigma_i \rightarrow -\sigma_i.$$

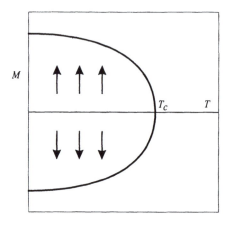

Fig. 12.1 The phase diagram for a uniaxial ferromagnetic system in the absence of an external magnetic field. For high temperatures the magnetization is zero; as the temperature is lowered below the critical temperature, T_c, the magnetization is either positive or negative for these are the states of lowest free energy. Whichever state it chooses, the symmetry of the system is broken.

Such a transformation reflects a symmetry of the Ising Hamiltonian. We can generate such a transformation by a parity change in which $z \rightarrow -z$ since this reverses the direction of all spins.

The Hamiltonian has this symmetry but the equilibrium state of the system may not. If the net magnetization is zero, a parity change leaves the system unaffected; if the magnetization is non-zero, a parity change alters the direction of spins and changes the sign of the magnetization, thereby altering the equilibrium state of the system. The low-temperature equilibrium state of the system no longer retains the full symmetry of the Hamiltonian. We say that there is *broken symmetry*.

In a continuous phase transition the system usually possesses a symmetry property which is present in the high-temperature phase, but which is lost at low temperatures. As we cool matter down it no longer retains the full symmetry of the basic laws of quantum mechanics, even though it must still obey these laws.

Following Landau (1937) we describe continuous phase transitions near the critical point in terms of an order parameter, ϕ. For temperatures below the critical temperature the order parameter is finite; as the critical point is approached the order parameter tends to zero; for temperatures above the critical temperature the order parameter is zero. The order parameter is the additional variable which we need to specify the system in the low-temperature phase of lower symmetry. For example, we must specify the magnetization in the Ising model as well as the temperature and number of particles in order to define the thermodynamic state of the system.

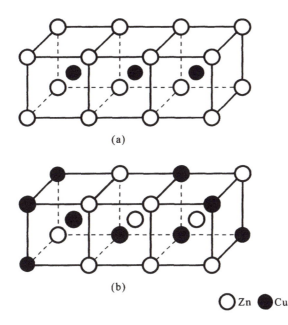

Fig. 12.2 The arrangement of Cu and Zn atoms can be ordered (a) or disordered (b).

There is a wide diversity of phase transitions which can be described in terms of an order parameter. A solid may undergo a structural phase transition involving displacements of atoms in the unit cell. For example, the structure could be simple-cubic at high temperatures and become tetragonal below a critical temperature. As the atoms become displaced the symmetry of the crystal is altered. The order parameter can be defined in terms of the displacement of the atom from its position in the high-temperature phase. Above the transition there is no displacement; below, the displacement is finite.

Another example is the order–disorder transition which occurs in body-centred cubic CuZn, known as β-brass. In the ordered phase at $T = 0$ all the Cu atoms are placed at the centre of the unit cell, the Zn atoms at the corners of the cell. All the atoms are neatly arranged and ordered. As the temperature is raised some of the Cu atoms replace the Zn atoms on the corners as shown in Fig. 12.2; the crystal becomes more disordered. Above a transition temperature, T_c, the atoms are arranged completely randomly on the two sites. The atoms are completely disordered. The order parameter can be defined in terms of the difference in fractional concentration of Cu and Zn on, say, the centre site. For temperatures above T_c the order parameter is zero, for lower temperatures it is finite.

All of the above involve a scalar order parameter, that is a single number. The order parameter can have two numbers associated with it. For example, the two numbers could be the real and imaginary parts of a complex order parameter.

In the case of ferromagnetism, the system could have a magnetization in the x–y plane with components M_x and M_y. The order parameter then consists of a vector made up of these two components. Some metals become superconducting for temperatures below a critical temperature. The characteristic feature of a *superconductor* is that its electrical resistivity is effectively zero. Similarly liquid ^{4}He at low temperatures can flow without resistance, one of the defining properties of a *superfluid*. In both of these cases we can describe the properties of the system by assuming that there is a complex order parameter $\Psi = \Psi' + i\Psi''$ where Ψ' is the real and Ψ'' is the imaginary part. A complex number can be treated as a two-component vector. The phase of the complex order parameter must be single valued, a requirement which is responsible for some of the unusual properties of superfluids and superconductors.

The order parameter can have three components: for an isotropic magnet there are three components of the magnetization which make up the order parameter. The more components the system has, the more complex the behaviour of the system. The number of components affects the thermal properties of the system close to the critical point.

12.4 Landau theory

The mean field theory of the Ising model yields a thermodynamic potential $(\Phi(\phi) = F_H)$ which can be written in terms of ϕ as

$$\Phi(\phi) = \Phi_0 - \eta\phi + \tfrac{1}{2}a\phi^2 + \tfrac{1}{4}b\phi^4 + \cdots$$

where $\eta = V\mu_0 H$ represents the external field. This is a power series expansion in the order parameter. In Landau theory we assume that near the transition the order parameter is small so that we may be justified in writing the thermodynamic potential $\Phi(\phi)$ as a power series in the order parameter. (For simplicity we consider a scalar order parameter, such as the z-component of magnetization in the uniaxial ferromagnet.) The expansion of $\Phi(\phi)$ must take account of any symmetry property that the system possesses. For example, we can have $\Phi(\phi) = \Phi(-\phi)$ in which case the expansion only includes even powers of ϕ; this is the case for a uniaxial ferromagnet in the absence of a magnetic field. The system (and hence the thermodynamic potential) is symmetric under the parity change, $z \to -z$, but the z-component of the magnetization changes sign. Hence there can be no terms which are odd powers of M_z in the expansion of $F_H(M_z)$.

When the system possesses this symmetry we can write

$$\Phi(\phi) = \Phi_0 + \frac{a}{2}\phi^2 + \frac{b}{4}\phi^4 + \frac{c}{6}\phi^6 + \cdots.$$

The quantity $a(T)$ is small near the critical temperature and varies with temperature. To illustrate the approach we only need to retain terms up to ϕ^6 as long as c is positive.

The thermodynamic potential is a minimum in equilibrium which means that $\Phi'(\phi) = 0$ and $\Phi''(\phi) > 0$. The former condition gives the equation of state

$$\Phi'(\phi) = 0 = \phi\left(a + b\phi^2 + c\phi^4\right).$$

We look for real solutions of this equation. Either we have the solution

$$\phi = 0$$

or

$$\phi^2 = \frac{b}{2c}\left(-1 \pm \sqrt{1 - 4ac/b^2}\right). \tag{12.4.1}$$

If $\Phi''(\phi) < 0$ the solutions are unstable; if $\Phi''(\phi) > 0$ the solutions are either stable or metastable. To find out whether it is stable or metastable we can plot a graph and look for a lower minimum.

If $b < 0$ with $c > 0$, ϕ is the real solution of

$$\phi^2 = (-b/2c)\left(1 + \sqrt{1 - 4ac/b^2}\right). \tag{12.4.2}$$

If $b > 0$ and a is small, ϕ is the real solution of

$$\phi^2 = \frac{b}{2c}\left(-1 + \sqrt{1 - 4ac/b^2}\right) \simeq -\frac{a}{b},$$

which is independent of c. These two solutions describe first- and second-order phase transitions.

12.4.1 Case I: $b > 0$, second-order transition

We start with the case where $b > 0$. The critical point occurs where $a = 0$. We can expand $a(T)$ in a power series in temperature; the lowest non-zero term in the expansion is

$$a(T) = \alpha(T - T_c)$$

with α a positive number. In the critical region, close to the transition, a is small. If a is positive the only possible solution is $\phi = 0$. If a is negative there are two additional solutions,

$$\phi = \pm\sqrt{\frac{-a}{b}} = \sqrt{\frac{\alpha(T_c - T)}{b}}.$$

Both of these solutions are stable when a is negative, and the solution with $\phi = 0$ is unstable. The two cases are illustrated in Fig. 12.3.

Suppose the system starts at a temperature above T_c so that $a(T)$ is positive, and the order parameter is zero. The system is cooled to a temperature just below T_c. Either the system evolves to a state with order parameter $\phi = +\sqrt{-a/b}$ or it evolves to a state with $\phi = -\sqrt{-a/b}$. Both states are equally stable. Whichever state is chosen breaks the underlying symmetry $\phi \rightarrow -\phi$.

If we now heat the system, ϕ evolves continuously to the value $\phi = 0$ at the transition temperature. ϕ is continuous at the transition; however, the specific heat is discontinuous there.

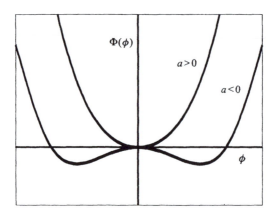

Fig. 12.3 The thermodynamic potential $\Phi(\phi) = a\phi^2/2 + b\phi^4/4$ with $b > 0$ for positive and negative values of a.

We can calculate the entropy of the system by differentiating $\Phi(\phi)$ with respect to temperature. For temperatures above T_c, ϕ is zero and we get

$$S = -\left(\frac{\partial \Phi(\phi)}{\partial T}\right) = -\left(\frac{\partial \Phi_0}{\partial T}\right).$$

The free energy for $T < T_c$ is

$$\Phi(\phi) = \Phi_0 - \frac{a^2}{2b} + \frac{a^2}{4b} = \Phi_0 - \frac{a^2}{4b} \tag{12.4.3}$$

so that the entropy is

$$S = -\left(\frac{\partial \Phi_0}{\partial T}\right) - \frac{\alpha^2(T_c - T)}{2b}.$$

As the temperature increases to T_c the entropy goes smoothly to $-(\partial\Phi_0/\partial T)$; for temperatures above T_c the entropy remains $-(\partial\Phi_0/\partial T)$. Therefore the entropy is continuous at the transition. However, the specific heat $T(\partial S/\partial T)_M$ is discontinuous at the transition. The discontinuity is

$$\Delta C_M = -\frac{\alpha^2}{2b} \tag{12.4.4}$$

at the transition temperature. Landau's theory predicts a discontinuity in the specific heat in a second-order phase transition.

In fact a discontinuity in the specific heat is not observed in practice. (A discontinuity is seen in the specific heat of conventional superconductors, but that

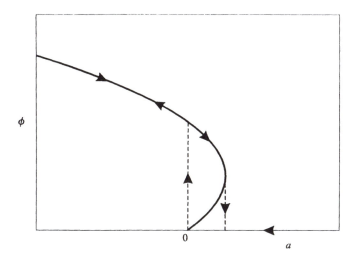

Fig. 12.4 Hysteresis at a first-order transition. Starting with $\phi = 0$ at high temperatures the system is cooled and follows the path indicated by the arrows. When $a = 0$
the state with $\phi = 0$ becomes unstable, and on further cooling the system jumps to the
branch with finite ϕ. On subsequent warming the system follows the arrows. It jumps
to the $\phi = 0$ state at a positive value of a.

is because the critical region is so narrow that it cannot be explored experimen-
tally.) Instead the specific heat usually diverges as $|T_c - T|$ to some power which
means that there is not a discontinuity in any of the derivatives of the thermodynamic potential; the observed transition is certainly not second order. It makes
more sense to describe such transitions as continuous since all derivatives of the
thermodynamic potential are continuous.

12.4.2 Case II: $b < 0$, first-order transition

When $b < 0$ and $c > 0$ the situation is a little more complex. Figure 12.4 shows
how $\Phi(\phi)$ varies with ϕ for different values of a. When a is very large there is only
one minimum at $\phi = 0$. For values of a less than $b^2/4c$ there are two minima:
the one at $\phi = 0$ is stable, the other given by the positive root of

$$\phi^2 = (-b/2c)\left(1 + \sqrt{1 - 4ac/b^2}\right)$$

is metastable. When $a = 0$ the two minima are equally stable. Finally when
$a < 0$ the minimum at $\phi = 0$ is metastable, the other is stable.

Suppose the system starts at high temperatures, where $\phi = 0$, and is cooled.
When the temperature is above T_c (so that a is positive) the stable state is
$\phi = 0$. As soon as the system is cooled below T_c this state becomes metastable.
If the temperature is cooled sufficiently the metastable state ($\phi = 0$) decays in
the stable state given by the positive root of eqn (12.4.2). We get one part of

the path shown in Fig. 12.4. Now let us warm the system. When a is positive the system becomes metastable and would prefer to exist in the $\phi = 0$ state. If we could heat the system to the point where $a = b^2/4c$, the second minimum would disappear. In practice the system decays before this point is reached due to fluctuations taking the system over the potential barrier. In this way we can go round a thermal cycle, cooling the system and then heating it causing the state to change repeatedly.

Notice that the state of the system for a given temperature close to the transition depends on where we have got to in this thermal cycle. We could end up in the stable state or the metastable state. The dependence of a system on its previous history is called *hysteresis*. Hysteresis is a characteristic property of a first-order phase transition.

Notice also that in a first-order phase transition there is a jump in the value of the order parameter at $a = 0$. Of course there is no jump in the thermodynamic potential at the phase transition since the two wells are equally deep there, by definition. But the variation of the depth of the well with temperature, that is

$$S = - \left(\frac{\partial \Phi(\phi)}{\partial T} \right)$$

is different for the two phases. There is a discontinuity in the entropy at the transition which means that there is a latent heat.

12.5   Symmetry-breaking field

Consider the uniaxial ferromagnet in the absence of an external magnetic field. The thermodynamic potential, $F_H(M_z)$, is an even function of M_z. Now let us add an external magnetic field along the z-axis. The energy of the system is shifted by $-\mu_0 H M_z$. The thermodynamic potential is no longer an even function of M_z. The magnetic field breaks the symmetry of the system, that is the invariance under the parity change $z \rightarrow -z$.

More generally, the thermodynamic potential in the presence of a symmetry-breaking field can be written as

$$\Phi(\phi) = \Phi_0 + \frac{a}{2}\phi^2 + \frac{b}{4}\phi^4 + \cdots - \eta\phi \tag{12.5.1}$$

where η represents the symmetry-breaking field. In the absence of this term the transition is first order if $b < 0$ and continuous if $b > 0$. We now show that a continuous transition is turned into a first-order phase transition by the symmetry-breaking term

If we plot graphs of $\Phi(\phi)$ as a function of ϕ we find that there can be two unequal minima, as shown in Fig. 12.5. When η is positive the lower minimum is for $\phi > 0$. Let us call this well 2. Suppose we start at high temperatures with a positive in well 2. As the temperature is lowered the depth of the well changes but there is no phase transition in which the system moves to the other well.

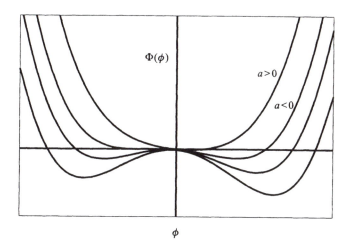

Fig. 12.5 Thermodynamic potential with a symmetry-breaking field. For high temperatures $a > 0$ and the minimum is for positive ϕ. As the temperature is lowered the value of ϕ grows. There is no sign of a phase transition; everything behaves continuously. However, if the symmetry-breaking field is reversed when $a < 0$ there is a discontinuity in equilibrium value of ϕ when the field is zero, so there is a first-order phase transition.

The system evolves smoothly. If we heat the system up we return smoothly to the starting point.

At $\phi = 0$ the derivative $\partial\Phi(\phi)/\partial\eta = -\phi$. Consider the low-temperature case where a is negative. Suppose η is positive and it evolves towards zero. In the limit that η tends to zero the value of ϕ tends to $\sqrt{-a/b}$. In contrast when η is negative the same limiting procedure leads to $\phi = -\sqrt{-a/b}$. It follows that the value of ϕ is discontinuous across the line in the phase diagram where $\eta = 0$, as shown in Fig. 12.6; the transition is first order.

12.6 Critical exponents

We define $\partial\eta/\partial\phi$ as the inverse of the susceptibility of the system, χ^{-1}.

$$\chi^{-1} = \alpha(T - T_c) + 3b\phi^2.$$

It is easy to evaluate the susceptibility in the limit that $\eta = 0$. For temperatures above T_c we find

$$\chi^{-1} = \alpha(T - T_c);$$

for temperatures below, we get

$$\chi^{-1} = 2\alpha(T_c - T).$$

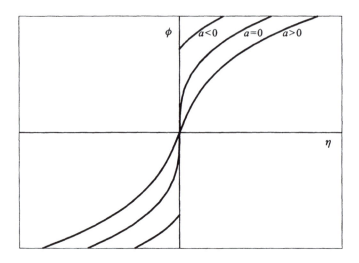

Fig. 12.6 A discontinuity in the order parameter as a function of η appears for temperatures below the phase transition.

In both cases the susceptibility diverges as $|T - T_c|^{-\gamma}$ with $\gamma = 1$ according to Landau theory. The quantity γ is called a *critical exponent*.

When the system is at the critical temperature, we have $a = 0$ and

$$\eta = b\phi^\delta$$

with $\delta = 3$ according to Landau theory. δ is another critical exponent.

Another critical exponent is β where

$$\phi = d(T_c - T)^\beta.$$

According to Landau theory $\beta = 1/2$.

The prediction of Landau theory is that these critical exponents β, δ, and γ are universal: all phase transitions for all systems have these values: $\beta = 1/2$, $\delta = 3$, and $\gamma = 1$. These critical exponents found by Landau theory do not agree with experiment in the region close to the critical point. According to experiment (and modern theory) all one-component systems have the same value of β: it is universal. Experiment gives $\beta = 0.325 \pm 0.005$ for all three-dimensional systems with a one-component order parameter. (The values we quote for the critical exponents are taken from Chapter 28 of Zinn-Justin 1996). If the number of components is different then the observed value of β changes. For example, for a three-dimensional system with a three-component order parameter, $\beta = 0.35 \pm 0.035$. Similarly, the other critical exponents do not agree with the predictions of Landau theory. Instead of $\gamma = 1$, measured values of γ are typically $\gamma = 1.24 \pm 0.01$ for a one-component order parameter, and $\gamma = 1.40 \pm 0.03$ for a three-component order parameter.

It might seem that we have been wasting our time using mean field theory as it makes predictions that are inconsistent with experiment for temperatures in the critical region. The reason for its failure, stated simply, is that we have ignored fluctuations. If the system is sufficiently far from the critical point—in practice this can be the majority of the phase diagram—mean field theory can work quite well. In systems where fluctuations are not very important mean field theory works extremely well. For example, mean field theory gives a very good description of the behaviour of conventional superconductors.

The advantage of using mean field theory is that it enables us to get simple solutions to complicated problems which are usually in reasonable qualitative agreement with experiment. We can use mean field theory to make predictions about the different phases which can appear, and to make an estimate of the energy scale involved, and hence the transition temperature.

To go beyond mean field theory requires material which lies outside any book entitled 'Introductory Statistical Mechanics'. The reader who wishes to go more deeply into current theories of the critical region is urged to read the lucid articles by Maris and Kadanoff (1978), or by Fisher (1983), or to consult more advanced textbooks, such as the one by Binney *et al.* (1992).

12.7 Problems

1. The energy associated with the magnetic coupling between spins is roughly

$$\frac{\mu_0 \mu^2}{4\pi r^3}.$$

The magnitude of the magnetic moments, μ, is of the order of a Bohr magneton, $\mu_B = e\hbar/2m_e \simeq 10^{-23}$ A m. If the atoms are about 0.2 nm apart, estimate the magnetic coupling and the corresponding transition temperature. Compare this temperature with the transition temperature of typical ferromagnets.

2. Suppose the thermodynamic potential of a system is

$$\Phi(\phi) = \frac{a}{2}\phi^2 + \frac{d}{6}\phi^6 - \eta\phi$$

with $a = \alpha(T - T_c)$ and with both α and d positive. Determine the critical exponents β, γ, and δ.

3. The van der Waals theory (Appendix F) predicts that the pressure for one mole of gas varies as

$$P(V,T) = \frac{RT}{V - b} - \frac{a}{V^2}.$$

At the critical point $(\partial P/\partial V)_T = 0$ and $(\partial^2 P/\partial V^2)_T = 0$. These two relations define the critical temperature, T_c, the critical pressure, P_c, and the critical volume, V_c. Show that we can expand about the critical point as

$$P = P_c + A(T - T_c)(V - V_c) + B(V - V_c)^3 + C(T - T_c)$$

and identify the coefficients A, B, and C. When $T = T_c$ show that

$$(P - P_c) = B(V - V_c)^\delta$$

and determine δ. The phase boundary line near the critical point is given by $P - P_c = C(T - T_c)$. When P lies on this line show that

$$(V - V_c) = A(T_c - T)^\alpha$$

and determine α.

4. The free-energy density in the van der Waals theory is given by (see Appendix F)

$$f(n) = -nk_BT\left(\ln\left[\left(\frac{2\pi mk_BT}{h^2}\right)^{3/2}\left(\frac{1}{n} - b\right)\right] + 1 + \frac{an}{k_BT}\right).$$

What is the significance of the condition $d^2 f/dn^2 = 0$? Show that the condition $d^2 f/dn^2 = 0$ gives the equation

$$n(1 - bn)^2 = \frac{k_BT}{2a}.$$

What is the value of n which gives the maximum value of the left-hand side of this equation? Hence determine T_c.

5. One way of describing an interacting gas of particles is by the lattice gas model. In this simple a volume V is divided up into atomic-size cubes of volume v_0; the cubes are stacked on top of each other so as to form a 'simple-cubic lattice'. Due to short-ranged repulsion between the atoms there can be no more than one gas atom in each of these $N_0 = V/v_0$ sites. The attractive interaction between atoms is thought to be short ranged so it is only significant between nearest-neighbour sites in the cubic lattice. We can model this by a Hamiltonian of the form

$$\hat{H} = -\varepsilon \sum_{(i,j)_{nn}} n_i n_j$$

where $(i,j)_{nn}$ refers to nearest-neighbours only. If the site i is occupied by one atom $n_i = 1$; if the site is empty $n_i = 0$. We place on the average $\overline{N}$ atoms on the N_0 sites so the average occupancy per site is $n = \overline{N}/N_0$.

(a) Replace $n_i n_j$ by

$$n_i n_j = (n_i - n + n)(n_j - n + n) \simeq n^2 + (n_i - n)n + (n_j - n)n$$

and show that the mean field Hamiltonian is

$$\hat{H}_{\mathrm{mf}} = \sum_{i=1}^{N_0} \left(3\varepsilon n^2 - 6\varepsilon n n_i\right).$$

(b) Show that the grand potential for a fixed value of n is

$$\Phi_G = 3\varepsilon n^2 N_0 - k_{\mathrm{B}}TN_0 \ln\left(1 + \mathrm{e}^{(6\varepsilon n+\mu)/k_{\mathrm{B}}T}\right).$$

(c) Calculate the pressure and average number of particles in the system for a fixed value of n; eliminate the chemical potential to get

$$P = -\frac{k_{\mathrm{B}}T}{v_0} \ln\left(1 - \frac{\bar{N}v_0}{V}\right) - 3\varepsilon v_0 \frac{\bar{N}^2}{V^2}.$$

(d) Next evaluate the isothermal compressibility $\chi = -V^{-1} \left(\partial V/\partial P\right)_T$ and show that it becomes negative when

$$T < 6\varepsilon v_0 \bar{N}\left(V - \bar{N}v_0\right)/k_{\mathrm{B}}V^2.$$

(e) Find the coordinates of the critical point and show that it corresponds to $n_c = 1/2$. What is the transition temperature?

6. A ferroelectric crystal has a free energy of the form

$$F = \tfrac{1}{2}\alpha(T - T_c)P^2 + \tfrac{1}{4}bP^4 + \tfrac{1}{6}cP^6 + DxP^2 + \tfrac{1}{2}Ex^2$$

where P is the electric polarization and x represents the strain applied to the crystal. Minimize the system with respect to x and determine the free energy at the minimum. Under what circumstances is there a first-order phase transition for this system?

7. Consider a two-component order parameter (ϕ_1, ϕ_2) with thermodynamic potential

$$\Phi\left(\phi_1, \phi_2\right) = \tfrac{1}{2}a_1\phi_1^2 + \tfrac{1}{2}a_2\phi_2^2 + \tfrac{1}{4}b_{11}\phi_1^4 + \tfrac{1}{4}b_{22}\phi_2^4 + \tfrac{1}{2}b_{12}\phi_1^2\phi_2^2.$$

Both a_1 and a_2 are negative and both b_{11} and b_{22} are positive. Imagine that ϕ_1 is zero. What is the equilibrium value of ϕ_2? Now imagine that ϕ_1 is extremely small: examine the stability of the system to changes in ϕ_1. Show that the system is unstable if a certain condition is met. Repeat the analysis for ϕ_2 zero and ϕ_1 non-zero. Show that both order parameters are non-zero if

$$b_{12} < b_{11}a_2/a_1; \qquad b_{12} < b_{22}a_1/a_2.$$

13

Ginzburg–Landau theory

There may always be another reality
To make fiction of the truth we think we've arrived at
Christopher Fry, A Yard of Sun, Act 2

The approach we have used so far has treated the order parameter as if it were constant throughout space, and ignored any spatial variations. Such an approach may be adequate for a single phase, but does not allow us to describe two phases coexisting in the same container. For example, in a liquid–gas transition, the liquid phase lies at the bottom of the container and the gas at the top, with an interface (meniscus) separating the two phases; the density changes smoothly with position near the interface. To describe the interface between the two phases we need to allow for a change in the density with position, something which is not included in Landau theory. More generally, the coexistence of two phases with order parameters ϕ_1 and ϕ_2 implies a region where the order parameter changes in space from one value to the other.

Can we extend the theory to allow for spatial variation in the order parameter, $\phi(\mathbf{r})$, and then determine the shape of the interface and the surface contribution to the free energy? Another question we want to answer concerns the size of fluctuations of the order parameter, something we have neglected. Are they sufficiently small so that we are justified in ignoring them, or do they dominate the properties of the system and need a careful treatment?

13.1   Ginzburg–Landau theory

To answer these questions we need to construct a more sophisticated model in which the order parameter is allowed to vary in space. The simplest such model is due to Ginzburg and Landau (1950). There are three main benefits of adopting their model: we can describe the interface between phases and estimate the surface free energy; we can investigate the importance of fluctuations; we find new physics: the model predicts the existence of topological singularities in the order parameter, such as quantized vortices in superfluid systems.

Ginzburg and Landau supposed the order parameter $\phi(\mathbf{r})$ varies with position and that the thermodynamic potential Φ depends on $\phi(\mathbf{r})$ and its derivatives as

$$\Phi[\phi(\mathbf{r})] = \int \mathrm{d}^3 r \, \left\{ f(\phi(\mathbf{r})) + \tfrac{1}{2}\lambda(\nabla\phi(\mathbf{r}))^2 \right\}. \tag{13.1.1}$$

(Normally we use an upper case symbol for the extensive quantity and a lower case symbol for the corresponding density; here we use f for the thermodynamic potential per unit volume as we have used the lower case Φ for the order parameter.) This proposition is much more sophisticated than the previous guess; we are now proposing that the thermodynamic potential which describes equilibrium states can be expressed as an *energy functional* of the corresponding order parameter and its derivatives.

The term $\frac{1}{2}\lambda(\nabla\phi(\mathbf{r}))^2$ must cause an increase in the thermodynamic potential whenever there is any spatial variation in the order parameter away from its equilibrium value. In other words fluctuations away from equilibrium must cost an increase in energy. If this statement were not true so that $\frac{1}{2}\lambda(\nabla\phi(\mathbf{r}))^2$ caused a decrease in the thermodynamic potential, the system would be highly unstable to spatial fluctuations of the order parameter; such fluctuations would grow rapidly until the system evolved into a new equilibrium state. It follows that λ must be positive.

The quantity $\frac{1}{2}\lambda(\nabla\phi(\mathbf{r}))^2$ in eqn (13.1.1) is called the *stiffness term*; λ is called the stiffness parameter.

Suppose the system starts out with a uniform order parameter which we denote as $\overline{\phi}$. Fluctuations in ϕ are written as a Fourier series

$$\phi(\mathbf{r}) = \overline{\phi} + \phi_1(\mathbf{r}) = \overline{\phi} + \sum_{\mathbf{k}} \phi_{\mathbf{k}}\, e^{i\mathbf{k}\cdot\mathbf{r}} \tag{13.1.2}$$

where $\mathbf{k}$ is the wave vector of the fluctuation of magnitude $\phi_{\mathbf{k}}$. If we substitute eqn (13.1.2) into eqn (13.1.1) we get

$$\begin{aligned}
\Phi[\phi(\mathbf{r})] &= \int \mathrm{d}^3\mathbf{r}\, \left\{ f(\overline{\phi}) + \phi_1(\mathbf{r})f' + \tfrac{1}{2}\phi_1^2(\mathbf{r})f'' + \cdots + \tfrac{1}{2}\lambda(\nabla\phi_1(\mathbf{r}))^2 \right\} \\
&= V\left(f(\overline{\phi}) + \tfrac{1}{2}\sum_{\mathbf{k}} |\phi_{\mathbf{k}}|^2 \left(f'' + \lambda k^2 \right) + \cdots \right).
\end{aligned} \tag{13.1.3}$$

If there is a single minimum, f'' is positive for all ϕ and the system is stable. Any large fluctuation away from the state of uniform ϕ decays rapidly until all that are left are thermal fluctuations. If there are two (or more minima), there must be a range (or ranges) of values of ϕ in which f'' is negative; the system is unstable if ϕ lies in a region where $f'' < 0$. The unstable system can decay into a state of lower thermodynamic potential, either directly (if the order parameter is not conserved) or by phase separation (if the order parameter is conserved).

If the order parameter is not conserved the system evolves towards the nearest minimum. Suppose there are just two minima: if the nearer minimum is the lower, the system is stable; if it is the higher, the system is metastable. The metastable state decays through nucleation of the more stable phase.

The situation is a little more complicated if the order parameter is conserved. The system described by eqn (13.1.1) is stable if there is a rise in free energy as the amplitude of the fluctuations increases. For a fluctuation of wave vector $\mathbf{k}$, the system is stable if $f'' + \lambda k^2 > 0$, unstable if $f'' + \lambda k^2 < 0$. For sufficiently large wave vectors the system is always stable. The real test of stability concerns the limit $k \to 0$. The system is unstable in this limit provided $f'' < 0$. Conversely any state with $f'' > 0$ is stable against small fluctuations. The value where $f'' = 0$ is called the *spinodal limit;* the corresponding value of ϕ is called the *spinodal point.* The spinodal point is the value of ϕ which separates the region of unstable values of the order parameter from metastable values of the order parameter.

If $f'' + \lambda k^2 < 0$ the system is unstable against fluctuations of wave vector $k < k_c$ with $k_c = \sqrt{-f''/\lambda}$. We can expect these fluctuations in ϕ to grow in magnitude until a new stable region is found. The inverse of k_c defines a characteristic length

$$\xi_c = k_c^{-1} = \sqrt{\lambda/|f''|} \tag{13.1.4}$$

called the *correlation length.* The system is stable against fluctuations for wave vectors greater than k_c. If desired we can take a thermal average over fluctuations involving these wave vectors, for such thermal averages are well defined. The process of doing this is called coarse graining. All information about the system on length-scales smaller than the correlation length is lost in this process.

What then is the significance of the correlation length? It is helpful to visualize the variation of the order parameter. One way of doing this is to attribute the least value of the magnitude of the order parameter with a white shade, the greatest value with a black shade with a uniform variation of greyness between these limiting values. If we were to look at representative samples of the system whose size is much smaller than the correlation length, the order parameter would have a uniform shade of grey across the sample. In contrast, in samples whose size is much larger than the correlation length, the order parameter would vary across the sample: there would be dark patches, stripes, and swirls forming distinct patterns. The correlation length sets the length-scale of these patterns.

13.2   Ginzburg criterion

The introduction of the stiffness parameter allows us to define the correlation length, the scale of length over which fluctuations are correlated in the system. Similarly the difference f_m between the maximum and minimum values of f defines a scale of energy density. The product of $f_m \xi_c^3$ is an energy which can be compared to $k_B T$ the thermal energy:

$$r(T) = \frac{f_m \xi_c^3}{k_B T} \tag{13.2.1}$$

The quantity $r(T)$ is a measure of the importance of fluctuations. The *Ginzburg criterion* (Ginzburg 1960) says that fluctuations are unimportant when $r(T) > 1$. In the limit that λ goes to infinity the correlation length diverges and $r(T)$ tends to infinity: all spatial fluctuations in the order parameter are ruthlessly removed because they cost an infinite amount of energy. Mean field theory is valid in this limit since the order parameter cannot vary in space.

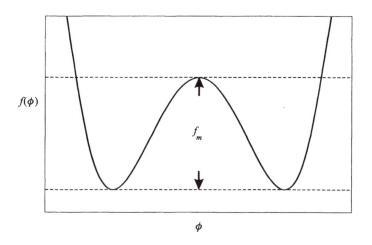

Fig. 13.1 The difference between the maximum and minimum values of $f(\phi)$ defines f_m.

Conversely, when $r(T)$ is small, mean field theory breaks down and fluctuations become important in determining the properties of the system. If we want to have a complete description of the system we must develop a theory which accounts properly for fluctuations, a topic which lies outside an introductory treatment of statistical physics.

We can estimate the magnitude of $r(T)$ using Landau theory. For a thermodynamic potential per unit volume of the form

$$f(\phi) = \frac{a\phi^2}{2} + \frac{b\phi^4}{4}$$

where the parameter a varies with temperature as $a = \alpha(T - T_c)$. The minimum of $f(\phi)$ for a which is negative occurs when $\phi^2 = -a/b$; for either of these values of ϕ we get

$$f_m = 0 - \left(-\frac{a^2}{4b}\right) = \frac{\alpha^2(T - T_c)^2}{4b}.$$

Also

$$f'' = a + 3b\phi^2 = -2a$$

so that

$$\xi_c = \sqrt{\frac{\lambda}{2\alpha(T_c - T)}}. \tag{13.2.2}$$

The correlation length diverges at the transition temperature. It follows that

$$r(T) = \frac{\alpha^{1/2}|T_c - T|^{1/2}\lambda^{3/2}}{8\sqrt{2}bk_BT} = r_0\,|1 - T/T_c|^{1/2} \tag{13.2.3}$$

with $r_0 = \alpha^{1/2}T_c^{1/2}\lambda^{3/2}/8\sqrt{2}bk_BT$.

Mean field theory can be a very good approximation over a large part of the phase diagram; it breaks down near the critical points in a region whose size depends on the quantity r_0. For example, superconductors of the conventional sort have a very large value of r_0 so that there is an extremely narrow critical region, so narrow that nobody has seen significant deviations from mean field theory. Perhaps we have to be about 1 μK from T_c to see deviations from mean field theory. In contrast, superfluid helium has a large region, say from 1.9 K to 2.172 K, where r_0 is less than one and mean field theory does not work. A proper description of fluctuation effects is essential in this region.

13.3 Surface tension

For a plane interface between two phases which are in equilibrium the profile $\phi(\mathbf{r})$ only varies as a function of the position z. Asymptotically $\phi(z)$ goes to ϕ_1^* for z tending to $-\infty$, and to ϕ_2^* for z tending to ∞. We can find the energy associated with the wall by minimizing the grand potential using the mathematical technique called the calculus of variations. The form of the result can be obtained more directly by using simple dimensional arguments.

The quantity f_m sets the scale of the energy density, whereas ξ_c sets the scale of length. A quantity with the dimensions of energy per unit area is $f_m\xi_c$. If we suppose that the relevant length-scale is the correlation length, it is reasonable to suppose on dimensional grounds that the surface tension is given by

$$\gamma = \kappa f_m\xi_c \tag{13.3.1}$$

where κ is a numerical factor of order one. For example, for the thermodynamic potential per unit volume of the form

$$f(\phi) = \frac{a\phi^2}{2} + \frac{b\phi^4}{4}$$

we have when a is negative

$$\gamma = \frac{\kappa}{\sqrt{2}}\frac{a^{3/2}\lambda^{1/2}}{b}.$$

The full calculation gives a numerical factor $\kappa = 4/3$:

$$\gamma = \frac{4/3}{2^{1/2}}\frac{a^{3/2}\lambda^{1/2}}{b}.$$

The shape of the interface is found to be

$$\phi(x) = \phi_2^* \tanh\left(x\sqrt{|a|/2\lambda}\right).$$

The width of the interface is of the order of the correlation length which makes perfect sense.

13.4   Nucleation of droplets

Consider a three-dimensional system such as a supercooled gas in a metastable state. The relevant thermodynamic potential is the grand potential; the corresponding energy density, Φ_G/V. But both V and Φ_G are extensive quantities, so that

$$\left(\frac{\partial \Phi_G}{\partial V}\right)_{\mu,T} = \frac{\Phi_G}{V} = -P.$$

It follows that the quantity $f(\phi)$ is just $-P(n)$. The supercooled gas would be more stable as a liquid, but to form the liquid phase it must create a small bubble of liquid which grows until the new equilibrium state is reached.

For simplicity we assume that the radius R of the bubble is much larger than the thickness ξ_c of the interface. When this is the case we can describe the surface energy as $4\pi\gamma R^2$. Let phase 1 (the gas) have pressure P_1 and phase 2 (the liquid) have a pressure $P_2 > P_1$ so that it is the more stable phase. The change in the thermodynamic potential of the system when there is a bubble of phase 2 in a matrix of phase 1 is

$$\Delta F = (P_1 - P_2)\frac{4\pi}{3}R^3 + 4\pi\gamma R^2.$$

The first term represents the decrease in grand potential on producing a bubble of the more stable (liquid) phase; the second term is the energy associated with the interface.

Let us denote $(P_2 - P_1)$ as ΔP. We only consider just the case where $\Delta P \ll P_m$, where P_m is the difference between the minimum and maximum in the grand potential per unit volume (the analogue of f_m). The quantity ΔF has a maximum as a function of R for a radius

$$(\Delta P)4\pi R^2 = 8\pi\gamma R$$

that is

$$R_C = \frac{2\gamma}{\Delta P}$$

with a corresponding barrier energy

$$\Delta E_C = \frac{16\pi\gamma^3}{3\Delta P^2}.$$

To nucleate the more stable phase the bubble has to overcome this barrier in the grand potential.

The same calculation can be done for other three-dimensional systems, leading to an energy barrier for nucleation of the form

$$\Delta E_C = \frac{16\pi\,(\kappa f_m \xi_c)^3}{3\Delta f^2}.$$

When $R > R_C$ the droplet grows, when $R < R_C$ the droplet shrinks. In order to nucleate the droplet there must be a fluctuation of sufficient energy to overcome this barrier. The probability that such a fluctuation will appear is proportional to

$$e^{-\Delta E_C/k_B T} = e^{-\left(16\pi\kappa^3 f_m^2/3\Delta f^2\right)\left(f_m \xi_c^3/k_B T\right)} = e^{-16\pi\kappa^3 f_m^2 r(T)/3\Delta f^2}.$$

The rate at which nucleation occurs depends on the creation of a fluctuation large enough to overcome the barrier, so it must be of the form

$$\nu = \nu_0\, e^{-16\pi\kappa^3 f_m^2 r(T)/3\Delta f^2}. \tag{13.4.1}$$

Even if the prefactor ν_0 is large, the exponential term can be sufficiently small that the nucleation rate is incredibly slow, so that nucleation is effectively unobservable. The system is then in a metastable state. To achieve this we need a large value of $r(T)$, so that fluctuations are small, or a small value of Δf compared to f_m. But remember the calculation is only valid if $\Delta f \ll f_m$; when this condition breaks down we need a new theory. A large value of $r(T)$ corresponds to the region where mean field theory works. The whole calculation is only self-consistent in the mean field regime with a sufficiently small value of Δf.

If the factor in the exponential is very large (say 1000), as is the case when Δf is tiny, the probability of nucleation is negligible. As Δf increases the probability of nucleation per second increases dramatically; for sufficiently large Δf, drops of the stable phase appear, grow, and when they fill the container the phase transition is completed. In practice, droplets of the more stable phase form on any impurities, surfaces, or defects rather than uniformly throughout the system. The reason is that the potential barrier is lowered by such imperfections. Our beautiful theory rarely fits experiment because of imperfections which are present.

In the critical region the value of $r(T)$ is small so it is easy to nucleate the new phase if we slowly cool through the phase transition. If the system is rapidly cooled we may be able to bypass the critical region and ensure that $\Delta E_C/k_B T$ is large at the end of the cooling. The rapid quenching through a first- or second-order transition is a topic which has recently become quite active in the field of superfluidity. It is known experimentally that, however careful one is to cool smoothly and slowly, it is extremely difficult to do so without producing topological defects, such as dislocation lines in a solid or vortex lines in a superfluid. It has been speculated that similar line defects (called cosmic strings) might have been produced during a phase transition during the early stages of the expansion of the Universe. (For a review see Leggett 1998.)

13.5 Superfluidity

If we put a large number of ^{4}He atoms in a container at a temperature above 4.2 K the atoms form a gas, but for lower temperatures they form a liquid which behaves in all respects as an ordinary liquid. This liquid phase is called helium I. However, when the temperature drops below 2.18 K the liquid develops spectacular properties: it flows through tiny capillaries without apparent friction; when placed in a beaker it flows up the sides and creeps over the rim; under certain conditions it forms a fountain when heated from below; when a ring-shaped vessel containing helium is rotated, the moment of inertia due to the liquid is much less than expected. This collection of effects is associated with what is called the *superfluid state* of liquid helium, known as helium II. The transition between the two states, helium I and helium II, is called the lambda transition, so named because the heat capacity plotted as a function of temperature resembles in shape the Greek letter λ.

It is generally believed that the phenomenon associated with superfluidity is directly connected with the Bose statistics of the ^{4}He atoms and is a manifestation of *Bose condensation*. Here we try to describe some of the essential properties of a superfluid using the notion of a *complex order parameter*. (For an introductory treatment of superfluid helium see the books by McClintock, *et al.* (1984), and by Wilks (1967).)

The simplest system to consider from a conceptual point of view is the flow of helium in an annular container, that is a hollow ring filled with the liquid. Suppose we suspend the ring from a torsion wire so that the axis of the ring is vertical. We then set the ring into rotation about the vertical axis. If the liquid inside the ring is a normal liquid, such as water, it will initially remain at rest when the rotation starts, but within a short time the liquid will come into equilibrium with the walls of the rotating ring and move with the same angular velocity. That is how a normal fluid responds. In contrast, if the same experiment is done on superfluid helium II at very low temperatures, part of the liquid appears to move with the walls, but part never comes into rotation no matter how long we wait. It is as if there were two liquids present: one behaves normally and moves with the walls, the other remains at rest with respect to the walls and experiences no frictional drag slowing it down. The part that behaves as if there were no friction is called the superfluid.

Because part of the liquid remains stationary, the moment of inertia of the ring and its contents is much less than expected. This effect is called non-classical rotational inertia. We can observe this effect directly by measuring the period of oscillation of the ring on the torsion wire. When the liquid acts in the normal manner the moment of inertia associated with the liquid is denoted as I_c. Now suppose the system becomes superfluid so the moment of inertia is less than I_c. The fraction of the fluid which is normal is associated with a density, ρ_n, which is defined as

$$\frac{\rho_n}{\rho} = \frac{I}{I_c} \tag{13.5.1}$$

where I is the measured moment of inertia and ρ is the mass density. ρ_n is called the *normal fluid density*. We also define the *superfluid density* as the remainder

$$\rho_s = \rho - \rho_n. \tag{13.5.2}$$

The superfluid density is zero when the liquid is in the normal state.

There are two aspects to an operational definition of superfluidity: the first aspect is that the moment of inertia is smaller than the classical value for a uniformly rotating liquid; the second aspect is that the liquid can exist indefinitely in a metastable state in which it flows persistently around the ring without decaying.

The conventional way of describing this mathematically is the two-fluid model (see Wilks 1967). For a normal liquid of mass density ρ the total momentum density is

$$\mathbf{g} = \rho \mathbf{v}_n \tag{13.5.3}$$

where $\mathbf{v}_n$ is the velocity of the walls of the container: a normal liquid is dragged along because of viscous effects so that when the system is in equilibrium there is no relative motion between the liquid and the walls. In contrast, for a superfluid

$$\mathbf{g} = \rho_s \mathbf{v}_s + \rho_n \mathbf{v}_n \tag{13.5.4}$$

where $\mathbf{v}_n$ is the velocity of the normal fluid which moves with the walls of the container and $\mathbf{v}_s$ is the velocity of the superfluid. If the two speeds are equal we have $\mathbf{g} = (\rho_s + \rho_n) \mathbf{v}_n$; but in general $\mathbf{v}_s \neq \mathbf{v}_n$ and the two fluids move at different velocities.

Why can superfluid helium apparently flow without friction in a variety of situations, such as in fine capillaries, in films, or around rings, where an ordinary liquid would not flow at all? The helium seems to be able to flow persistently around the ring without decaying due to friction. What stops the decay of the current? The answer is that the state in which the liquid flows around a ring is metastable.

13.6 Order parameter

Suppose the free-energy density in the rest frame of the walls of the container is taken to be of the form

$$\Phi[\Psi(\mathbf{r})] \;=\; \int d^3r \left\{ f(\Psi(\mathbf{r})) + \tfrac{1}{2}\lambda |\nabla\Psi(\mathbf{r},t)|^2 \right\}, \tag{13.6.1}$$

$$f(\Psi) \;=\; f_n + \frac{a}{2}|\Psi|^2 + \frac{b}{4}|\Psi|^4 . \tag{13.6.2}$$

with $a > 0$ for temperatures above the critical temperature T_c and negative for temperatures below T_c. The stiffness term must be positive for stability so, since the order parameter is a complex quantity, we must have the square of the modulus of $\nabla\Psi(\mathbf{r}, t)$ in eqn (13.6.1). We write $\lambda = \hbar^2/m$ with m left undefined for the present. Following the Londons (1935) we suppose that the superfluid state can be described by a complex order parameter

$$\Psi(\mathbf{r}) = A(\mathbf{r}) e^{iS(\mathbf{r})}$$

with $A(\mathbf{r})$ and $S(\mathbf{r})$ real functions of position. The free-energy density is then

$$f = f_n + \frac{a}{2}A^2 + \frac{b}{4}A^4 + \frac{\hbar^2}{2m}\left(\nabla A(\mathbf{r}, t)\right)^2 + \frac{\hbar^2}{2m}A^2(\mathbf{r}, t)\left(\nabla S(\mathbf{r})\right)^2.$$

Suppose that both $\nabla S(\mathbf{r}) = 0$ and $\nabla A(\mathbf{r}) = 0$ so that $S(\mathbf{r})$ and $A(\mathbf{r})$ are uniform everywhere. When this is the case the free-energy density is

$$f = f_n + \frac{a}{2}A^2 + \frac{b}{4}A^4. \tag{13.6.3}$$

If a is negative the order parameter has a magnitude given by

$$A = \sqrt{-a/b}.$$

The order parameter is finite below T_c and represents the superfluid state.

Any state which has a phase varying with position must have an energy which is larger than the state described by eqn (13.6.3). A simple calculation shows that the free-energy density is increased by

$$\Delta f = \frac{\hbar^2}{2m}\left(\frac{-a}{b}\right)(\nabla S)^2 + \text{order}\left(\nabla S^4\right) \simeq \frac{\hbar^2}{2m}A^2(\nabla S)^2.$$

We can interpret this extra energy as the kinetic-energy density associated with uniform flow of the liquid.

How then do we describe currents? The simplest way to do this is to pretend that $\Psi(\mathbf{r}, t)$ behaves as if it were a quantum mechanical wave function. If this were the case

$$\rho(\mathbf{r}, t) = |\Psi(\mathbf{r}, t)|^2$$

would represent a probability density; the corresponding probability current would be

$$\mathbf{j} = -\frac{i\hbar}{2m}\left(\Psi^*\nabla\Psi - \Psi\nabla\Psi^*\right).$$

If $\Psi = A(\mathbf{r}, t)e^{iS(\mathbf{r}, t)}$ did indeed represent a single-particle wavefunction then we could write the superfluid mass density as

$$\rho_s(\mathbf{r}, t) = mA^2(\mathbf{r}, t)$$

and the corresponding superfluid mass current as

$$\mathbf{j}_s = -\frac{i\hbar}{2} \left(\Psi^* \nabla \Psi - \Psi \nabla \Psi^* \right) = A^2(\mathbf{r}, t) \hbar \nabla S.$$

By taking the ratio of the two we get the superfluid velocity

$$\mathbf{v}_s = \frac{\mathbf{j}_s}{\rho_s} = \frac{\hbar}{m} \nabla S. \tag{13.6.4}$$

Hence the kinetic-energy density is increased by

$$\Delta f = \frac{1}{2} \rho_s(\mathbf{r}, t) \mathbf{v}_s^2.$$

This term represents the kinetic-energy density associated with the motion of the superfluid.

Why then is superfluid flow stable against decay? The reason is that superfluid flow is a metastable state which cannot decay because of a topological constraint on the complex order parameter. The line integral

$$\int_A^B \nabla S.\mathrm{dr} = \int_A^B \mathrm{d}S = S_B - S_A$$

equals the difference in phase between the end-points. If we make the end-points coincide the path of the line integral becomes a closed contour. The line integral around a closed contour could be zero, or it could be an integral number times 2π

$$\oint_C \nabla S.\mathrm{dr} = n 2\pi$$

where n is called the winding number.

To simplify matters let us consider a ring geometry. Imagine a long, thin, hollow cylinder of helium of diameter d which is bent into a circle of radius $R \gg d$, as shown in Fig. 13.2. We can have states of the system with different winding numbers: $0, \pm 1, \pm 2$, and so on. If the phase is constant everywhere so that the gradient is zero we get the $n = 0$ state; if the phase equals the angle of rotation θ (in radians) around the ring then $n = 1$; if the phase equals 2θ then $n = 2$; and so on. Each of the states with $n \neq 0$ has a larger kinetic-energy density than the $n = 0$ state. They are metastable states.

If the system reaches equilibrium in one of the states with non-zero value of n there is a finite current in the ring. Although this state is metastable, the decay rate is usually so slow as to be unobservable. The only way it can decay is if we can unwind the phase somehow. The way it does this is called phase slippage. The process of phase slippage involves distorting the order parameter leading to an increase in the energy of the system. In effect there is an energy barrier to overcome. It is the presence of the energy barrier which stabilizes the state with a non-zero value of n.

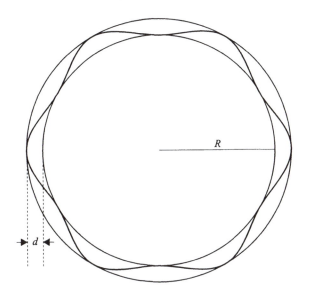

Fig. 13.2 The annulus (or ring), which is made by bending a thin cylinder into a circle of radius R, has a circular cross-section of diameter $d \ll R$. The complex order parameter must be single valued. Nevertheless, the phase can vary by $2\pi n$ (n is an integer) on going once around the ring. The state shown has $n = 6$: we can fit in six wavelengths in the circumference of the ring.

13.7 Circulation

The topological constraint is usual described in terms of the circulation of the fluid,

$$K = \oint \mathbf{v}_S.\mathrm{d}\mathbf{r}.$$

Equation (13.6.4) implies that for any point where $\rho(\mathbf{r},t)$ is non-zero (so that $\mathbf{v}_S$ is well-defined)

$$\mathrm{curl} \mathbf{v}_S = \frac{\hbar}{m}(\mathrm{curl}\nabla S) = 0.$$

The use of Stokes' theorem gives for any closed contour

$$K = \oint \mathbf{v}_S.\mathrm{d}\mathbf{r} = \int\int \mathrm{d}\mathbf{A}.\mathrm{curl} \mathbf{v}_S = 0.$$

It seems that we must have zero circulation in the liquid. However, this conclusion is only valid if the closed contour encloses a region where ρ_s is non-zero everywhere so that $\mathbf{v}_S$ is well-defined throughout. When this is not the case we can do the integral another way:

$$K = \oint_C \mathbf{v}_S.\mathrm{d}\mathbf{r} = \frac{\hbar}{m} \oint_C \nabla S.\mathrm{d}\mathbf{r} = \frac{\hbar}{m} \oint_C \mathrm{d}S.$$

On going around any closed contour, C, the phase of the order parameter can change by an integral number times 2π. Hence

$$K = \oint_C \mathbf{v}_S.\mathrm{d}\mathbf{r} = \frac{\hbar}{m} 2\pi n. \qquad (13.7.1)$$

The circulation is quantized in units of h/m.

If we can shrink the contour to a point with the phase well-defined everywhere we say the system is *simply connected*. In a simply connected geometry with finite $\rho(\mathbf{r}, t)$ everywhere, Stokes' theorem guarantees that $n = 0$ since $\mathrm{curl}\,\mathbf{v}_S = 0$ everywhere. For a ring, however, the contour that goes around the ring cannot be shrunk to a point. Any contour which cannot be shrunk to a point is called *multiply connected*. In a multiply connected region, such as a ring, eqn (13.7.1) is non-trivial; it expresses a topological constraint on the system. The integer number n tells us how many times the phase rotates by 2π on going round the contour C. The circulation in a multiply connected region is quantized in units of h/m.

13.8   Vortices

A rotating bucket of fluid seems to be a simply connected region such that each contour can be shrunk to zero without difficulty. This is the case if the super-fluid density is finite everywhere. However, a rotating bucket of a superfluid can become multiply connected by the introduction of vortex lines running parallel to the axis of rotation (Yarmchuk *et al.* 1979). A *vortex line* is a line singularity on which the superfluid density vanishes. The vortex line cannot just end: either it closes on itself forming a vortex ring, or it joins onto the wall of the container.

The line integral of $\mathbf{v}_S.\mathrm{d}\mathbf{r}$ for a contour which passes around one of these vortex lines gives $\pm h/m$, so that $n = \pm 1$. If $n = 1$ the vortex has positive cir-culation, if $n = -1$ it has negative circulation. States of higher n are unstable and decay into singly quantized vortices of circulation $\pm h/m$. The circulation of a vortex line has been determined in two ways. Vinen (1961) studied the vibra-tions of a fine wire stretched in the liquid parallel to the axis of rotation. The vibrational modes of the wire were affected by the circulation of the fluid around it which enabled the circulation to be measured. Rayfield and Reif (1964) mea-sured the circulation by studying the energy–momentum relationship of vortex rings. Such rings can be created by means of positive or negative ions travelling through the fluid. By measuring the speed of the ring as a function of the energy of the ring, they were able to obtain the circulation. For both experiments the measured values of the circulation show that the mass m is the bare mass of the helium atom.

13.9 Charged superfluids

We can think of superconductivity as the superfluidity of charged particles. For example, superconductivity in metals involves the pairing of electrons; the pair has a mass $m^* = 2m$ and charge $e^* = 2e$. For a superfluid system made of charged particles the definition of superfluid velocity needs to be generalized. The relation between momentum and velocity for a classical particle of charge q is

$$\mathbf{v} = \frac{1}{m^*}\left(\mathbf{p}+q\mathbf{A}\right).\tag{13.9.1}$$

In a quantum treatment we have

$$\mathbf{v}\Psi = \frac{1}{m^*}\left(\frac{\hbar}{i}\nabla+q\mathbf{A}\right)\Psi.$$

Suppose that the order parameter, $\Psi(\mathbf{r}) = A\,e^{iS(\mathbf{r})}$, has an amplitude which is independent of position. If we use the analogy between the wavefunction and the order parameter we are led to the superfluid velocity for particles of charge $q = -e^*$ when

$$\mathbf{v}_s\Psi = \frac{1}{m^*}\left(\frac{\hbar}{i}\nabla+q\mathbf{A}\right)\left(A\,e^{iS}\right) = \frac{\hbar}{m^*}\left(\nabla S - \frac{e^*}{\hbar}\mathbf{A}\right)\Psi.$$

Instead of having simply $\mathbf{v}_s = \hbar\nabla S/m^*$, as we might expect, there is an extra term involving the vector potential. The presence of this term leads to magnetic effects.

It is straightforward to modify the other equations. We get for example

$$\mathrm{curl}\,\mathbf{v}_S = \frac{\hbar}{m^*}\left(\mathrm{curl}\left(\nabla S - \frac{e^*}{\hbar}\mathbf{A}\right)\right) = -\frac{e^*}{m^*}\mathrm{curl}\mathbf{A} = -\frac{e}{m}\mathbf{B}\tag{13.9.2}$$

where we have assumed that the charge is $2e$ and the mass is $2m$. Equation (13.9.2) is known as the *London equation* (London and London 1935). The current $\mathbf{j} = n_s\mathbf{v}_S$ where n_s is the superfluid number density. When n_s is uniform in space, the curl of the London equation gives

$$\mathrm{curl}\,\mathbf{j} = -\frac{n_s e\mathbf{B}}{m}.$$

When this equation is combined with Maxwell's equation

$$\mathrm{curl}\,\mathbf{B} = e\mu_0\mathbf{j}\tag{13.9.3}$$

we get

$$\nabla^2\mathbf{B} = \frac{\mathbf{B}}{\lambda_L^2}\tag{13.9.4}$$

with

$$\lambda_L^2 = \frac{m}{Ne^2\mu_0}.$$

Suppose we apply a magnetic induction field B_e along the y-direction in the half space $x < 0$ just outside and parallel to a flat surface of a superconductor. The solution of eqn (13.9.4) in this case is

$$B_y(x) = B_e \exp(-x/\lambda_L)$$

where x is the distance inside the superconductor measured from the surface. The distance λ_L, the *London penetration depth*, tells us the length-scale over which the magnetic induction decays to zero. The reason for the decay is the presence of currents in the superconductor which screen out the magnetic induction field: this is called the *Meissner effect* (Meissner and Ochesenfeld 1933). A long rod of superconductor lined up to lie parallel to an external field $\mathbf{B}_e$ is a perfect diamagnet because it excludes the field. (For different geometries the demagnetizing field leads to more complex structures called the intermediate state.)

This analysis is only correct if the London equation is applicable, which is the case when $\lambda_L \gg \xi_C$. Such a 'London superconductor' is in fact the exception which is found above all in very impure materials, 'dirty superconductors'. The Meissner effect is then complicated by the penetration of flux in the shape of quantized filaments, the analogues of vortex lines in superfluid liquid helium. Conventional superconductors, such as Hg, Pb, Sn, et cetera, correspond to $\xi_C \gg \lambda_L$; in these materials, Pippard superconductors (Pippard 1953), the theory of the Meissner effect has to be reworked.

13.10   Quantization of flux

Let us consider a ring of superconductor of thickness $\gg \lambda_L$ and take the contour, C, to lie well inside of the superconductor where the magnetic induction is zero. In the region where the magnetic induction is zero, the current, $\mathbf{j} = n_s \mathbf{v}_S$, is zero according to eqn (13.9.3); hence the superfluid velocity is zero on the contour. Thus

$$0 = \oint \mathbf{v}_s.\mathrm{dl} = \oint \frac{\hbar}{m^*}\left(\nabla S - \frac{e^*}{\hbar}\mathbf{A}\right).\mathrm{dl}.$$

It follows that

$$\oint \nabla S.\mathrm{dl} = \oint \frac{e^*}{\hbar}\mathbf{A}.\mathrm{dl} = \frac{e^*}{\hbar}\int\int \mathrm{dS}.\mathbf{B} .$$

Now the line integral $\nabla S.\mathrm{dl}$ is an integer times 2π. Hence the magnetic flux through any area enclosed by the contour is

$$\int\int \mathrm{dS}.\mathbf{B} = \frac{\hbar}{e^*} 2\pi n.$$

The magnetic flux is quantized in units of $h/e^* = \phi_0$ where $\phi_0 = h/2e = 2 \times 10^{-15}$ T m^{-2} is the *elementary flux quantum*. The quantization of magnetic flux through the ring was confirmed in the experiment of Deaver and Fairbank (1961).

13.11 Problems

1. When a certain material undergoes an order–disorder transition the free-energy density, $f[\phi(\mathbf{r})]$, for temperatures somewhat below the transition temperature, T_c, may be written as

$$f[\phi(\mathbf{r})] = -\frac{\alpha}{2}(T_c - T)\phi^2 + \frac{b}{4}\phi^4 + \frac{\lambda}{2}(\nabla\phi)^2$$

where $\phi(x, y, z)$ is a scalar order parameter and α, b, and λ are constants. In a uniformly ordered sample at temperature T, what is the value of the order parameter ϕ_0 which minimizes the free energy? We would like to know the shape of the domain wall between two large regions where $\phi = +\phi_0^*(T)$ and $\phi = -\phi_0^*(T)$ respectively. Consider an order parameter $\phi(x)$ which is just a function of x, and which tends to $\pm\phi_0^*(T)$ as x tends to $\pm\infty$ with the transition region centred on $x = 0$. Calculus of variations gives the equation

$$\lambda\frac{\partial^2\phi}{\partial x^2} = \frac{\delta f[\phi(\mathbf{r})]}{\delta\phi} = -\alpha(T_c - T)\phi + b\phi^3.$$

Show by substitution that $\phi(x)$ can be written as

$$\phi(x) = \phi_0^* \tanh(x/l)$$

and obtain a value for l.

2. Show that the condition

$$R_C = \frac{2\gamma}{\Delta P} \gg \xi_c$$

implies that $P_m \gg \Delta P$.

3. If the thermodynamic potential is of the form

$$\Phi(\phi) = \frac{a}{2}\phi^2 + \frac{c}{3}\phi^3 + \frac{b}{4}\phi^4$$

show that the order parameter is non-zero when $c^2 > 4ab$. Find the spinodal points.

4. We can define a correlation function as

$$G(\mathbf{r}, \mathbf{r}') = \langle\delta\phi(\mathbf{r})\,\delta\phi(\mathbf{r}')\rangle.$$

Use the equipartition theorem to show that

$$G(\mathbf{r}, \mathbf{r}') = \sum_{\mathbf{k}} \frac{k_B T}{V(f'' + \lambda k^2)} e^{i\mathbf{k}\cdot(\mathbf{r}-\mathbf{r}')}.$$

Given that

$$\int_0^\infty \frac{x\sin(ax)}{x^2+\beta^2}\,dx := \frac{\pi}{2}\,e^{-a\beta};\ \text{for}\ a>0\ \text{and}\ \mathrm{Re}(\beta)>0$$

show that for a three-dimensional system which is stable (so that $f'' > 0$) the sum over k, when replaced by an integral, gives

$$G(\mathbf{r},\mathbf{r}') \sim \exp(-\,|\mathbf{r}-\mathbf{r}'|\,/\xi_c).$$

Prove this and obtain an expression for the correlation length ξ_c.
5. Calculate the temperature dependence of the quantity

$$r(T) = f_m\xi_c^d/k_{\mathrm B}T$$

for a two-dimensional $(d=2)$ and a four-dimensional system $(d=4)$ when

$$f(\phi) = \frac{a\phi^2}{2} + \frac{b\phi^4}{4} = \frac{\alpha(T-T_c)\phi^2}{2} + \frac{b\phi^4}{4}.$$

Obtain a formula for the energy barrier, ΔE_C, for nucleation of a new phase in a both a two- and a four-dimensional system.
6. For a charged superfluid the Ginzburg–Landau form of the free energy including the energy of the magnetic induction field $\mathbf{B} = \mathrm{curl}\mathbf{A}$ is

$$\Phi[\Psi(\mathbf{r})] = \int \left(\frac{a}{2}|\Psi(\mathbf{r})|^2 + \frac{b}{4}|\Psi(\mathbf{r})|^4 + \frac{\lambda}{2}\left|\nabla\Psi(\mathbf{r}) - \frac{ie^*}{\hbar}\mathbf{A}(\mathbf{r})\Psi(\mathbf{r})\right|^2 \right) d^3\mathbf{r}$$

$$+ \int \frac{1}{2\mu_0}\,(\mathrm{curl}\,\mathbf{A})^2\,d^3\mathbf{r}.$$

Minimize the total free energy by varying both the order parameter $\Psi(\mathbf{r})$ and the vector potential $\mathbf{A}(\mathbf{r})$ using the calculus of variations to get the two equations

$$a\Psi(\mathbf{r}) + b|\Psi(\mathbf{r})|^2\,\Psi(\mathbf{r}) + \lambda\left(-i\nabla - \frac{e^*}{\hbar}\mathbf{A}(\mathbf{r})\right)^2\Psi(\mathbf{r}) = 0$$

and

$$\mathbf{j} - \frac{ie^*\lambda}{2\hbar}\,(\Psi^*(\mathbf{r})\nabla\Psi(\mathbf{r}) - \Psi(\mathbf{r})\nabla\Psi^*(\mathbf{r})) - \frac{(e^*)^2\,\lambda\,|\Psi(\mathbf{r})|^2\,\mathbf{A}(\mathbf{r})}{\hbar^2} = 0.$$

These are the Ginzburg–Landau equations; they tell us how the order parameter and the current vary in the presence of a magnetic field.

A

Thermodynamics in a magnetic field

Let us start with two of Maxwell's equations:

$$\text{curl}\,\mathbf{H} = \mathbf{J} + \frac{\partial \mathbf{D}}{\partial t} \tag{A.1}$$

and

$$-\,\text{curl}\,\mathbf{E} = \frac{\partial \mathbf{B}}{\partial t} \tag{A.2}$$

where $\mathbf{J}$ is the electrical current density, $\mathbf{E}$ is the electric field, $\mathbf{D}$ is the electric induction, $\mathbf{H}$ is the magnetic field, and $\mathbf{B}$ is the magnetic induction. We take the scalar product of eqn (A.1) with $\mathbf{E}$ and take the scalar product of eqn (A.2) with $\mathbf{H}$; adding the resulting equations together gives

$$\mathbf{E}.\text{curl}\,\mathbf{H} - \mathbf{H}.\text{curl}\,\mathbf{E} = \mathbf{J}.\mathbf{E} + \left(\mathbf{E}.\frac{\partial \mathbf{D}}{\partial t} + \mathbf{H}.\frac{\partial \mathbf{B}}{\partial t} \right). \tag{A.3}$$

According to a standard vector identity, the expression on the left-hand side of eqn (A.3) is equal to $-\text{div}(\mathbf{E} \times \mathbf{H})$. Therefore we can write

$$-\,\mathbf{J}.\mathbf{E} = \text{div}(\mathbf{E} \times \mathbf{H}) + \left(\mathbf{E}.\frac{\partial \mathbf{D}}{\partial t} + \mathbf{H}.\frac{\partial \mathbf{B}}{\partial t} \right). \tag{A.4}$$

We integrate both sides of this equation over the volume of the system and use Gauss' theorem for the term involving $\text{div}(\mathbf{E} \times \mathbf{H})$. For the small time interval δt, this procedure gives

$$-\int \mathbf{J}.\mathbf{E}\,\delta t\;\mathrm{d}V = \oint \mathbf{E} \times \mathbf{H}.\mathrm{d}\mathbf{S}\,\delta t + \int (\mathbf{E}.\mathrm{d}\mathbf{D} + \mathbf{H}.\mathrm{d}\mathbf{B})\,\mathrm{d}V \tag{A.5}$$

where we have defined the quantities $\mathrm{d}\mathbf{D}$ and $\mathrm{d}\mathbf{B}$ by

$$\mathrm{d}\mathbf{D} = \left(\frac{\partial \mathbf{D}}{\partial t} \right)\delta t \tag{A.6a}$$

$$\mathrm{d}\mathbf{B} = \left(\frac{\partial \mathbf{B}}{\partial t} \right)\delta t. \tag{A.6b}$$

Equations (A.4) and (A.5) are two forms of Poynting's theorem which expresses the law of energy conservation in electrodynamics.

For a set of electrons each of charge $-e$ we can write $\mathbf{J}$ as $-n_e e\mathbf{v}$, where n_e is the number density of electrons and $\mathbf{v}$ is their average velocity. The quantity $\mathbf{J}.\mathbf{E} = -n_e e\mathbf{E}.\mathbf{v}$. But $-e\mathbf{E}$ is the force on the electrons, so the quantity $\mathbf{J}.\mathbf{E}\,\delta t$ is the work done by the field on the electrons per unit volume; when written as $-\mathbf{J}.\mathbf{E}\,\delta t$ it is the work done *by* the electrons.

This work can have two effects: the electrons can do work by radiating electromagnetic energy through the bounding surface (this is represented by the term involving a surface integral on the right-hand side of eqn (A.5)); the work can also increase the energy stored by the electromagnetic field (this is represented by the term involving a volume integral on the right-hand side of eqn (A.5)). Provided all changes are done sufficiently slowly that no energy is radiated away, we are only concerned with the second term. It then follows that the electromagnetic work done on the system by the moving charges is

$$\mathrm{d}W_{\mathrm{emag}} = \int (\mathbf{E}.\mathrm{d}\mathbf{D} + \mathbf{H}.\mathrm{d}\mathbf{B})\,\mathrm{d}V. \tag{A.7}$$

For simplicity suppose that there is no electric field present. The change in the free energy, F, of the system in the presence of a magnetic field is

$$\mathrm{d}F = -S\,\mathrm{d}T - P\,\mathrm{d}V + \int \mathbf{H}.\mathrm{d}\mathbf{B}\,\mathrm{d}V. \tag{A.8}$$

The free energy here is expressed in terms of the thermodynamic variables T, V, and $\mathbf{B}$. The difficulty is that we cannot do experiments with constant $\mathbf{B}$, something which cannot be measured or specified, so this is not a convenient choice of thermodynamic variables.

We can determine the magnetic field, $\mathbf{H}$, in materials of simple shapes such as thin cylindrical rods placed coaxially in long solenoidal coils. In this case the magnetic field is exactly equal to the magnetic field generated by the coil. If we can generate a constant current in the coil then we create a constant magnetic field, $\mathbf{H}$. It makes more sense then to specify the magnetic field rather than the magnetic induction, $\mathbf{B}$.

For a long thin rod placed coaxially along a cylindrical coil the proper choice of thermodynamic potential is

$$F_H = F - \int \mathbf{H}.\mathbf{B}\,\mathrm{d}V. \tag{A.9}$$

Then

$$\mathrm{d}F_H = -S\,\mathrm{d}T - P\,\mathrm{d}V - \int \mathbf{B}.\mathrm{d}\mathbf{H}\,\mathrm{d}V. \tag{A.10}$$

We can then work out the thermodynamics keeping the magnetic field constant.

Suppose the volume of a long cylindrical rod is held constant and that both **B** and **H** are uniform and act along the z direction, parallel to the rod. In this case we can put $\mathbf{B} = \mu_0 \left(\mathbf{H} + \mathbf{M}\right)$ with **M** the magnetization and find

$$\mathrm{d}F_H = -S\,\mathrm{d}T - V\mu_0\left(H\mathrm{d}H + \mathbf{M}.\mathrm{d}\mathbf{H}\right),$$

where V is the volume of the rod. The change in free energy associated with the magnetic field, $V\mu_0 H\mathrm{d}H$, is independent of the magnetization of the rod; this term can be omitted. We then find that the change in the free energy is

$$\mathrm{d}F_H = -S\,\mathrm{d}T - V\mu_0\mathbf{M}.\mathrm{d}\mathbf{H}, \tag{A.11}$$

from which it follows that

$$\mathbf{M} = -\frac{1}{V\mu_0}\left(\frac{\partial F_H}{\partial \mathbf{H}}\right)_T. \tag{A.12}$$

The magnetic susceptibility for an isotropic medium is therefore

$$\chi_M = \left(\frac{\partial M}{\partial H}\right)_T = -\frac{1}{V\mu_0}\left(\frac{\partial^2 F_H}{\partial H^2}\right)_T. \tag{A.13}$$

The whole calculation has been based on the idea that the sample is a long thin rod. If the sample has a different shape (say it is spherical) a demagnetizing field $\mathbf{H}_d$ is created which gives a contribution to the total magnetic field $\mathbf{H} = \mathbf{H}_j + \mathbf{H}_d$; the total magnetic field is not just the field $\mathbf{H}_j$ created by the current in the coil. We can keep the current in the coil fixed but that does not correspond to constant **H** because the demagnetizing field can change as the sample becomes magnetized. This makes the choice of the proper thermodynamic potential more complicated for geometries other than a long cylindrical rod.

B

Useful integrals

The integral

$$I_0 = \int_0^\infty e^{-x^2} dx \tag{B.1}$$

can be calculated by a trick. Write I_0 as

$$I_0 = \int_0^\infty e^{-y^2} dy$$

and then multiply the two expressions for I_0 together; this gives

$$I_0^2 = \int_0^\infty \int_0^\infty e^{-x^2} e^{-y^2} dx\, dy.$$

If we now convert to polar coordinates (r, θ) instead of Cartesian coordinates (x, y) we get

$$I_0^2 = \int_0^{\pi/2} d\theta \int_0^\infty e^{-r^2} r\, dr = \frac{\pi}{4}.$$

Hence

$$I_0 = \frac{\pi^{1/2}}{2}.$$

There are many integrals which can be solved now that we know I_0. Let us define a set of integrals as

$$I_{2n} = \int_0^\infty x^{2n} e^{-x^2} dx \tag{B.2}$$

with n an integer. For example

$$I_2 = \int_0^\infty x^2 e^{-x^2} dx,$$

an integral which can be done by parts. We get

$$I_2 = \frac{I_0}{2}.$$

The integrals I_4, I_6, I_8, and the rest can be done by iteration. For I_{2n} we can show by integration by parts that

$$I_{2n} = \left(\frac{2n-1}{2}\right) I_{2n-2},$$

and by iterating we get

$$I_{2n} = \left(\frac{2n-1}{2}\right) \times \left(\frac{2n-3}{2}\right) \times \cdots \times \frac{1}{2} I_0. \tag{B.3}$$

For example $I_8 = 105\pi^{1/2}/32$.

The class of integrals

$$I_{2n+1} = \int_0^\infty x^{2n+1} e^{-x^2} dx$$

with n integer can be done in the same way. We use integration by parts to get the relation

$$I_{2n+1} = n \, I_{2n-1}.$$

Hence by iterating we get

$$I_{2n+1} = n! I_1.$$

But

$$I_1 = \int_0^\infty x \, e^{-x^2} dx = \frac{1}{2},$$

hence

$$I_{2n+1} = \frac{n!}{2}. \tag{B.4}$$

For example $I_5 = 2!/2 = 1$.

A more elegant way of writing I_{2n+1} is in terms of the *gamma function* which is defined by

$$\Gamma(m+1) = \int_0^\infty y^m e^{-y} dy . \tag{B.5}$$

By substituting $y = x^2$ we get

$$\Gamma(m+1) = 2 \int_0^\infty x^{2m+1} e^{-x^2} dx = 2I_{2m+1}.$$

The gamma function is defined for all positive values of m, not just integers or rational numbers.

Another set of integrals that is useful is

$$J_m = \int_0^\infty \frac{x^m}{e^x - 1} \, dx.$$

The internal energy associated with black body radiation involves J_3. These integrals can be evaluated by expanding the denominator as a power series in e^{-x} to give

$$
\begin{aligned}
J_m &= \int_0^\infty x^m \left(e^{-x} + e^{-2x} + e^{-3x} + \cdots\right) dx \\
&= \Gamma(m+1) \sum_{n=1}^\infty \frac{1}{n^{m+1}}. \tag{B.6}
\end{aligned}
$$

This sum can easily be found using a pocket calculator. For example, by summing up a few terms we get for $m = 3$

$$
\begin{aligned}
J_m &= 3! \left(1 + \frac{1}{16} + \frac{1}{81} + \frac{1}{256} + \cdots\right) \\
&= 6.493939 = \frac{\pi^4}{15}.
\end{aligned}
$$

The sum

$$\zeta(m) = \sum_{n=1}^\infty \frac{1}{n^m} \tag{B.7}$$

is called the Reimann zeta function. An elegant way of writing J_m is

$$J_m = \Gamma(m+1)\zeta(m+1).$$

Another related integral occurs in the determination of the critical temperature for the onset of Bose condensation in a non-interacting Bose gas (eqn (10.5.3)). It is

$$\int_0^\infty \frac{x^2}{e^{x^2} - 1} \, dx = I_2 \sum_{n=1}^\infty \frac{1}{n^{3/2}}. \tag{B.8}$$

Here we have expanded $\left(e^{x^2} - 1\right)^{-1}$ as a power series in e^{-x^2}. The sum, which can be evaluated using a pocket calculator, is 1.158.

Sometimes summations can be approximated by an integral with little error. Suppose that the function $f(n)$ varies continuously with n. The area under the curve of $f(n + x)$ with x going from $-1/2$ to $1/2$ is

$$\int_{-1/2}^{1/2} f(n+x) \, dx = \int_{-1/2}^{1/2} \left\{ f(n) + x \frac{df(n)}{dn} + \frac{x^2}{2} \frac{d^2 f(n)}{dn^2} + \cdots \right\} dx$$

$$= f(n) + \left[\frac{x^2}{2} \frac{df(n)}{dn} \right]_{-1/2}^{1/2} + \left[\frac{x^3}{6} \frac{d^2 f(n)}{d^2 n} \right]_{-1/2}^{1/2}$$

$$= f(n) + \frac{1}{24} \frac{d^2 f(n)}{d^2 n}.$$

If $d^2 f(n)/d^2 n$ is very small compared to $f(n)$ then the integral can be taken to be $f(n)$. Hence we can replace $f(n)$ by

$$f(n) \simeq \int_{n-1/2}^{n+1/2} f(x)\, dx.$$

The sum of $f(n)$ with n integer can be approximated by an integral

$$\sum_{n=0}^{\infty} f(n) \approx \int_{-1/2}^{\infty} f(x)\, dx. \qquad \text{(B.9)}$$

Suppose $f(n) = (2n+1)\, e^{-\alpha n(n+1)}$. Then

$$\frac{d^2 f(n)}{d^2 n} = \{ \alpha^2 (2n+1)^2 - 6\alpha \} f(n).$$

For $\alpha \ll 1$ the approximation is valid and we get

$$\sum_{n=0}^{\infty} (2n+1)\, e^{-\alpha n(n+1)} \approx \int_{-1/2}^{\infty} (2x+1)\, e^{-\alpha x(x+1)}\, dx$$

$$\approx \frac{1}{\alpha}.$$

Provided α is very small the sum can be replaced by an integral with little error.

C

The quantum treatment of a diatomic molecule

Consider two particles of masses m_1 and m_2 held together by an attractive central potential energy $V(r)$. The Hamiltonian which describes the motion of about the centre of mass is

$$\hat{H} = -\frac{\hbar^2}{2\mu}\nabla^2 + V(r) \tag{C.1}$$

where μ is the reduced mass given by

$$\frac{1}{\mu} = \frac{1}{m_1} + \frac{1}{m_2}. \tag{C.2}$$

In spherical coordinates the Laplacian operator is

$$\nabla^2 = \frac{\partial^2}{\partial r^2} + 2\frac{\partial}{\partial r} + \frac{1}{r^2}\left(\frac{\partial^2}{\partial\theta^2} + \cot\theta\frac{\partial}{\partial\theta} + \frac{1}{\sin^2\theta}\frac{\partial^2}{\partial\phi^2}\right). \tag{C.3}$$

We can represent the square of the orbital angular momentum operator as

$$\hat{l}^2 = -\left(\frac{\partial^2}{\partial\theta^2} + \cot\theta\frac{\partial}{\partial\theta} + \frac{1}{\sin^2\theta}\frac{\partial^2}{\partial\phi^2}\right). \tag{C.4}$$

The eigenstates of $\hat{l}^2$, which are called spherical harmonic functions $Y_{l,m}(\theta,\phi)$, satisfy the eigenvalue equations

$$\hat{l}^2 Y_{l,m}(\theta,\phi) = l(l+1)Y_{l,m}(\theta,\phi) \tag{C.5}$$

and

$$\hat{l}_z Y_{l,m}(\theta,\phi) = mY_{l,m}(\theta,\phi). \tag{C.6}$$

The single-particle wavefunction can be written as

$$\Phi(r,\theta,\phi) = \frac{R(r)Y_{l,m}(\theta,\phi)}{r} \tag{C.7}$$

where the radial wavefunction $R(r)$ satisfies the differential equation

$$-\frac{\hbar^2}{2\mu}\frac{d^2R(r)}{dr^2} + \frac{\hbar^2 l(l+1)R(r)}{2\mu r^2} + V(r)R(r) = \varepsilon R(r). \tag{C.8}$$

This equation is similar to a one-dimensional Schrödinger equation in which the effective potential is

$$U(r) = V(r) + \frac{\hbar^2 l(l+1)}{2\mu r^2}. \tag{C.9}$$

The second term on the right-hand side of eqn (C.9) can be thought of as a centrifugal potential energy.

For a molecule, such as carbon monoxide, the potential $V(r)$ has a deep minimum at a separation $r = d$. To a good approximation we can treat the centrifugal potential energy as if the particles were a constant distance d apart. Then

$$U(r) \approx V(r) + \frac{\hbar^2 l(l+1)}{2\mu d^2}. \tag{C.10}$$

The second term on the right-hand side is the kinetic energy of rotation about the centre of mass. We can treat μd^2 as the moment of inertia, I, of the molecule. We rewrite eqn (C.8) as

$$-\frac{\hbar^2}{2\mu}\frac{d^2R(r)}{dr^2} + \frac{\hbar^2 l(l+1)R(r)}{2\mu d^2} + V(r)R(r) \approx \varepsilon R(r). \tag{C.11}$$

The term $\hbar^2 l(l+1)/2\mu d^2$ just shifts the energy by a constant amount and so it can be removed. The total energy eigenvalue is written as

$$\varepsilon = \varepsilon_n + \frac{\hbar^2 l(l+1)}{2\mu d^2}. \tag{C.12}$$

The solution of the resulting equation then yields a set of eigenstates $R_{n,l}(r)$ and eigenvalues ε_n which correspond to vibrational motion in the potential well. The total energy is the sum of rotational and vibrational energies. For r close to d, the potential varies as $(r-d)^2$, so the vibrational state is that of a simple harmonic oscillator. The energy eigenvalue, $\varepsilon_n = \hbar\omega(n+1/2)$, only depends on n in this approximation. The total energy eigenvalue is

$$\varepsilon_{n,l} = \hbar\omega(n+1/2) + \frac{\hbar^2 l(l+1)}{2\mu d^2}. \tag{C.13}$$

The splitting of the energy levels of rotation is equal to $\hbar^2(l+1)/\mu d^2$ between the l and $(l+1)$ states. The splitting can be observed by spectroscopic techniques. For carbon monoxide this splitting in energy is consistent with the observed spectra if $\hbar/2\pi\mu d^2 = 115.2\,\text{GHz}$.

There are additional complications when the particles in the molecule are identical, as happens in molecular hydrogen. Suppose the two identical atoms are in a state of definite orbital angular momentum, such as

$$\Phi_{n,l,m}(r,\theta,\phi) = \frac{R_{n,l}(r)Y_{l,m}(\theta,\phi)}{r} \tag{C.14}$$

where r is the distance of one atom from the centre of mass. If now $\mathbf{r}$ is changed to $-\mathbf{r}$ the two particles are exchanged. How does the wavefunction change under this transformation?

In spherical coordinates the transformation is equivalent to

$$(r,\theta,\phi) \rightarrow (r, \pi - \theta, \pi + \phi). \tag{C.15}$$

Under this transformation the radial function is unchanged whereas the spherical harmonic function transforms as

$$Y_{l,m}(\pi - \theta, \pi + \phi) \rightarrow (-1)^l \, Y_{l,m}(\theta,\phi). \tag{C.16}$$

In other words the wavefunction changes sign on interchange of particles if l is an odd integer but keeps the same sign if l is an even integer. For example,

$$\begin{aligned}
Y_{1,1}(\theta,\phi) &= A\sin(\theta)\,e^{i\phi} \rightarrow A\sin(\pi - \theta)\,e^{i\pi}\,e^{i\phi} \\
&= -A\sin(\theta)\,e^{i\pi}\,e^{i\phi} = -Y_{1,1}(\theta,\phi).
\end{aligned}$$

The wavefunction changes sign for $l = 1, m = 1$.

D

Travelling waves

D.1 Travelling waves in one dimension

In one dimension a travelling wave is of the form

$$\phi(x) = A\,e^{i(kx - \omega t)} \tag{D.1}$$

where A is a constant, k is the wave vector, and ω is the angular frequency. Normally we treat classical waves as real quantities and describe them using sine and cosine functions, but in quantum mechanics we use a complex wavefunction.

According to de Broglie, the momentum associated with such a travelling wave in one dimension is given by

$$p = \frac{h}{\lambda} = \hbar k \tag{D.2}$$

where $\hbar = h/2\pi$ and $k = 2\pi/\lambda$. In one dimension the momentum, $\hbar k$, is a simple scalar quantity.

A problem arises when we try to normalize travelling waves. The wavefunction must satisfy the condition

$$\int |\phi(x)|^2\,dx = 1. \tag{D.3}$$

What are the limits on the integral? Suppose that the range of integration is from $-\infty$ to ∞. Then

$$\int_{-\infty}^{\infty} |\phi(x)|^2\,dx \;=\; \int_{-\infty}^{\infty} |A|^2\,dx$$

$$=\; 2\,|A|^2\,\infty = 1.$$

The integral is then infinite, which is nonsense.

Instead, let us localize the particle in the region $0 < x < L$, for then we can normalize the wavefunction. With this restriction we find that $A = L^{-1/2}$ and the wavefunction is finite.

We have solved one problem but have generated another: if we normalize over a finite range we need to specify the boundary conditions on the wavefunction. We need to specify the boundary conditions at $x = 0$ and at $x = L$ in such a way that we are left with a travelling wave solution. It is no good demanding that the wavefunction vanishes at $x = 0$ and at $x = L$, for if we were to make such a choice then we would set up standing waves which do not have a well-defined value of the momentum. Which boundary conditions should we choose?

The solution is to impose periodic boundary conditions. The wave is reflected at the end of the box in a strange way such that

$$\phi(x = L) = \phi(x = 0). \tag{D.4}$$

This implies that the wave repeats periodically, and so it can be written as a Fourier series

$$\phi(x) = \sum_n c_n e^{2\pi i n x/L} \tag{D.5}$$

where n is an integer. It then follows that $\phi(x)$ satisfies the periodic boundary condition

$$\phi(L_x) = \sum_n c_n e^{2\pi i n L/L} = \sum_n c_n = \phi(0) \tag{D.6}$$

since $e^{2\pi i n} = 1$.

The function $\phi(x)$ can be expanded in a complete set of normalized eigenfunctions; we choose them to be eigenfunctions of the momentum operator. Normalized eigenfunctions of the momentum operator which satisfy the boundary conditions are

$$\phi_k(x) = \frac{e^{2\pi i n x/L}}{L^{1/2}} = \frac{e^{ikx}}{L^{1/2}} \tag{D.7}$$

where $k = 2\pi n/L$. The wavefunctions are eigenfunctions of the momentum operator, $\hat{p}_x = -i\hbar\partial/\partial x$, with momentum eigenvalues $\hbar k = nh/L$. The momentum eigenvalues are quantized. Successive eigenvalues are spaced apart by h/L. The quantization of momentum eigenvalues is a consequence of the periodic boundary conditions.

To calculate the density of states in **k** space in one dimension we need the number of states in the range k to $k + dk$. In this range there are $(L/2\pi)\,dk$ states so the density of states is

$$D(k)\,dk = \frac{L_x}{2\pi}\,dk. \tag{D.8}$$

Notice that the range of k values extends from $-\infty$ to ∞ for travelling waves.

D.2   Travelling waves in three dimensions

The same technique can be used for travelling waves in three dimensions. We choose a rectangular box of lengths $L_x, L_y,$ and L_z. The periodic boundary conditions for the wavefunction $\phi(x, y, z)$ are

$$\phi(0,0,0) = \phi(L_x,0,0) = \phi(0, L_y, 0) = \phi(0, 0, L_z). \tag{D.9}$$

The wavefunction is periodic in all three directions. The wavefunction can be written as a triple Fourier series

$$\phi(x, y, z) = \sum_{n_1} c_{n_1, n_2, n_3} \, e^{2\pi i n_1 x/L_x} e^{2\pi i n_2 y/L_y} e^{2\pi i n_3 z/L_z} \tag{D.10}$$

with n_1, n_2, and n_3 integers. The normalized wavefunctions are

$$\phi_{n_1, n_2, n_3}(x, y, z) = \frac{e^{i(k_x x + k_y y + k_z z)}}{(L_x L_y L_z)^{1/2}} \tag{D.11}$$

where now

$$k_x = \frac{2\pi n_1}{L_x}, \qquad k_y = \frac{2\pi n_2}{L_y}, \qquad k_z = \frac{2\pi n_3}{L_z}. \tag{D.12}$$

We can construct a vector **k** as

$$\mathbf{k} = \hat{\mathbf{i}} k_x + \hat{\mathbf{j}} k_y + \hat{\mathbf{k}} k_z \tag{D.13}$$

where $\hat{\mathbf{k}}$ is a unit vector in the z direction. The position vector **r** is given by

$$\mathbf{r} = \hat{\mathbf{i}} x + \hat{\mathbf{j}} y + \hat{\mathbf{k}} z \tag{D.14}$$

so that the wavefunction can be written as

$$\phi_k(x, y, z) = \frac{e^{i\mathbf{k}.\mathbf{r}}}{(L_x L_y L_z)^{1/2}}. \tag{D.15}$$

This wavefunction is an eigenfunction of the momentum operator with eigenvalue $\hbar\mathbf{k}$.

We can construct a set of points in **k** space with particular values of $k_x, k_y,$ and k_z to represent these single-particle states. Each point in **k** space occupies a volume

$$\Delta V_k = \frac{(2\pi)^3}{L_x L_y L_z}.$$

The density of single-particle states in **k** space is

$$\frac{1}{\Delta V_k} = \frac{L_x L_y L_z}{(2\pi)^3} = \frac{V}{(2\pi)^3}.$$

We also want $D(k)$, the number of states in the range k to $k + dk$. The volume of **k** space in this range is $4\pi k^2 \, dk$. The density of points in **k** space is $V/(2\pi)^3$ so $D(k)$ is

$$D(k) \, dk = \frac{V}{(2\pi)^3} 4\pi k^2 \, dk = \frac{V k^2}{2\pi^2} dk. \tag{D.16}$$

E

Partial differentials and thermodynamics

The ability to manipulate *partial differentials* is a valuable skill when analysing thermodynamic problems. Here we review four different aspects: mathematical relations, Maxwell relations, the $T\,\mathrm{d}S$ equations, and relations that come from the fact that the entropy is proportional to the number of particles.

E.1 Mathematical relations

If three variables x, y, and z, are related in such a way that $F(x, y, z) = 0$ we can, in principle, write one variable as a function of the other two. For example, we could write $x = x(y, z)$ or $z = z(x, y)$. Differentiating $x = x(y, z)$ by parts gives

$$\mathrm{d}x = \left(\frac{\partial x}{\partial y}\right)_z \mathrm{d}y + \left(\frac{\partial x}{\partial z}\right)_y \mathrm{d}z \qquad (\text{E1})$$

where the terms in brackets are partial derivatives of x. Differentiating $z = z(x, y)$ by parts gives

$$\mathrm{d}z = \left(\frac{\partial z}{\partial x}\right)_y \mathrm{d}x + \left(\frac{\partial z}{\partial y}\right)_x \mathrm{d}y. \qquad (\text{E2})$$

If we substitute the expression for $\mathrm{d}z$ back into eqn (E1) we get

$$
\begin{aligned}
\mathrm{d}x &= \left(\frac{\partial x}{\partial y}\right)_z \mathrm{d}y + \left(\frac{\partial x}{\partial z}\right)_y \left(\left(\frac{\partial z}{\partial x}\right)_y \mathrm{d}x + \left(\frac{\partial z}{\partial y}\right)_x \mathrm{d}y\right) \\
&= \left(\frac{\partial x}{\partial z}\right)_y \left(\frac{\partial z}{\partial x}\right)_y \mathrm{d}x + \left(\left(\frac{\partial x}{\partial y}\right)_z + \left(\frac{\partial x}{\partial z}\right)_y \left(\frac{\partial z}{\partial y}\right)_x\right) \mathrm{d}y.
\end{aligned}
$$

This relationship is always true no matter what values are taken for $\mathrm{d}x$ and $\mathrm{d}y$. If we take the value $\mathrm{d}y = 0$, and $\mathrm{d}x \neq 0$, we have

$$1 = \left(\frac{\partial x}{\partial z}\right)_y \left(\frac{\partial z}{\partial x}\right)_y,$$

which can be written as

$$\left(\frac{\partial x}{\partial z}\right)_y = \frac{1}{\left(\frac{\partial z}{\partial x}\right)_y}. \tag{E3}$$

Equation (E3) is called the reciprocal theorem.

If we take $dx = 0$, and $dy \neq 0$, we have

$$\left(\frac{\partial x}{\partial y}\right)_z + \left(\frac{\partial x}{\partial z}\right)_y \left(\frac{\partial z}{\partial y}\right)_x = 0$$

which can be expressed, using the reciprocal theorem, as

$$\left(\frac{\partial x}{\partial z}\right)_y \left(\frac{\partial z}{\partial y}\right)_x \left(\frac{\partial y}{\partial x}\right)_z = -1. \tag{E4}$$

Equation (E4) is called the reciprocity relation.

E.2 Maxwell relations

In Chapter 1 we showed that any small change in the function of state $G = g(x, y)$ is given by the equation

$$dG = \left(\frac{\partial g(x, y)}{\partial x}\right)_y dx + \left(\frac{\partial g(x, y)}{\partial y}\right)_x dy \tag{1.4.7}$$

which can be rewritten as:

$$dG = A(x, y)\,dx + B(x, y)\,dy, \tag{1.4.8}$$

where $A(x, y) = (\partial g(x, y)/\partial x)_y$ and $B(x, y) = (\partial g(x, y)/\partial y)_x$.

Suppose we want to differentiate $G = g(x, y)$ with respect to x and y; if G is an analytic function of x and y the order of differentiation does not matter so that

$$\left(\frac{\partial}{\partial x}\left(\frac{\partial G}{\partial y}\right)_x\right)_y = \left(\frac{\partial}{\partial y}\left(\frac{\partial G}{\partial x}\right)_y\right)_x.$$

It follows that

$$\left(\frac{\partial B(x, y)}{\partial x}\right)_y = \left(\frac{\partial A(x, y)}{\partial y}\right)_x. \tag{1.4.10}$$

For example, when we write $U(S, V)$ as a differential, we can put $S = x$ and $V = y$ and get

$$dU(S, V) = \left(\frac{\partial U}{\partial S}\right)_V dS + \left(\frac{\partial U}{\partial V}\right)_S dV = T\,dS - P\,dV.$$

Then we find

$$\left(\frac{\partial T}{\partial V}\right)_S = -\left(\frac{\partial P}{\partial S}\right)_V. \tag{E5}$$

This is the first of four *Maxwell relations*. The other Maxwell relations can be found from expressions for the enthalpy, the Helmholtz free energy, and the Gibbs free energy. Using eqns (1.7.2) and (2.5.10) we obtain the relationship

$$dH(S, P) = \left(\frac{\partial H}{\partial S}\right)_P dS + \left(\frac{\partial H}{\partial P}\right)_S dP = T\,dS + V\,dP,$$

from which we get the second Maxwell relation

$$\left(\frac{\partial T}{\partial P}\right)_S = \left(\frac{\partial V}{\partial S}\right)_P. \tag{E6}$$

The Helmholtz free energy is

$$dF(T, V) = \left(\frac{\partial F}{\partial T}\right)_V dT + \left(\frac{\partial F}{\partial V}\right)_T dV = -S\,dT - P\,dV \tag{2.6.9}$$

from which we get the third Maxwell relation

$$\left(\frac{\partial S}{\partial V}\right)_T = \left(\frac{\partial P}{\partial T}\right)_V. \tag{E7}$$

Finally, the Gibbs free energy is

$$dG(T, P) = \left(\frac{\partial G}{\partial T}\right)_P dT + \left(\frac{\partial G}{\partial P}\right)_T dP = -S\,dT + V\,dP \tag{2.6.11}$$

from which we get the fourth Maxwell relation

$$-\left(\frac{\partial S}{\partial P}\right)_T = \left(\frac{\partial V}{\partial T}\right)_P. \tag{E8}$$

If the work is not $-P\,dV$ but, say, $F\,dl$ for a stretched string, we can either repeat the analysis or replace $-P$ with F, and replace V with l everywhere in the above relations. The first Maxwell relation, for example, becomes

$$\left(\frac{\partial T}{\partial l}\right)_S = \left(\frac{\partial F}{\partial S}\right)_l.$$

E.3   TdS relations

We can write the entropy as a function of T and V so that

$$dS(T, V) = \left(\frac{\partial S}{\partial T}\right)_V dT + \left(\frac{\partial S}{\partial V}\right)_T dV.$$

By using the third Maxwell relation and eqn (2.5.11) we get

$$T\,dS = T\left(\frac{\partial S}{\partial T}\right)_V dT + T\left(\frac{\partial P}{\partial T}\right)_V dV,$$

$$= C_V \, dT + T \left(\frac{\partial P}{\partial T} \right)_V dV.$$

This is the first $T \, dS$ relation.

The second $T \, dS$ relation is found by writing the entropy as a function of T and P.

$$dS(T, P) = \left(\frac{\partial S}{\partial T} \right)_P dT + \left(\frac{\partial S}{\partial V} \right)_T dP.$$

Using the fourth Maxwell relation and eqn (2.5.12) we get

$$T \, dS = T \left(\frac{\partial S}{\partial T} \right)_P dT - T \left(\frac{\partial V}{\partial T} \right)_P dP,$$
$$= C_P \, dT - T \left(\frac{\partial V}{\partial T} \right)_P dP.$$

E.4 Extensive quantities

The entropy is an *extensive quantity*, which means that it is proportional to the number of particles, N, in the system. Other extensive quantities for a pure system are U, V, and N. For a mixture

$$N = \sum_i N_i$$

where N_1 is the number of particles of species 1, N_2 is the number of particles of species 2, and so on. In this case all the N_i are taken to be extensive, that is proportional to N.

Let us write the entropy, $S(U, V, N_1, N_2, \ldots)$, as a function of all the extensive quantities. Suppose we could increase N by a small fraction so that

$$N \Rightarrow N(1 + \epsilon) = N + \epsilon N = N + dN.$$

Thus if ϵ is very small $dN = \epsilon N$ represents the 'small' change in the number of particles. Similarly

$$U \Rightarrow U(1 + \epsilon) = U + \epsilon U = U + dU,$$
$$V \Rightarrow V(1 + \epsilon) = V + \epsilon V = V + dV,$$
$$S \Rightarrow S(1 + \epsilon) = S + \epsilon S = S + dS.$$

We can get another expression for dS. As the number of particles increases the entropy must change as

$$S(U, V, N_1, N_2, \ldots) \Rightarrow S(U(1 + \epsilon), V(1 + \epsilon), N_1(1 + \epsilon), N_2(1 + \epsilon), \ldots)$$
$$= S(U + dU, V + dV, N_1 + dN_1, N_2 + dN_2, \ldots).$$

By making a Taylor expansion to first order in ϵ we find that the change in entropy is

$$
\begin{aligned}
\mathrm{d}S &= \left(\frac{\partial S}{\partial U}\right)_{V,N_i} \mathrm{d}U + \left(\frac{\partial S}{\partial V}\right)_{U,N_i} \mathrm{d}V + \sum_i \left(\frac{\partial S}{\partial N_i}\right)_{U,V,N_j} \mathrm{d}N_i \\
&= \left(\frac{\partial S}{\partial U}\right)_{V,N_i} U\epsilon + \left(\frac{\partial S}{\partial V}\right)_{U,N_i} V\epsilon + \sum_i \left(\frac{\partial S}{\partial N_i}\right)_{U,V,N_j} N_i\epsilon.
\end{aligned}
$$

Equating $\mathrm{d}S = \epsilon S$ to the expression above and cancelling ϵ gives

$$
S = U\left(\frac{\partial S}{\partial U}\right)_{V,N_i} + V\left(\frac{\partial S}{\partial V}\right)_{U,N_i} + \sum_i N_i \left(\frac{\partial S}{\partial N_i}\right)_{U,V,N_j}.
$$

When we express these partial derivatives in terms of thermodynamic quantities we get

$$
TS = U + PV - \sum_i \mu_i N_i.
$$

This relation comes from the requirement that the entropy is extensive. For a pure system it gives

$$
\mu N = U + PV - TS = G.
$$

The chemical potential is the Gibbs free energy per particle. Then by taking differentials of this equation we get the *Gibbs–Duhem relation*

$$
N \, \mathrm{d}\mu = -S \, \mathrm{d}T + V \, \mathrm{d}P.
$$

For a two-component system we get the *Euler equation*

$$
\mu_1 N_1 + \mu_2 N_2 = U + PV - TS = G.
$$

This says that for a pure system the Gibbs free energy is the sum of the chemical potentials of all the constituent particles. The same result is found when there are more than two constituents.

F
Van der Waals equation

In this appendix we use the techniques of classical statistical mechanics, as described in section 5.14, to derive the partition function of the van der Waals gas. There we argued that the partition function for a single particle in one dimension is given by

$$Z = \frac{1}{h} \int_{-\infty}^{\infty} \mathrm{d}x \int_{-\infty}^{\infty} \mathrm{d}p_x \, e^{-E(p_x, x)/k_B T}. \tag{5.14.1}$$

If the particle moves in three dimensions the partition function becomes

$$Z = \frac{1}{h^3} \int \mathrm{d}^3\mathbf{r} \int \mathrm{d}^3\mathbf{p} \, e^{-E(\mathbf{p}, \mathbf{r})/k_B T}, \tag{F.1}$$

where $\mathrm{d}^3\mathbf{r}$ represents integration over all space, and $\mathrm{d}^3\mathbf{p}$ represents integration over all possible values of momentum.

We adapt this approach to derive an approximate expression for the free energy of a gas of 'hard sphere' molecules which interact with each other via a

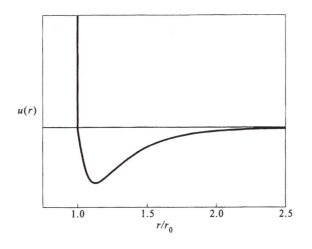

Fig. F.1 The model potential with a hard core.

potential of the form

$$u(r) \;=\; \infty \qquad \text{when } r < r_0 \qquad\qquad\qquad\qquad \text{(F.2)}$$
$$u(r) \;<\; 0 \qquad \text{when } r > r_0. \qquad\qquad\qquad\qquad \text{(F.3)}$$

For r less that r_0 the potential is infinite which means that the particles cannot penetrate inside this region; for r greater than r_0 the potential is negative indicating an attractive interaction between the particles. By making a few simple approximations we can generate the Helmholtz free energy which gives rise to van der Waals theory of gases, a theory which predicts the liquid–gas phase transition.

Consider a monatomic gas made up of N identical particles each of mass m. The gas is kept at a high temperature T in a volume V. We can calculate the thermodynamics of such a gas in a classical approximation from the partition function

$$Z_N = \frac{1}{N! h^{3N}} \int \cdots \int d^3 p_1 \, d^3 p_2 \cdots d^3 p_N \, d^3 r_1 \, d^3 r_2 \cdots d^3 r_N \, e^{-H/k_B T} \qquad \text{(F.4)}$$

with H the classical Hamiltonian

$$\hat{H} = \sum_{i=1}^{N} \frac{p_i^2}{2m} + U(\mathbf{r}_1, \mathbf{r}_2, \ldots, \mathbf{r}_N). \qquad\qquad \text{(F.5)}$$

$U(\mathbf{r}_1, \mathbf{r}_2, \ldots, \mathbf{r}_N)$ is the potential energy of the particles, $\sum_{i=1}^{N} p_i^2/2m$ the sum of the kinetic energies. The factor $(N!)^{-1}$ arises because the particles are indistinguishable.

Integration over each momentum variable is best done in Cartesian coordinates; for example, the x-component of momentum involves the integral

$$\int_{-\infty}^{\infty} dp_x \, e^{-p_x^2/2mk_B T} = (2\pi m k_B T)^{1/2} \,.$$

In this way we get

$$Z_N = \frac{1}{N!} \left(\frac{2\pi m k_B T}{h^2} \right)^{3N/2} \int \cdots \int d^3 r_1 \, d^3 r_2 \cdots d^3 r_N \, e^{-U(\mathbf{r}_1, \mathbf{r}_2, \ldots, \mathbf{r}_N)/k_B T}.$$
$$\text{(F.6)}$$

In the limit that the ratio $U(\mathbf{r}_1, \mathbf{r}_2, \ldots, \mathbf{r}_N)/k_B T$ tends to zero we recover the expression for the partition function of an ideal gas. We want to do better than this and take account of the interaction between particles.

Let us suppose that the potential energy is the sum of the interaction potentials between pairs of atoms:

$$U(\mathbf{r}_1, \mathbf{r}_2, \ldots, \mathbf{r}_N) = \frac{1}{2} \sum_{i \neq j} u\left(r_{ij}\right) \qquad\qquad \text{(F.7)}$$

where $r_{ij} = |\mathbf{r}_i - \mathbf{r}_j|$ and $u(r)$ is given by eqn (F.2).

We want to evaluate the multiple integral

$$I = \int \cdots \int d^3\mathbf{r}_1 \, d^3\mathbf{r}_2 \cdots d^3\mathbf{r}_N \, e^{-\sum_{i \neq j} u(r_{ij})/2k_B T} \qquad \text{(F.8)}$$

To make progress we must make approximations. Let us suppose that the particles are distributed randomly throughout the volume and that none lie within r_0 of any other particle. If $r < r_0$ the potential is infinite so that no particles can enter the region. The volume excluded around each particle is $\alpha = 4\pi r_0^3/3$ so we can write

$$I = (V - (N-1)\alpha)(V - (N-1)\alpha) \cdots (V - \alpha) V \simeq \left(V - \frac{N\alpha}{2}\right)^N. \qquad \text{(F.9)}$$

This is a crude approximation although it is exact to first order in α. We can describe $V_0 = N\alpha/2 = bN$ as an excluded volume.

Such a simple model is too crude, for it ignores the attractive part of the interaction for $r > r_0$. The remaining part of the integral over real space is done for $r > r_0$ to define an average attractive potential

$$U_0 = 4\pi \int_{r_0}^{\infty} dr \, r^2 u(r) \, n(r) \qquad \text{(F.10)}$$

which is negative. The quantity $n(r)$ is the density of the remaining particles. The simplest approximation is to suppose the particles to be uniformly distributed in space so that $n(r) = N/V$. Thus we can write

$$U_0 = -2aN/V, \qquad \text{(F.11)}$$

$$a = -2\pi \int_{r_0}^{\infty} dr \, r^2 u(r). \qquad \text{(F.12)}$$

The overall partition function is

$$Z_N = \frac{1}{N!} \left(\frac{2\pi m k_B T}{h^2}\right)^{3N/2} \left[(V - V_0) \, e^{-U_0/2k_B T}\right]^N$$

$$= \frac{1}{N!} \left(\frac{2\pi m k_B T}{h^2}\right)^{3N/2} \left[(V - bN) \, e^{aN/V k_B T}\right]^N \qquad \text{(F.13)}$$

so that the free energy is

$$
\begin{aligned}
F &= -k_B T \ln(Z_N) \\
&= -N k_B T \left(\ln \left[\left(\frac{2\pi m k_B T}{h^2} \right)^{3/2} \left(\frac{V}{N} - b \right) \right] + 1 + \frac{aN}{V k_B T} \right). \quad \text{(F.14)}
\end{aligned}
$$

From F we get all the thermodynamics of the gas. For example, the expression for the pressure of a van der Waals gas is

$$
P = - \left(\frac{\partial F}{\partial V} \right)_{T,N} = \frac{N k_B T}{V - bN} - \frac{aN^2}{V^2}. \quad \text{(F.15)}
$$

Suppose the system is held at constant pressure; the thermodynamic potential is the Gibbs free energy $G = F + PV$. The minimum in the Gibbs free energy as a function of volume occurs where

$$
\left(\frac{\partial G}{\partial V} \right)_{T,N} = V \left(\frac{\partial P}{\partial V} \right)_{T,N}. \quad \text{(F.16)}
$$

The minimum in the Gibbs free energy occurs where $(\partial P / \partial V) = 0$. The critical point occurs when $(\partial^2 G / \partial V^2) = 0$ which means

$$
\left(\frac{\partial^2 G}{\partial V^2} \right)_{T,N} = \left(\frac{\partial P}{\partial V} \right)_{T,N} + V \left(\frac{\partial^2 P}{\partial V^2} \right)_{T,N} = 0. \quad \text{(F.17)}
$$

Since $(\partial P / \partial V) = 0$ we must have $(\partial^2 P / \partial V^2) = 0$. Thus the critical point for the van der Waals gas is found from the two conditions

$$
\left(\frac{\partial P}{\partial V} \right)_{T,N} = -\frac{N k_B T}{(V - bN)^2} + \frac{2aN^2}{V^3} = 0, \quad \text{(F.18)}
$$

$$
\left(\frac{\partial^2 P}{\partial V^2} \right)_{T,N} = \frac{2N k_B T}{(V - bN)^3} - \frac{6aN^2}{V^4} = 0. \quad \text{(F.19)}
$$

These equations have solutions

$$
T_c = \frac{8a}{27 b k_B}; \quad V_c = 3bN; \quad P_c = \frac{a}{27 b^2}.
$$

The ratio

$$
\frac{P_c V_c}{N k_B T_c} = \frac{3}{8} = 0.375. \quad \text{(F.20)}
$$

Some typical values of P_c, V_c, and T_c for simple gases are shown in Table F.1. The ratio N/V_c defines a critical density n_c.

Table F.1 Values of the pressure, molar volume, and temperature at the critical point for simple gases.

Gas	T_c (K)	V_c (m^3 mol^{-1})	P_c (atm)	$P_c V_c / RT_c$
^{4}He	5.2	57.8×10^{-6}	2.26	0.30
Ne	44.4	41.7×10^{-6}	26.9	0.31
H$_2$	33.1	65.0×10^{-6}	12.8	0.31
O$_2$	154.4	74.4×10^{-6}	49.7	0.29

We can rewrite eqn (F.14) in terms of n_c as

$$\frac{3bF}{V k_B T} = -\frac{n}{n_c} \left(\ln \left[\left(\frac{2\pi m k_B T_c b^{2/3}}{h^2} \right)^{3/2} \left(\frac{T}{T_c} \right)^{3/2} \left(\frac{3n_c}{n} - 1 \right) \right] + 1 + \frac{9n/n_c}{8T/T_c} \right).$$

$$(F.21)$$

It is then fairly easy to plot $3bF/V k_B T$ as a function of n/n_c for any particular temperature. For temperatures below T_c we can make a double tangent construction to the curve.

The values of V_c give an estimate of the size of the particles since $V_c = 3Nb = N2\pi r_0^3$. From Table F.1 we can see that V_c is roughly 6×10^{-5} m^3 mol^{-1} for all the particles listed; this corresponds to an interatomic spacing r_0 of about 2.5×10^{-10} m which seems reasonable as an estimate of an atomic size. The quantity T_c is a measure of the strength of the attractive interaction between particles; it is weakest for helium which is an inert gas, and much stronger for oxygen.

As an exercise we leave you to show that the internal energy of a van der Waals gas is (using $\beta = 1/k_B T$)

$$U = -\frac{\partial}{\partial \beta} (\ln(Z_N)) = \frac{3}{2} N k_B T - \frac{aN^2}{V}, \qquad (F.22)$$

that the entropy is

$$S = -\left(\frac{\partial F}{\partial T} \right)_{V,N} = N k_B \left[\frac{5}{2} + \ln \left(\frac{V}{N} - b \right) + \frac{3}{2} \ln \left(\frac{2\pi m k_B T}{h^2} \right) \right], \qquad (F.23)$$

and that the chemical potential is

$$\mu = \left(\frac{\partial F}{\partial N} \right)_{U,V} = \frac{F}{N} + k_B T \frac{V}{V - bN} - \frac{aN}{V}$$

$$= k_B T \left[\left(\frac{n_c}{n} \right) \frac{3bF}{V k_B T} + \frac{1}{1 - (n/3n_c)} - \frac{9nT_c}{8n_c T} \right]. \qquad (F.24)$$

G

Answers to problems

Chapter 1

1. $P = \frac{R\theta}{(V-b)} - \frac{a}{V^2}$.

3. (a) exact since

$$\frac{\partial\,(10y + 6z)}{\partial z} = \frac{\partial\,(6y)}{\partial y},$$

(b) not exact since

$$\frac{\partial\,(3y^2 + 4yz)}{\partial z} \neq \frac{\partial\,(2yz + y^2)}{\partial y},$$

(c) not exact.

4. (a) $\beta = \frac{1}{2}(3\alpha - 1)$.

5. When we differentiate we get

$$dx = (2yz^3)e^{y^2 z}dy + (2z + y^2 z^2)e^{y^2 z}dz.$$

Divide this by $z\,e^{y^2 z}$ and we get

$$đx' = (2yz^2)dy + (2 + y^2 z)\,dz.$$

This does not satisfy the condition of an exact differential since

$$\frac{\partial\,(2yz^2)}{\partial z} \neq \frac{\partial\,(2 + y^2 z)}{\partial y}.$$

7. The work done in a reversible adiabatic process is

$$W = -\int_{V_1}^{V_2} P\,dV = -\int_{V_1}^{V_2} \alpha V^{-\gamma}dV = \frac{\alpha}{\gamma - 1}\left(V_2^{(1-\gamma)} - V_1^{(1-\gamma)}\right).$$

8. 1.09 mm.

9. The change in the internal energy is the sum of the work done by the mass on the piston and the heat released by the resistor:

$$\Delta U = Mgh + I^2 Rt = 0.396 \text{ J}.$$

10. Isobaric does most work, adiabatic does least.
11. (a) 33×10^6 J, (b) 1 hr 50 min.
12. (b) 51×10^{15} J $= 14.3$ TWh.

Chapter 2

1. $C_P = aT = T(\partial S/\partial T)_P$. It follows that $(\partial S/\partial T)_P = a$ and so $S = aT + S(0)$. Since $S(0) = 0$ we have $S = aT = 0.43 \text{ J K}^{-1}$.
2. 0.42.
3. The efficiency is $\eta = 1 - \frac{50+273}{250+273} = \frac{200}{523}$.
4. 0.37.
5. $Q_2 = Q_1 T_2/T_1 = 0.667 \text{ kJ}$; $W = 0.333 \text{ kJ}$.
6. Pressure from 1 atm to 2.5 atm; volume from 6.3 litres to 2.5 litres.
7. The entropy change of the water is given by

$$\Delta S = \int_{273}^{373} \frac{dQ}{T} = \int_{273}^{373} \frac{C \, dT}{T} = C \ln \left(\frac{373}{273} \right).$$

The entropy change of the heat bath is $-Q/373$ where Q is equal to the energy given to the water: $Q = C100$. Therefore the total entropy change of the universe is

$$\Delta S = -\frac{C100}{373} + C \ln \left(\frac{373}{273} \right) = 184.9 \text{ J K}^{-1}.$$

8. (a) $0.044C$, (d) tends to zero.
9. The internal energy of each block is CT where T is the temperature. The total internal energy is conserved which means that when they reach equilibrium both objects go to a temperature of 323 K. Thus

$$\Delta S = \int_{273}^{323} \frac{C \, dT}{T} + \int_{373}^{323} \frac{C \, dT}{T} = C \ln \left(\frac{323^2}{273 \times 373} \right) = C \times 0.02425.$$

10. $aT + bT^3/3$.
11. Since $C = bT = dU/dT$ we have $U = U_0 + bT^2/2$. The total internal energy is constant which implies that $2T_f^2 = (100^2 + 200^2)$ and so $T_f = 158.1$ K. The entropy change is found to be

$$\Delta S = \int_{100}^{T_f} \frac{bT \, dT}{T} + \int_{200}^{T_f} \frac{bT \, dT}{T} = b(2T_f - 300) = 16.2b \text{ J K}^{-1}.$$

12. Go for design 3.

13. 5 grams of ice per second are frozen; the temperature of the ice is 273 K. The room is at 303 K. The heat released by the ice is $Q_2 = -5 \times 320\,\mathrm{J\,s^{-1}} = -1600\,\mathrm{J\,s^{-1}}$. The efficiency of the Carnot engine is

$$\eta = 1 - \frac{T_2}{T_1} = 1 - \frac{273}{303} = \frac{W}{Q_1}.$$

Hence

$$\frac{303}{30} = \frac{Q_1}{W} = \frac{W - Q_2}{W} = 1 - \frac{Q_2}{W}$$

or

$$\frac{Q_2}{W} = 1 - \frac{303}{30} = -\frac{273}{30}.$$

But $Q_2 = -1600\,\mathrm{J\,s^{-1}}$ so

$$W = \frac{1600 \times 30}{273} = 175.8\,\mathrm{J\,s^{-1}}.$$

14. $\Delta S = 4 \times 10^{-4}\,\mathrm{J\,K^{-1}}$.

15. In equilibrium the spring is extended by a distance Mg/K. It starts with an extension $Mg/2K$ so it has a distance $Mg/K - Mg/2K = Mg/2K$ to fall. The gravitational potential energy released as the mass falls a distance $Mg/2K$ is $M^2g^2/2K$. Some of this energy is absorbed as potential energy of the spring. The absorbed energy is

$$\int_{Mg/2K}^{Mg/K} Kx\,\mathrm{d}x = \left[\frac{Kx^2}{2}\right]_{Mg/2K}^{Mg/K} = 3M^2g^2/8K.$$

The rest of the energy, $M^2g^2/8K$, is released as heat into the surroundings. The increase in entropy of the universe is $M^2g^2/8KT$.

16. $0.127\,\mathrm{J\,K^{-1}}$.

17. The charge on the capacitor is $Q = CV$ where V is the voltage and C the capacitance. The work done by the battery is $QV = CV^2$ and the energy stored in the capacitor is $CV^2/2$. Hence an energy $CV^2/2$ is released as heat, giving an entropy change of $CV^2/2T$.

19. The heat capacity is independent of temperature, so the entropy change is

$$\Delta S = C \int_{T_1}^{T_f} \frac{\mathrm{d}T}{T} + C \int_{T_2}^{T_f} \frac{\mathrm{d}T}{T}.$$

The maximum work occurs when the process is reversible so $\Delta S = 0$. Hence

$$C \ln\left(\frac{T_f^2}{T_1 T_2}\right) = 0,$$

and so $T_f = (T_1 T_2)^{1/2}$. But the internal energy is $U = CT$. The initial internal energy is $U_i = C(T_1 + T_2)$ and the final internal energy is $U_f = 2CT_f$. The work done is the change in the internal energy

$$\Delta W = C\left\{T_1 + T_2 - 2(T_1 T_2)^{1/2}\right\}.$$

Chapter 3

1. To get a six we can have the arrangements $\{1,5\}, \{2,4\}, \{3,3\}, \{4,2\}, \{5,1\}$. Thus there are five ways out of 36 possible outcomes. So $p_6 = 5/36$. To get seven there are six ways so that $p_7 = 6/36$. The total probability is $11/36$.
2. 0.65.
3. Prob(ace of spades) $= 1/52$. Prob(another ace) $= 3/51$, that is three aces left out of the 51 cards that remain. Therefore total probability is $(1/52) \times (3/51) = 1/884$.
4. 20.
5. We can choose five objects out of twelve in

$$W = 12 \times 11 \times 10 \times 9 \times 8 = 95\,040$$

ways if the order of the choice is important.
Or we can choose in

$$W = 12 \times 11 \times 10 \times 9 \times 8/(5 \times 4 \times 3 \times 2 \times 1) = 792$$

ways if the order of the choice is unimportant.
6. (a) $1/250\,000$, (b) $1/200$.
7. The number of ways is $W = 12!/5!4!3! = 27\,720$.
8. (a) 0.25, (b) 0.0625, (c) 0.0026, (d) 0.6962.
9. The probability that two people do not have the same birthday is $p_2 = 364/365$. The probability that three people do not have the same birthday is $p_3 = (364/365) \times (363/365)$. The probability that four people do not have the same birthday is $p_4 = (364/365) \times (363/365) \times (363/365)$. The probability that five people do not have the same birthday is

$$p_5 = \frac{364}{365} \times \frac{363}{365} \times \frac{362}{365} \times \frac{361}{365}.$$

10. (a) p^n, (b) $p^{n-1}(1-p)$, (c) $(n-1)p^{n-2}(1-p)^2$.
11. There are eight coins and we want at least six heads. We can have eight heads, or seven heads and a tail, or we can have six heads and two tails.

(a) Eight heads $p_8 = 1/2^8$.
(b) Seven heads $p_7 = (1/2^8) \times 8!/7!1! = 8/2^8$ as there are eight ways of choosing one head out of eight with the rest tails.

(c) Six heads

$$p_6 = \frac{1}{2^8} \frac{8!}{6!2!} = \frac{28}{2^8},$$

as there are 28 ways of choosing two heads out of eight with the rest tails.

The total probability is the sum of these

$$p = \frac{(1+8+28)}{256} = \frac{37}{256}.$$

12. 0.265.

13. From the way the game is played, it is clear that Alice has twice the chance of winning as Bernard, who has twice the chance of winning as Charles. Thus

$$p_A = 2p_B = 4p_C.$$

The total probability is one which means that

$$1 = p_A + p_B + p_C = 4p_C + 2p_C + p_C.$$

Thus

$$p_C = 1/7, \quad p_B = 2/7, \quad p_A = 4/7.$$

14. (a) 0.683, (b) 0.00006.

15. The average value of x is

$$\bar{x} = -20 \times \frac{3}{10} - 10 \times \frac{2}{10} + 30 \times \frac{5}{10} = 7.$$

The standard deviation, Δx, is found from the equation:

$$(\Delta x)^2 = (-27)^2 \times \frac{3}{10} + (-17)^2 \times \frac{2}{10} + (23)^2 \times \frac{5}{10}$$
$$= (23.26)^2.$$

Hence $\Delta x = 23.26$.

16. 0.32.

17. Five chips chosen from 100, out of which 95 are good.
The probability that the first is good is 95/100, the probability that the first and the second are both good is $(95/100) \times (94/100)$, and so on until the probability that the first five are good is

$$\frac{95}{100} \times \frac{94}{99} \times \frac{93}{98} \times \frac{92}{97} \times \frac{91}{96} = 0.7696.$$

18. $(P_H)^L (1 - P_H)$.

19. Each die has six faces all equally probable. We can get all the faces to be the same in six ways. The total number of ways of arranging three dice is 6^3. Hence

$$p = \frac{6}{6^3} = \frac{1}{36}.$$

If the sum of faces is 11, then there are several ways to get this total:
if {6,1,4} then 6 permutations, i.e. (614), (641), (416), (461), (164), (146);
if {6,2,3} then 6 permutations,
if {5,3,3} then 3 permutations;
if {5,4,2} then 6 permutations,
if {5,5,1} then 3 permutations;
if {4,4,3} then 3 permutations.
The grand total is 27 so that

$$p = \frac{27}{6^3} = \frac{1}{8}.$$

20. $14\,414\,400$

21. Minimum has three m's, two i's, one n, and one u.

$$W = \frac{7!}{3!2!1!1!} = 420.$$

The number of ways of writing 'mmm****' where * can be (i, n, u) is $4!/(2!1!1!) = 12$. Thus

$$p = \frac{12}{420} = \frac{1}{35}.$$

Chapter 4

1. 10^{-6} J is taken from a system at $300\,\text{K}$ so

$$\Delta S = -\frac{10^{-6}}{300}\,\text{J K}^{-1}.$$

The heat is added to a system at $299\,\text{K}$, so

$$\Delta S = \frac{10^{-6}}{299}\,\text{J K}^{-1}.$$

The total change in the entropy of the two systems taken together is

$$\Delta S_T = \frac{10^{-6}}{299} - \frac{10^{-6}}{300} = 1.1 \times 10^{-11}\,\text{J K}^{-1}.$$

By using Boltzmann's hypothesis we get

$$\ln\left(\frac{W_f}{W_i}\right) = \frac{1.1 \times 10^{-11}}{1.38 \times 10^{-23}} = 8 \times 10^{11}.$$

Thus

$$W_f = W_i \, e^{8 \times 10^{11}}.$$

2. 4.9×10^{41}.

3. 10^{-7} J is added at 30 K. The entropy change is

$$\Delta S = \frac{10^{-7}}{30} = 3.33 \times 10^{-9} \, \mathrm{J\,K^{-1}}.$$

Hence

$$\ln \left(\frac{W_f}{W_i} \right) = \frac{3.33 \times 10^{-9}}{1.38 \times 10^{-23}} = 2.4 \times 10^{14}.$$

Thus

$$W_f = W_i \, e^{2.4 \times 10^{14}}.$$

4. 0.35 eV.

5. $S = k_B \ln(A) + k_B \gamma (VU)^{1/2}$ so that

$$\frac{1}{T} = \left(\frac{\partial S}{\partial U} \right)_V = \frac{k_B \gamma V^{1/2}}{2 U^{1/2}}$$

and

$$T = \frac{2 U^{1/2}}{k_B \gamma V^{1/2}}.$$

When $U = 0$ the temperature is zero.

6. $U = -(2CT)^{-1}$.

7. Using the techniques given in section 4.5 we get

$$\Delta V_A^2 = - \frac{k_B}{(\partial^2 S / \partial V_A^2)_{U_A}}.$$

The total entropy is $S = S_A + S_B$; hence

$$\left(\frac{\partial S}{\partial V_A} \right)_{U_A} = \frac{(P_A - P_B)}{T},$$

and

$$\left(\frac{\partial^2 S}{\partial V_A^2} \right)_{U_A} = \frac{1}{T} \left(\left(\frac{\partial P_A}{\partial V_A} \right)_{U_A} + \left(\frac{\partial P_B}{\partial V_B} \right)_{U_B} \right).$$

Thus

$$\Delta V_A^2 = -k_B T \left(\left(\frac{\partial P_A}{\partial V_A} \right)_{U_A} + \left(\frac{\partial P_B}{\partial V_B} \right)_{U_B} \right)^{-1}.$$

8. $N!/(n!(N-n)!$

9. M atoms are taken from N in

$$W = \frac{N!}{(N-M)!M!}$$

ways. M atoms are placed on N interstices in

$$W = \frac{N!}{(N-M)!M!}$$

ways. Therefore

$$W_T = \left(\frac{N!}{(N-M)!M!}\right)^2.$$

The internal energy is $U = M\epsilon$ so that

$$\frac{1}{T} = \frac{k_B}{\epsilon}\left(\frac{\partial \ln(W)}{\partial M}\right).$$

But

$$\ln(W) = 2\{N\ln(N) - M\ln(M) - (N-M)\ln(N-M)\}$$

so that we get

$$\frac{1}{T} = \frac{2k_B}{\epsilon}\{-\ln(M) + \ln(N-M)\}.$$

Inverting this equation gives

$$\frac{M}{N} = \frac{1}{e^{\epsilon/2k_B T} + 1}.$$

10. $T = U^{1/4}/\sigma V^{1/4}$.

11. Use

$$\frac{1}{T} = \left(\frac{\partial S}{\partial U}\right)_{V,N} = \frac{Nk_B}{U}.$$

The chemical potential is

$$\mu = -T\left(\frac{\partial S}{\partial N}\right)_{U,V} = k_B T \ln\left(\frac{N}{A}\right) + k_B T \ln\left(\frac{2\pi\hbar^2}{mk_B T}\right).$$

12. Maximum in the entropy when $n = 1/2$.

13. Using

$$S = k_B \ln(W)$$

and

$$\frac{1}{T} = \left(\frac{\partial S}{\partial U}\right)_V$$

gives

$$\frac{1}{T} = k_B \frac{\partial}{\partial U}\left(\ln((N-1+U/\hbar\omega)!) - \ln((N-1)!) - \ln((U/\hbar\omega)!)\right)_V$$

$$= \frac{k_B}{\hbar\omega} \ln \left(\frac{N - 1 + U/\hbar\omega}{U/\hbar\omega} \right).$$

By rearranging this equation we get

$$U = \frac{(N - 1)\hbar\omega}{e^{\hbar\omega/k_B T} - 1}.$$

For large N the average energy per oscillator is roughly equal to

$$\frac{U}{N} \simeq \frac{U}{N - 1} = \frac{\hbar\omega}{e^{\hbar\omega/k_B T} - 1}.$$

14. $\Delta S_{2\text{level}} = k_B \ln(n_2/n_1)$; $\Delta S_{\text{bath}} = (\varepsilon_2 - \varepsilon_1)/T$; reversible process has no change in the total entropy.

Chapter 5

1. The partition function is

$$Z = e^0 + e^{-1} + e^{-2} = 1.503.$$

The probability of being in the ground state is

$$p_1 = \frac{e^0}{Z} = 0.665.$$

The probability of being in the first excited state is

$$p_2 = \frac{e^{-1}}{Z} = 0.245.$$

The probability of being in the second excited state is

$$p_3 = \frac{e^{-2}}{Z} = 0.090.$$

2. 0.375.
3. $\Delta/k_B T = 1.6 \times 10^{-20}/(1.38 \times 10^{-23} \times 300) = 3.865.$
   The probability of the system being in the upper state is

$$p_1 = \frac{e^{-\Delta/k_B T}}{Z} = \frac{e^{-\Delta/k_B T}}{1 + e^{-\Delta/k_B T}} = 0.0205.$$

The probability of the system being in the upper states is 0.25 if

$$0.25 = \frac{1}{4} = \frac{e^{-\Delta/k_B T}}{1 + e^{-\Delta/k_B T}}.$$

Hence $e^{\Delta/k_B T} = 3$ or $\Delta/k_B T = \ln(3)$ which means that

$$T = \frac{\Delta}{k_B \ln(3)} = 1055 \text{ K.}$$

4. 0.233.

5. $Z = 1 + 3 e^{-1} + 5 e^{-2} = 2.780$.
The probability of being in the lowest state is

$$p_0 = \frac{1}{Z} = 0.360;$$

the probability of being in the next states is

$$p_1 = \frac{3 e^{-1}}{Z} = 0.397;$$

the probability of being in the top states is

$$p_2 = \frac{5 e^{-2}}{Z} = 0.243.$$

The average energy is

$$\bar{E} = (0 \times p_0 + 100 \times p_1 + 200 \times p_2) k_B = 88.3 k_B.$$

6. $P = P_{00} + P_{10} + P_{20} + P_{01}$ where

$$P_{n_1 n_2} = e^{-(n_1 + 2n_2)\hbar\omega/k_B T}(1 - e^{-\hbar\omega/k_B T})(1 - e^{-2\hbar\omega/k_B T}).$$

7. The free energy is

$$F = -k_B T \ln(Z) = -a k_B T^4 V.$$

The pressure is

$$P = -\left(\frac{\partial F}{\partial V}\right)_T = a k_B T^4.$$

The entropy is

$$S = -\left(\frac{\partial F}{\partial T}\right)_V = 4 a k_B T^3 V.$$

Since $F = U - TS$ we have

$$U = F + TS = -a k_B T^4 V + 4 a k_B T^4 V = 3 a k_B T^4 V.$$

8. $(1 + e^{-\hbar\omega/k_B T})^{-1}$.

9. The energy levels of the hydrogen atom are $\varepsilon_n = -13.6/n^2$ in eV. Since $k_B T = 2/3\,\text{eV}$

$$\frac{\varepsilon_n}{k_B T} = -\frac{3 \times 13.6}{2n^2} = -\frac{3\alpha}{2n^2}.$$

The ratio of the occupancies of $n = 3$ to $n = 1$ levels depends on the Boltzmann factor and on the degeneracies of the two levels. Thus

$$\frac{N_3}{N_1} = \frac{2 \times 3^2 \times e^{\alpha/6}}{2 \times 1^2 \times e^{3\alpha/2}}$$

$$= 9\,e^{-4\alpha/3} = 1.2 \times 10^{-7}.$$

10. 0.696.

11. The probability, p_0, of being in the ground state is 0.63; $E_0 = 0.0023\,\text{eV}$. The ratio

$$\frac{p_n}{p_0} = e^{-(E_n - E_0)/k_B T}$$

so

$$\frac{p_1}{p_0} = e^{-(E_1 - E_0)/k_B T} = \frac{0.23}{0.63} = 0.365.$$

Hence

$$-\left(\frac{E_1 - E_0}{k_B T}\right) = \ln(0.365).$$

But $E_1 - E_0 = 0.0086\,\text{eV}$ so that

$$T = -\frac{(E_1 - E_0)}{k_B \ln(0.365)} = 99.0\,\text{K}.$$

Repeat for other ratios and we get

$$T = -\frac{(E_2 - E_0)}{k_B \ln(0.1349)} = 99.6\,\text{K},$$

and

$$T = -\frac{(E_3 - E_0)}{k_B \ln(0.0492)} = 99.4\,\text{K}.$$

13. The partition function is

$$Z = g_0\,e^{-\varepsilon_0/k_B T} + g_1\,e^{-\varepsilon_1/k_B T} = 3\,e^{-\varepsilon_0/k_B T} + 2\,e^{-\varepsilon_1/k_B T},$$

with $\varepsilon_0 = 0$ and $\varepsilon_1 = \Delta$. Thus

$$Z = 3 + 2\,e^{-\Delta/k_B T},$$

where $\Delta/k_B T = 11.25$ at a temperature of 1000 K; therefore

$$Z = 3 + 2\,\mathrm{e}^{-11.25} = 3.000026.$$

If the temperature is 3000 K then

$$Z = 3 + 2\,\mathrm{e}^{-11.25/3} = 3.047.$$

14. $Z_1 = 1 + \mathrm{e}^{-\epsilon/k_\mathrm{B}T}\left(\mathrm{e}^{\mu B/k_\mathrm{B}T} + \mathrm{e}^{-\mu B/k_\mathrm{B}T}\right)$ from which we get Z_N and the thermodynamics.

15. If $n_1 = n_2 = 0; n_3 = 2$ then the degeneracy is three. The energy level is $7\hbar\omega/2$. But when $n_1 = n_2 = 1; n_3 = 0$ the energy level is $7\hbar\omega/2$ also, and the degeneracy is three. The total degeneracy when $E = 7\hbar\omega/2$ is six. Repeat for the next level of energy $E = 9\hbar\omega/2$ and the degeneracy is ten. The ratio of the number in level of energy $9\hbar\omega/2$ to the number in level of energy $7\hbar\omega/2$ is

$$\frac{N_{9/2}}{N_{7/2}} = \frac{10}{6}\,\mathrm{e}^{-\hbar\omega/k_\mathrm{B}T} > 1$$

if

$$\ln\left(\frac{10}{6}\right) - \frac{\hbar\omega}{k_\mathrm{B}T} > 0,$$

so

$$k_\mathrm{B}T > \frac{\hbar\omega}{\ln(5/3)}.$$

17. The pressure is

$$P = k_\mathrm{B}T\left(\frac{\partial \ln(Z)}{\partial V}\right)_T = \frac{Nk_\mathrm{B}T}{V - Nb} - \frac{N^2 a^2}{V^2}.$$

Rearranging this gives something close to van der Waals equation:

$$\left(P + \frac{N^2 a^2}{V^2}\right)(V - Nb) = Nk_\mathrm{B}T.$$

The internal energy is given by eqn (5.6.7)

$$\bar{U} = k_\mathrm{B}T^2\left(\frac{\partial \ln(Z)}{\partial T}\right)_V = N\left(\frac{3k_\mathrm{B}T}{2} - \frac{Na^2}{V}\right).$$

18. 1.6×10^{-10} N.

19. The energy levels are

$$\varepsilon_i = \frac{\hbar^2\pi^2}{2mL^2}\left(n_1^2 + n_2^2 + n_3^2\right) = \alpha\left(n_1^2 + n_2^2 + n_3^2\right)$$

with

$$\alpha = \frac{\hbar^2 \pi^2}{2mL^2} = 6.02 \times 10^{-34} \text{ J}$$

for $L = 0.01$ m. The states can be characterized by the set of numbers (n_1, n_2, n_3) with $n_i \geq 1$. The degeneracy is found from the number of arrangements of these numbers that can be made giving the same value of $(n_1^2 + n_2^2 + n_3^2)$; e.g. the state $(2, 1, 1)$ is degenerate with $(1, 2, 1)$, and with $(1, 1, 2)$. In this way we get the quantum states:

State	Energy	Degeneracy
$(1, 1, 1)$	3α	1
$(2, 1, 1)$	6α	3
$(2, 2, 1)$	9α	3
$(2, 2, 2)$	12α	1
$(3, 1, 1)$	11α	3
$(3, 2, 1)$	14α	6

20. Yes.

21. The algebra needs some care, but the main steps are these:

$$-\frac{\partial \ln(Z)}{\partial \beta} = \frac{\sum_i E_i e^{-\beta E_i}}{\sum_i e^{-\beta E_i}} = \sum_i p_i E_i$$

and

$$\frac{\partial^2 \ln(Z)}{\partial \beta^2} = \frac{\sum_i E_i^2 e^{-\beta E_i}}{\sum_i e^{-\beta E_i}} - \left(\frac{\sum_i E_i e^{-\beta E_i}}{\sum_i e^{-\beta E_i}}\right)^2.$$

22. 1.416×10^{-23} J K^{-1}.

23. For an LC circuit the total voltage is zero across the two elements. Hence

$$L\frac{dI}{dt} + \frac{Q}{C} = 0$$

where I is the current and Q the charge on the capacitor. Since $I = dQ/dt$ this becomes the differential equation

$$L\frac{d^2Q}{dt^2} + \frac{Q}{C} = 0.$$

This is the equation of a simple harmonic oscillator with angular frequency $\omega = (LC)^{-1/2}$. If we multiply the equation for simple harmonic motion by dQ/dt and integrate with respect to time we get

$$\frac{L}{2}\left(\frac{dQ}{dt}\right)^2 + \frac{Q^2}{2C} = \text{constant.}$$

The term $L(dQ/dt)^2/2$ is analogous to the kinetic energy of a mechanical harmonic oscillator, the term $Q^2/2C$ is analogous to the potential energy.

Both terms are quadratic in the 'coordinate' and so have $k_BT/2$ of internal energy associated with them according to the equipartition theorem. Hence

$$\sqrt{I^2} = \sqrt{\frac{k_BT}{L}}.$$

24. $6k_B$; $6.5k_B$.

25. In two dimensions there are two degrees of freedom for translation, one for rotation, and two for vibration (there is only one mode of vibration). In all there are five degrees of freedom.

(a) $C_V = \frac{N\Delta^2/k_BT^2}{\cosh^2(\Delta/k_BT)}$.

(b) C_V varies as T.

Chapter 6

1. $Z_1 = e^{\varepsilon/k_BT} + 1 + e^{-\varepsilon/k_BT}$ for one spin; then for N spins

$$Z_N = \left(e^{\varepsilon/k_BT} + 1 + e^{-\varepsilon/k_BT}\right)^N,$$

and

$$F = -Nk_BT\ln\left(1 + 2\cosh\left(\varepsilon/k_BT\right)\right).$$

2. $-3Nk_BT\ln(e^{\varepsilon/2k_BT} + e^{-\varepsilon/2k_BT})$.

3. The energy levels for fermions are 0 (one state), ε (two states), 2ε (two states), 3ε (one state).

$$Z = 1 + 2e^{-\varepsilon/k_BT} + 2e^{-2\varepsilon/k_BT} + e^{-3\varepsilon/k_BT}.$$

If the particles are bosons then there are states where both particles are in the same single-particle state. Hence there are the additional energy levels of energies: 0 (two states), 2ε (one state), 4ε (one state). The partition function is

$$Z = 3 + 2e^{-\varepsilon/k_BT} + 3e^{-2\varepsilon/k_BT} + e^{-3\varepsilon/k_BT} + e^{-4\varepsilon/k_BT}.$$

4. 44; 5; 1.

5. (a) The simple, 'classical' way is to work out Z_1^2. This gives

$$Z_2 = Z_1^2 = \left(1 + e^{-\varepsilon/k_BT} + e^{-2\varepsilon/k_BT} + e^{-3\varepsilon/k_BT}\right)^2.$$

(b) The allowed states for Fermi particles have energy $(0, \varepsilon)$, $(0, 2\varepsilon)$, $(0, 3\varepsilon)$, $(\varepsilon, 2\varepsilon)$, $(\varepsilon, 3\varepsilon)$, and $(2\varepsilon, 3\varepsilon)$. Hence the partition function is

$$Z = e^{-\varepsilon/k_BT} + e^{-2\varepsilon/k_BT} + 2e^{-3\varepsilon/k_BT} + e^{-4\varepsilon/k_BT} + e^{-5\varepsilon/k_BT}.$$

(c) The allowed states for Bose particles have energy $(0, 0)$, $(0, \varepsilon)$, $(0, 2\varepsilon)$, $(0, 3\varepsilon)$, $(\varepsilon, \varepsilon)$, $(\varepsilon, 2\varepsilon)$, $(\varepsilon, 3\varepsilon)$, $(2\varepsilon, 2\varepsilon)$, $(2\varepsilon, 3\varepsilon)$, and $(3\varepsilon, 3\varepsilon)$. Hence the partition function is

$$Z = 1 + e^{-\varepsilon/k_BT} + 2e^{-2\varepsilon/k_BT} + 2e^{-3\varepsilon/k_BT} + 2e^{-4\varepsilon/k_BT} + e^{-5\varepsilon/k_BT} + e^{-6\varepsilon/k_BT}.$$

6. $n = 3/(3 + \exp(\varepsilon/k_BT))$.
7. The partition function of one spin is

$$Z_1 = e^{\mu B/k_BT} + e^{-\mu B/k_BT}.$$

The partition function for N spins is

$$Z_N = \left(e^{\mu B/k_BT} + e^{-\mu B/k_BT} \right)^N.$$

$$F = -k_BT \ln \left\{ \left(e^{\mu B/k_BT} + e^{-\mu B/k_BT} \right)^N \right\}$$

$$= -N\mu B - Nk_BT \ln \left(1 + e^{-2\mu B/k_BT} \right).$$

The entropy is

$$S = -\left(\frac{\partial F}{\partial T} \right) = Nk_B \left(\ln \left(1 + e^{-2\mu B/k_BT} \right) + \frac{2\mu B}{k_BT} \frac{e^{-2\mu B/k_BT}}{1 + e^{-2\mu B/k_BT}} \right).$$

The entropy per particle is a function of $\mu B/k_BT$ only. If B is reduced by a factor of 10^3 from 10 to 10^{-2} T then the spin temperature goes down from 1 to 0.001 K. This process is called *adiabatic demagnetization*.

8. $\varepsilon_{n_1 n_2} = \hbar\omega(n_1 + n_2 + 1)$;
   bosons

$$Z = \frac{e^{-\hbar\omega/k_BT}}{(1 - e^{-\hbar\omega/k_BT})(1 - e^{-2\hbar\omega/k_BT})};$$

fermions

$$Z = \frac{e^{-2\hbar\omega/k_BT}}{(1 - e^{-\hbar\omega/k_BT})(1 - e^{-2\hbar\omega/k_BT})}.$$

9. The Helmholtz free energy is

$$F = -k_BT \ln(Z_N) = -k_BT \ln \left(\frac{Z_a^{N_a} Z_b^{N_b}}{N_a! N_b!} \right).$$

If the particles were identical then $Z_a = Z_b$, $N = N_a + N_b$, and the free energy would be

$$F_{id} = -k_BT \ln \left(\frac{Z_a^N}{N!} \right).$$

The difference in the free energies if $Z_a = Z_b$, even though the particles are not identical, is

$$\Delta F = -k_BT \ln \left(\frac{Z_a^N}{N_a! N_b!} \right) + k_BT \ln \left(\frac{Z_a^N}{N!} \right)$$

$$= -k_{\mathrm{B}}T \ln \left(\frac{N!}{N_a! N_b!}\right)$$

$$= -k_{\mathrm{B}}T \left\{(N_a + N_b) \ln (N) - N_a \ln (N_a) - N_b \ln (N_b)\right\}.$$

Hence the mixing entropy which arises because the particles are not identical is

$$\Delta S_{\mathrm{mixing}} = k_{\mathrm{B}} \left\{(N_a + N_b) \ln (N) - N_a \ln (N_a) - N_b \ln (N_b)\right\}$$

$$= -k_{\mathrm{B}} \left\{N_a \ln \left(\frac{N_a}{N}\right) + N_b \ln \left(\frac{N_b}{N}\right)\right\}.$$

Chapter 7

1. In two dimensions the density of states in energy is

$$D(\varepsilon)d\varepsilon = \frac{Ak\, dk}{2\pi\, d\varepsilon} d\varepsilon.$$

For free particles the dispersion relation is $\varepsilon(k) = \hbar^2 k^2 / 2m$ and so $d\varepsilon/dk = \hbar^2 k/m$. Hence

$$D(\varepsilon) = \frac{Am}{2\pi \hbar^2}.$$

2. $u_{\mathrm{rms}} = 1390 \mathrm{\ m\,s}^{-1}$; $3.1 \times 10^{-10} \mathrm{\ m}$; $1130 \mathrm{\ m\,s}^{-1}$.
3. We have

$$\frac{d\varepsilon}{dk} = \tfrac{3}{2}\alpha k^{1/2}.$$

In three dimensions the density of states in energy is

$$D(\varepsilon) = \frac{Vk^2}{2\pi^2} \frac{2}{3\alpha k^{1/2}}.$$

But since $k^{3/2} = \varepsilon/\alpha$ we get

$$D(\varepsilon) = \frac{V\varepsilon}{3\pi^2 \alpha^2}.$$

4. $1350 \mathrm{\ m\,s}^{-1}$.
5. The root mean square speed of hydrogen molecules is given by

$$u_{\mathrm{rms}} = \sqrt{\overline{u^2}} = \sqrt{\frac{3k_{\mathrm{B}}T}{m}} = 1935 \mathrm{\ m\,s}^{-1}.$$

The mean speed is given by

$$\bar{u} = \sqrt{\frac{8k_{\mathrm{B}}T}{\pi m}} = 1782 \mathrm{\ m\,s}^{-1}.$$

6. $Z_N = \left(V k_B^3 T^3 / 2\pi^2 \hbar^3 c\right)^N / N!.$

7. The partition function is

$$Z = \int_0^\infty D(\varepsilon) \, e^{-\varepsilon/k_B T} d\varepsilon$$

and $\varepsilon = \hbar c k$ for a highly relativistic particle. In one dimension the density of states in energy is $D(\varepsilon) = L/\pi\hbar c.$

$$Z = \frac{L}{\pi\hbar c} \int_0^\infty e^{-\varepsilon/k_B T} d\varepsilon = \frac{L k_B T}{\pi\hbar c}.$$

8. (b) $2\,093 \text{ m s}^{-1}$; (d) $k_B T$; (e) $A(P/k_B T)\bar{u} = 1.3 \times 10^{17} \text{ s}^{-1}.$

9. Given that $\varepsilon = \hbar c k$, the partition function is

$$Z = \int_0^\infty D(\varepsilon) e^{-\varepsilon/k_B T} d\varepsilon$$

$$= \frac{A}{2\pi\hbar^2 c^2} \int_0^\infty \varepsilon \, e^{-\varepsilon/k_B T} d\varepsilon$$

$$= \frac{A k_B^2 T^2}{2\pi\hbar^2 c^2}.$$

10. In terms of the escape velocity, $v_E = \sqrt{2g R_E}$, the fraction is

$$f = \frac{\int_{v_E}^\infty u^2 \exp(-mu^2/2k_B T) \, du}{\int_0^\infty u^2 \exp(-mu^2/2k_B T) \, du}$$

11. Using the density of states in **k** space for three dimensions gives

$$Z = \int_0^\infty D(\varepsilon) \, e^{-\varepsilon/k_B T} d\varepsilon$$

$$= \frac{V}{2\pi^2} \int_0^\infty \left(\frac{1}{3\alpha}\right) e^{-\varepsilon/k_B T} d\varepsilon$$

$$= \frac{V k_B T}{6\alpha\pi^2}.$$

12. The frequency distribution is

$$\exp(-(mc^2(\nu - \nu_0)^2/2k_B T \nu_0^2).$$

13. The average speed is

$$\bar{u} = \frac{\int_0^\infty u n(u) \, du}{\int_0^\infty n(u) \, du}$$

where $n(u)$ is the Maxwell distribution in speed. For a free particle of mass m in two dimensions we find

$$n(u) = \frac{NAm^2}{2\pi Z\hbar^2} u e^{-mu^2/2k_BT}.$$

Thus

$$\bar{u} = \frac{\int_0^\infty u^2 e^{-mu^2/2k_BT} du}{\int_0^\infty u e^{-mu^2/2k_BT} du}.$$

After putting $x^2 = mu^2/2k_BT$ we get

$$\bar{u} = \left(\frac{2k_BT}{m}\right)^{1/2} \frac{\int_0^\infty x^2 e^{-x^2} dx}{\int_0^\infty x e^{-x^2} dx}.$$

The integrals over x are standard: the one in the numerator is $\pi^{1/2}/4$, and the one in the denominator is $1/2$. As a result we get

$$\bar{u} = \left(\frac{\pi k_BT}{2m}\right)^{1/2}.$$

14. $(2m/\pi k_BT)^{1/2}$.
15. The angular speed is $\omega = \pi 10^4/60 \,\mathrm{rad\,s^{-1}}$. The temperature is $273 + 830 = 1103\,\mathrm{K}$. The mass of an atom is $209 \times 1.66 \times 10^{-27}\,\mathrm{kg}$, so the r.m.s. speed of each atom is $(3k_BT/m)^{1/2} = 363\,\mathrm{m\,s^{-1}}$.
    The displacement is $\theta r = \omega\, dr/u_{\mathrm{rms}} = 0.0289\,\mathrm{m}$.
    If we consider bismuth molecules, the mass is increased by a factor of 2, so the speed is reduced by a factor of $2^{1/2}$. The displacement is $0.0408\,\mathrm{m}$.
16. (c) If $v_{\max} = 3333\,\mathrm{m\,s^{-1}}$ then $T = 443\,\mathrm{K}$.
17. The distribution of speeds of particles which are incident on the hole is $un(u)$ since the faster particles arrive at the hole more quickly than the slower ones. In three dimensions

$$un(u) = \frac{NVm^3}{2\pi^2 Z\hbar^3} u^3 e^{-mu^2/2k_BT}.$$

The average kinetic energy of particles which leave the hole is given by

$$\overline{\frac{mu^2}{2}} = \frac{m}{2} \frac{\int_0^\infty u^3 n(u)\, du}{\int_0^\infty un(u)\, du}$$

$$= \frac{m}{2}\left(\frac{2k_BT}{m}\right) \frac{\int_0^\infty x^5 e^{-x^2} dx}{\int_0^\infty x^3 e^{-x^2} dx}.$$

The integrals can be done by parts or by looking them up in Appendix B. The final result is that the kinetic energy is $2k_BT$.

Chapter 8

1. The dispersion relation for ripplons is

$$\omega(k) = \sqrt{\frac{\gamma_s k^3}{\rho}}.$$

The thermal part of the internal energy for surface waves on liquid helium is

$$\bar{U} = \frac{A}{2\pi} \int_0^\infty \frac{\hbar\omega(k)}{e^{\hbar\omega(k)/k_B T} - 1} k \, dk.$$

It helps to change to a variable $z^{3/2} = \hbar\gamma_s^{1/2} k^{3/2}/\rho^{1/2} k_B T$. Then

$$\bar{U} = \frac{A k_B T}{2\pi} \left(\frac{\rho^{1/2} k_B T}{\hbar \gamma_s^{1/2}} \right)^{4/3} \int_0^\infty \frac{z^{5/2}}{e^{z^{3/2}} - 1} dz.$$

2. $u(E) = E^2/(2\pi\hbar^2 c^2 (e^{E/k_B T} - 1))$; for 'hole' of width d the rate of radiation is $dcu(E)/\pi$.

3. The energy density as a function of wavelength is

$$u(\lambda) = \frac{8\pi hc}{\lambda^5 \left(e^{hc/\lambda k_B T} - 1 \right)}.$$

The maximum occurs when the derivative of $u(\lambda)$ with respect to λ is zero.

$$\frac{du(\lambda)}{d\lambda} = 8\pi hc \left\{ \frac{hc\, e^{hc/\lambda k_B T}}{\lambda^7 k_B T \left(e^{hc/\lambda k_B T} - 1 \right)^2} - \frac{5}{\lambda^6 \left(e^{hc/\lambda k_B T} - 1 \right)} \right\} = 0.$$

Rearranging this gives

$$\frac{hc}{\lambda k_B T} = \frac{5 \left(e^{hc/\lambda k_B T} - 1 \right)}{e^{hc/\lambda k_B T}} = 5 \left(1 - e^{-hc/\lambda k_B T} \right).$$

We solve this equation by iteration starting with $hc/\lambda k_B T = 5$; after a few iterations we get $hc/\lambda k_B T = 4.965$. Thus

$$T = \frac{hc}{4.965 \lambda k_B} = 6\,037\,\text{K}.$$

4. $\Delta E^2 = 4\pi^2 V k_B^5 T^5 / 15 \hbar^3 c^3$.

5. The rate of heat radiated is

$$\frac{dQ}{dt} = 4\pi R_s^2 \sigma T^4 = 3.95 \times 10^{26}\,\text{W}.$$

The energy per unit area on Earth is

$$\frac{dQ}{dt} = \frac{3.95 \times 10^{26}}{4\pi \left(1.5 \times 10^{11}\right)^2} = 1.397 \, \text{kW m}^{-2}.$$

6. $E/k_{\mathrm{B}}T \simeq 2.85$ so the energy is 1.26×10^{-21} J at 30 K.

7. Volume of Universe, V, is $4\pi R_U^3/3$. The number of photons is

$$N = \frac{V}{\pi^2} \int_0^\infty \frac{k^2}{\left(e^{\hbar c k/k_{\mathrm{B}}T} - 1\right)} dk$$

$$= \frac{4R_U^3}{3\pi} \left(\frac{k_{\mathrm{B}}T}{\hbar c}\right)^3 \int_0^\infty \frac{x^2}{\left(e^x - 1\right)} dx.$$

The integral over x is $2(1 + 1/2^3 + \cdots) = 2.4$ so the answer is 2.3×10^{87}.
The energy of the photons is

$$U = \frac{V}{\pi^2} \int_0^\infty \frac{\hbar c k^3}{\left(e^{\hbar c k/k_{\mathrm{B}}T} - 1\right)} dk$$

$$= \frac{4R_U^3 k_{\mathrm{B}}T}{3\pi} \left(\frac{k_{\mathrm{B}}T}{\hbar c}\right)^3 \int_0^\infty \frac{x^3}{\left(e^x - 1\right)} dx.$$

The integral over x is about $6(1 + 1/2^4 + 1/3^4 + \cdots)$, so the final answer is 2.6×10^{65} J.

9. Spin waves in three dimensions have a thermal part of the internal energy

$$U = \frac{V}{2\pi^2} \int_0^\infty \frac{\hbar \alpha k^4}{\left(e^{\hbar \alpha k^2/k_{\mathrm{B}}T} - 1\right)} dk$$

$$= \frac{\hbar \alpha V}{2\pi^2} \left(\frac{k_{\mathrm{B}}T}{\hbar \alpha}\right)^{5/2} \int_0^\infty \frac{x^4}{\left(e^{x^2} - 1\right)} dx,$$

so the internal energy is proportional to $T^{5/2}$.

10. (c) $3k_{\mathrm{B}}T/16(F/L)$.

11. The free energy is

$$F = \frac{k_{\mathrm{B}}TV}{2\pi^2} \int_0^\infty k^2 \ln\left(1 - e^{-\hbar\omega(k)/k_{\mathrm{B}}T}\right) dk.$$

The integral over k is separated out into a part that runs from 0 to 0.8 Å^{-1} and a part that runs around the roton minimum. The first term becomes

$$F_{\mathrm{phonon}} = \frac{k_{\mathrm{B}}TV}{2\pi^2} \int_0^{k_{\mathrm{max}}} k^2 \ln\left(1 - e^{-\hbar s k/k_{\mathrm{B}}T}\right) dk$$

$$= \frac{k_B T V}{2\pi^2} \left(\frac{k_B T}{\hbar s}\right)^3 \int_0^\infty x^2 \ln\left(1 - e^{-x}\right) dx,$$

the second becomes

$$F_{\text{roton}} = \frac{k_B T V}{2\pi^2} \int_{-\infty}^\infty (k_0 + \xi)^2 \ln\left(1 - e^{-\Delta/k_B T} e^{-\hbar^2 \xi^2 / 2\mu k_B T}\right) d\xi$$

from which it is easy to get the desired result by expanding the $\ln(1 - x)$ function.

Chapter 9

1. $3\mu_{O_2} = 2\mu_{O_3}$.
2. $k_B T \ln(\Pi) = ST - \bar{U} + w_z L_z$.
3. If n_D is the number density of donors then the number density of free electrons is $n_e = n_D + n_h$ where n_h is the number density of holes. The chemical potential of electrons is

$$\mu_e = E_g + k_B T \ln\left(n_e / n_{Q'_e}\right).$$

The chemical potential of holes is

$$\mu_h = k_B T \ln\left(n_h / n_{Q'_h}\right).$$

In chemical equilibrium the sum of these chemical potentials is zero,

$$E_g + k_B T \ln\left(n_e / n_{Q'_e}\right) + k_B T \ln\left(n_h / n_{Q'_h}\right) = 0,$$

from which it follows that

$$n_e n_h = n_e \left(n_e - n_D\right) = n_{Q'_e} n_{Q'_h} e^{-E_g / k_B T} = 4 n_{Q_e} n_{Q_h} e^{-E_g / k_B T}.$$

This is a simple quadratic equation for n_e which has to be solved. If the number density of donors, n_D, is zero we get $n_e^0 = n_h^0$. The solution is then

$$n_e^0 = 2 \sqrt{n_{Q_e} n_{Q_h}} \, e^{-E_g / 2 k_B T}.$$

When there are donors present the number density of electrons is

$$n_e = \frac{n_D}{2} + \sqrt{\frac{1}{4} (n_D)^2 + (n_e^0)^2}.$$

5. The reaction

$$p + e^- \rightleftharpoons H$$

occurs at 4000 K and reaches chemical and thermal equilibrium. The condition for equilibrium is

$$\mu_e + \mu_p = \mu_H$$

where the chemical potentials are

$$\mu_p = k_B T \ln \left(n_p/n_{Q'_p} \right) = k_B T \ln \left(\frac{n_p}{2} \left(\frac{2\pi\hbar^2}{m_p k_B T} \right)^{3/2} \right)$$

for the proton,

$$\mu_e = k_B T \ln \left(n_e/n_{Q'_e} \right) = k_B T \ln \left(\frac{n_e}{2} \left(\frac{2\pi\hbar^2}{m_e k_B T} \right)^{3/2} \right)$$

for the electron, and

$$\mu_H = k_B T \ln \left(n_H/n_{Q'_H} \right) + \Delta = k_B T \ln \left(\frac{n_H}{4} \left(\frac{2\pi\hbar^2}{m_H k_B T} \right)^{3/2} \right) + \Delta$$

for the hydrogen atom, provided we ignore all the excited states of the hydrogen atom, such as the 2p, 3s, et cetera, and concentrate on the orbital ground state only. Notice the atom has a spin degeneracy in its ground state of four corresponding to the spin states of both the electron and the proton. The condition for chemical equilibrium then gives

$$n_e n_p = n_H \left(\frac{m_e k_B T}{2\pi\hbar^2} \right)^{3/2} e^{\Delta/k_B T}.$$

The number density of protons and electrons must be equal to preserve overall neutrality. But the gas is half-ionized so that

$$n_e = n_p = n_H.$$

Hence

$$n_e = \left(\frac{m_e k_B T}{2\pi\hbar^2} \right)^{3/2} e^{\Delta/k_B T}.$$

The binding energy $\Delta = -13.6\,\text{eV}$ and at 4000 K, $\Delta/k_B T = -39.45$. Using the values for the fundamental constants we get

$$n_e = 4.5 \times 10^9 \, \text{m}^{-3}.$$

7. The grand partition function is

$$\Xi = \sum_{N=0}^{\infty} \sum_i e^{-\{E_i(N)-\mu N\}/k_B T}$$

where $E_i(N)$ represents the energy level of a quantum state with N particles. Let us put $\lambda = e^{\mu/k_B T}$ and call

$$Z_N = \sum_i e^{-E_i(N)/k_B T}.$$

Then we can write

$$\Xi = \sum_{N=0}^{\infty} \lambda^N Z_N.$$

But the partition function for N particles is roughly

$$Z_N = \frac{Z_1^N}{N!}$$

where Z_1 is the single-particle partition function. Hence

$$\Xi = \sum_{N=0}^{\infty} \frac{\lambda^N Z_1^N}{N!} = e^{\lambda Z_1}.$$

9. The total energy for N attached atoms is $E = N\epsilon$. The number of ways of creating a state of this energy (the degeneracy of the state) is

$$W = \frac{M!}{N!(M-N)!}.$$

The grand partition function involves a sum over all values of N for these quantum states:

$$\Xi = \sum_{N=0}^{M} \frac{M!}{N!(M-N)!} e^{-N(\epsilon-\mu)/k_B T}.$$

But the binomial expansion of $\left(1 + e^{-(\epsilon-\mu)/k_B T}\right)^M$ gives this series. Hence

$$\Xi = \left(1 + e^{-(\epsilon-\mu)/k_B T}\right)^M.$$

Chapter 10

1. In one dimension we get the total number of particles with both spin-up and spin-down

$$N = 2 \sum_k n(k) = \frac{2L}{\pi} \int_0^{k_F} dk = \frac{2Lk_F}{\pi}.$$

Hence

$$k_F = \frac{\pi N}{2L}.$$

In two dimensions we get for the total number of particles with both spin up and spin down

$$N = 2 \sum_k n(k) = \frac{A}{\pi} \int_0^{k_F} k \, dk = \frac{Ak_F^2}{2\pi}.$$

Hence

$$k_F = \sqrt{\frac{2\pi N}{A}}.$$

2. 5.4×10^{33} Pa.

3. The number density of protons and electrons are equal for charge neutrality. Therefore the Fermi spheres for both protons and electrons have the same Fermi wave vector. When the particles are non-relativistic the Fermi temperature is $T_F = \hbar^2 k_F^2 / 2mk_B$, so the difference between their Fermi temperatures arises from the different masses. Protons in the star have a Fermi temperature which is lower by the ratio of the electron to the proton mass:

$$\frac{m_e}{m_p} = \frac{0.911 \times 10^{-30}}{1.66 \times 10^{-27}}.$$

In section 10.4.3 we showed that the Fermi temperature of the electrons is 2.8×10^8 K; it follows that the Fermi temperature of the protons is 1.5×10^5 K. The system is degenerate if T_F is much greater than the temperature $T = 10^7$ K. But 1.5×10^5 K is much less than 10^7 K, so the protons are not degenerate.

4. $NE_F/2A$.

5. The total number of particles is given by

$$\bar{N} = 2 \sum_k \left(1 + e^{(\varepsilon(k)-\mu)/k_B T}\right)^{-1}$$

$$= 2 \sum_k e^{-(\varepsilon(k)-\mu)/k_B T} \left(1 - e^{-(\varepsilon(k)-\mu)/k_B T} + \cdots\right).$$

Turn this into an integral and we get

$$\bar{N} = \frac{V}{\pi^2} \int_0^\infty k^2 \left(e^{-(\varepsilon(k)-\mu)/k_B T} - e^{-2(\varepsilon(k)-\mu)/k_B T} + \cdots \right) dk.$$

Now the integrals can be converted into standard integrals given in Appendix B. We get

$$\bar{N} = \frac{V e^{\mu/k_B T}}{4\pi^{3/2}} \left(\frac{2mk_B T}{\hbar^2} \right)^{3/2} \left(1 - \frac{e^{\mu/k_B T}}{2^{3/2}} + \cdots \right).$$

Rearranging this equation gives

$$e^{\mu/k_B T} = \frac{4\pi^{3/2}\bar{N}}{V} \left(\frac{\hbar^2}{2mk_B T} \right)^{3/2} \left(1 + \frac{e^{\mu/k_B T}}{2^{3/2}} + \cdots \right)$$

$$= \frac{4\pi^{3/2}\bar{N}}{V} \left(\frac{\hbar^2}{2mk_B T} \right)^{3/2} \left(1 + \frac{4\pi^{3/2}\bar{N}}{2^{3/2}V} \left(\frac{\hbar^2}{2mk_B T} \right)^{3/2} + \cdots \right).$$

It is then possible to take logarithms and get an equation for μ :

$$\frac{\mu}{k_B T} = \ln \left(\frac{4\pi^{3/2}\bar{N}}{V} \left(\frac{\hbar^2}{2mk_B T} \right)^{3/2} \right)$$

$$+ \ln \left(1 + \frac{4\pi^{3/2}\bar{N}}{2^{3/2}V} \left(\frac{\hbar^2}{2mk_B T} \right)^{3/2} + \cdots \right).$$

The last term on the right-hand side is the correction to the formula for μ for a classical gas.

6. $\bar{U} = 3\hbar c k_F/4.$

7. Using $E_F = \hbar^2 (3\pi^2 n)^{2/3}/2m_e$ with $n = 2.6 \times 10^{28}$ m^{-3} gives the value of $E_F = 5.13 \times 10^{-19}$ J or 3.2 eV. At 300 K the molar specific heat is given by $C = \pi^2 N_A k_B^2 T/2E_F = 0.33$ J K^{-1} mol^{-1}.

8. $\chi = 1.23 \times 10^{-5}.$

9. The nuclear magnetic moment is μ_N. To be completely polarized at $T = 0$, the down-spin state with spin energy $+\mu_N B$ must lie above the Fermi energy of all the occupied up-spin states. These have a maximum energy of

$$\frac{\hbar^2 k_{F\uparrow}^2}{2m} - \mu_N B$$

where $k_{F\uparrow}$ is the Fermi wave vector of the up-spin states. Hence we get the condition

$$\mu_N B > \frac{\hbar^2 k_{F\uparrow}^2}{2m} - \mu_N B. \tag{I}$$

To calculate the Fermi wave vector for up-spin states we use

$$N = \sum_{k\uparrow} n(k) = \frac{V}{2\pi^2} \int_0^{k_{F\uparrow}} k^2 dk = \frac{V k_{F\uparrow}^3}{6\pi^2}$$

so that

$$k_{F\uparrow}^3 = 6\pi^2 n.$$

Notice that this is not the same formula for k_F^3 as for the unpolarized system since only the up states are included in the determination of $k_{F\uparrow}$. Rearranging inequality (I) gives

$$\left(\frac{4m\mu_N B}{\hbar^2}\right)^{3/2} > k_{F\uparrow}^3 = 6\pi^2 n.$$

The maximum density is

$$n_{max} = \frac{1}{6\pi^2} \left(\frac{4m\mu_N B}{\hbar^2}\right)^{3/2}.$$

10. (a) $\mu = k_B T \ln(1 - \exp(-2\pi\hbar^2 N/Amk_B T))$.
11. If

$$\hbar k_F = \hbar(3\pi^2 n)^{1/3} = 0.1 m_e c,$$

then by rearranging we get

$$n = \frac{1}{3\pi^2} \left(\frac{0.1 m_e c}{\hbar}\right)^3 = 5.88 \times 10^{32}\,\mathrm{m}^{-3}.$$

The pressure is given by $P = 2n E_F/5$. Now the Fermi energy is

$$E_F = \frac{(0.1 m_e c)^2}{2m_e} = 4.09 \times 10^{-16}\,\mathrm{J}.$$

Hence the pressure is $9.6 \times 10^{16}\,\mathrm{N\,m}^{-2}$.
12. (c) $T_F = 6.0\,\mathrm{K}$.
13. If there are only neutrons in the star, the mass per fermion is $m' = m_n$ instead of $m' = 2m_n$ when the star is made of helium. The Chandrasekhar mass scales as $(m')^{-2}$ so it is larger by a factor of 4. The value quoted in the text of $2.9 \times 10^{30}\,\mathrm{kg}$ therefore increases to $1.16 \times 10^{31}\,\mathrm{kg}$. Stars with masses beyond this value will collapse into a black hole provided they do not radiate mass away when they collapse.
14. We have ignored other contributions to the energy so experiment and theory disagree.
15. We put $\mu/k_B T = 0$ into the Fermi–Dirac distribution function. The number of particles is

$$N = \frac{V}{\pi^2} \int_0^\infty \frac{k^2}{e^{\hbar c k/k_B T} + 1} dk = \frac{V}{\pi^2} \left(\frac{k_B T}{\hbar c}\right)^3 \int_0^\infty \frac{x^2}{e^x + 1} dx.$$

The energy of these particles is

$$U = \frac{V}{\pi^2} \int_0^\infty \frac{\hbar c k^3}{e^{\hbar c k/k_B T} + 1} dk = \frac{V k_B T}{\pi^2} \left(\frac{k_B T}{\hbar c}\right)^3 \int_0^\infty \frac{x^3}{e^x + 1} dx.$$

16. $E_F = 4.0 \times 10^{-15}$ J $(r = 2 \times 10^7 \text{m})$; $E_F = 5.1 \times 10^{-11}$ J $(r = 10^4 \text{m})$.

17. The Fermi wave vector is $(3\pi^2 n)^{1/3} = 4.99 \times 10^9 \text{ m}^{-1}$.
    The Fermi energy of electrons is 1.52×10^{-19} J $= 0.95$ eV.
    If we change the particles from electrons to neutrons the Fermi energy is decreased by the ratio of the masses $0.911 \times 10^{-30}/1.66 \times 10^{-27}$ and becomes $E_F = 5.2 \times 10^{-4}$ eV.

18. 0.81 K.

19. The mass of the Sun is 2×10^{30} kg, and each hydrogen has one electron. The mass of the hydrogen is 1.67×10^{-27} kg so the number of electrons in the Sun is $2 \times 10^{30}/1.67 \times 10^{-27} = 1.2 \times 10^{57}$.
    The volume of the star is $4\pi R^3/3 = 3.35 \times 10^{22} \text{ m}^3$. The average number density is $N/V = 3.58 \times 10^{34} \text{ m}^{-3}$. (We assume that the density is uniform in the interests of simplicity.)
    The Fermi energy is $\hbar^2 k_F^2/2m_e$ if the particles are not relativistic and this gives $E_F = 6.35 \times 10^{-15}$ J $= 39\,600$ eV.
    To be non-relativistic $\hbar k_F$ must be less than $m_e c = 2.73 \times 10^{-22}$ kg m s^{-1}. But $\hbar k_F = \hbar (3\pi^2 N/V)^{1/3} = 1.075 \times 10^{-22}$ kg m s^{-1} so the particles are non-relativistic.
    The Fermi temperature of the electrons is 4.6×10^8 so at $T = 10^7$ K the electrons are degenerate. The Fermi temperature of the nucleons is smaller by the factor $m_p/m_e = 1/1830$, that is $T_F = 2.5 \times 10^5$ K. The nucleons are non-degenerate.

Chapter 11

1. The gas has a molar volume, V_G, which is much larger than the molar volume of the liquid, V_L. Therefore

$$\frac{dP}{dT} \simeq \frac{L_{GL}}{TV_G}.$$

Using the equation of state of an ideal gas we get

$$\frac{dP}{dT} \simeq \frac{PL_{GL}}{RT^2}$$

with solution for

$$P = P_0 \exp(-L_{GL}/RT).$$

3. The grand potential obeys the relation

$$P = -\left(\frac{\partial \Phi_G}{\partial V}\right)_{T,\mu} \tag{9.9.7}$$

so that

$$\int_{V_1^*}^{V_2^*} P\,dV = -\int_{V_1^*}^{V_2^*} \left(\frac{\partial \Phi_G}{\partial V}\right) dV = -\Phi_G(V_2^*) + \Phi_G(V_1^*).$$

When the system is in equilibrium between the two phases with molar volumes V_1^* and V_2^* the grand potentials are equal, $\Phi_G(V_2^*) = \Phi_G(V_1^*)$.

4. 3.3 K km^{-1}.

5. Differentiation gives

$$\frac{d^2 F}{dX^2} = 2Nzv + Nk_BT\left(\frac{1}{X} + \frac{1}{1-X}\right) = 0.$$

Since v is negative we can rearrange this into the required form. The largest value of $X(1-X)$ occurs when $X = 1/2$ so we get

$$T_{max} = \frac{z|v|}{2k_B}.$$

6. $\phi_2^* = \phi_a/\sqrt{3}$.

Chapter 12

1. $T_c < 1$ mK whereas the observed transition temperature in Fe, Co, or Ni is about 1000 K.

2. $\beta = 1/4$, $\delta = 5$, $\gamma = 1$.

3. Do a Taylor expansion about V_c and T_c and get $A = \partial^2 P/\partial V \partial T$, $B = \frac{1}{6}\partial^3 P/\partial V^3$, $C = \partial P/\partial T$, and evaluate these at the critical point; $\delta = 3$, $\alpha = 2$.

4. Maximum transition temperature when $n = 1/3b$, so

$$T_c = \frac{8a}{27bk_B}$$

5. To get the mean field Hamiltonian remember there are six nearest-neighbours for a simple cubic lattice, but we need to divide by two to avoid counting the same interaction twice. For N_0 sites we get

$$\Xi = (1 + e^{(6\varepsilon n + \mu)/k_BT})^{N_0} e^{-3\varepsilon n^2 N_0/k_BT}$$

from which we can get the grand potential. The average number of particles is

$$\overline{N} = N_0 \frac{e^{(6\varepsilon n + \mu)/k_B T}}{1 + e^{(6\varepsilon n + \mu)/k_B T}}$$

which fixes μ.

6. First order if $b - 2D^2/E$ is negative, otherwise second order.
7. With $\phi_2^2 = -a_2/2b_{22}$ we get

$$\Phi(\phi_1, \phi_2) = \left(\frac{a_1}{2} - \frac{a_2 b_{12}}{2b_{22}}\right)\phi_1^2 - \frac{a_2^2}{4b_{22}} + \frac{b_{11}}{4}\phi_1^4 .$$

This is stable if

$$\frac{a_1}{2} - \frac{a_2 b_{12}}{4b_{22}}$$

is positive, that is

$$|a_1| < \frac{|a_2| b_{12}}{b_{22}}.$$

Hence the system is unstable if

$$b_{12} < \frac{b_{22} a_1}{a_2},$$

and a finite ϕ_1 results. Repeating the argument for zero ϕ_2 and finite ϕ_1 gives

$$b_{12} < \frac{b_{11} a_2}{a_1}.$$

Chapter 13

1. The length l is $\sqrt{-2\lambda/\alpha\,(T_c - T)}$.
2. Use an expression for γ.
3. $\phi = 0$ or $\phi = (-c \pm \sqrt{c^2 - 4ab}))/2b$; spinodal points where $\phi = (-2c \pm \sqrt{4c^2 - 12ab}))/6b$.
4. $|\phi_k|^2 = k_B T/(V(f'' + \lambda k^2))$. When f'' is positive the correlation function in three dimensions is

$$G(\mathbf{r}, \mathbf{r}') = \frac{k_B T}{4\pi^3 \lambda\,|\mathbf{r} - \mathbf{r}'|} \int_0^\infty k\,\frac{\sin(k\,|\mathbf{r} - \mathbf{r}'|)}{V\,(f''/\lambda + k^2)}\,dk = \frac{k_B T\,e^{-|\mathbf{r}-\mathbf{r}'|\sqrt{f''/\lambda}}}{8\pi^2\lambda\,|\mathbf{r} - \mathbf{r}'|}.$$

Hence we have $\xi_c = \lambda/f''$.
5. In two dimensions $r(T) \sim |T_c - T|$ and in four dimensions $r(T) \sim |T_c - T|^0$.
6. To derive these we need to use the techniques of calculus of variations and note that

$$\mathrm{curl}\,\mathbf{B} = \mathrm{curl}\,\mathrm{curl}\,\mathbf{A} = \mu_0 \mathbf{j}.$$

H

Physical constants

atomic mass unit	a.m.u	$1.6605 \times 10^{-27}\,\mathrm{kg}$
Avogadro's number	N_A	$6.0221 \times 10^{23}\,\mathrm{mol}^{-1}$
Bohr magneton	μ_B	$9.2740 \times 10^{-24}\,\mathrm{J\,T}^{-1}$
Bohr radius	a_0	$5.2918 \times 10^{-11}\,\mathrm{m}$
Boltzmann's constant	k_B	$1.3807 \times 10^{-23}\,\mathrm{J\,K}^{-1}$
charge on proton	e	$1.6022 \times 10^{-19}\,\mathrm{C}$
Dirac's constant	$\hbar$	$1.0546 \times 10^{-34}\,\mathrm{J\,s}$
electron mass	m_e	$9.1094 \times 10^{-31}\,\mathrm{kg}$
universal gas constant	R	$8.3145 \times \mathrm{J\,K}^{-1}\,\mathrm{mol}^{-1}$
gravitational constant	G	$6.6726 \times 10^{-11}\,\mathrm{m}^3\,\mathrm{kg}^{-1}\,\mathrm{s}^{-2}$
molar volume at STP		$22.42 \times 10^{-3}\,\mathrm{m}^3$
permeability of free space	μ_0	$4\pi \times 10^{-7}\,\mathrm{H\,m}^{-1}$
permittivity of free space	$\varepsilon_0 = 1/\mu_0 c^2$	$8.8542 \times 10^{-12}\,\mathrm{F\,m}^{-1}$
Planck's constant	h	$6.6262 \times 10^{-34}\,\mathrm{J\,s}^{-1}$
proton mass	m_p	$1.6726 \times 10^{-27}\,\mathrm{kg}$
speed of light	c	$2.9979 \times 10^8\,\mathrm{m\,s}^{-1}$
standard atmosphere		$1.0133 \times 10^5\,\mathrm{Pa}$
Stefan constant	σ	$5.6705 \times 10^{-8}\,\mathrm{W\,m}^{-2}\,\mathrm{K}^{-4}$

Bibliography

Anderson, M.H., Ensher, J.R., Matthews, M.R., Wieman, C.E., and Cornell, E.A. (1995). Observation of Bose–Einstein condensation in a dilute atomic vapor. *Science* **269**, 198–201.

Binney, J.J., Dowrick, N.J., Fisher, A.J., and Newman, M.E.J. (1992). *The theory of critical phenomena.* Oxford University Press.

Bose, S.N. (1924). Plancks Gesetz und Lichtquantenhypothese. *Zeitschrift für Physik* **26**, 178–81.

Deaver, D.S. and Fairbank, W.M. (1961). Experimental evidence for quantized flux in superconducting cylinders. *Phys. Rev. Lett.* **7**, 43.

Debye, P. (1912). Zur Theorie der spezifischen Wärmen. *Annalen der Physik* **39**, 789–839.

Dirac, P.A.M. (1926). On the theory of quantum mechanics. *Proc. Roy. Soc.* **112A**, 661–77.

Ebner, C. and Edwards, D.O.E. (1970). The low temperature thermodynamic properties of superfluid solutions of ^{3}He in ^{4}He. *Physics Reports* **2**, 77–154.

Edwards, D.O.E. and Balibar S. (1989). Calculation of the phase diagram of ^{3}He–^{4}He solid and liquid mixtures. *Phys. Rev.* **39** 4083–97.

Edwards, D.O.E. and Pettersen, M.S. (1992). Lectures on the properties of liquid and solid ^{3}He–^{4}He mixtures at low temperatures. *J. Low Temp. Phys.* **87**, 473–523.

Einstein, A. (1905). Über einen die Erzeugung und Verwandlung des Lichtes betreffenden heuristischen Gesichtspunkt. *Annalen der Physik* **17**, 132–48.

Einstein, A. (1907). Die Plancksche Theorie der Strahlung und die Theorie der spezifischen Wärme. *Annalen der Physik* **22**, 180–90.

Einstein, A. (1924). *Sitzungsberichte, Preussische Akademie der Wissenschaften* 3.

Einstein, A. (1925). *Sitzungsberichte, Preussische Akademie der Wissenschaften* 261.

Fermi, E. (1926a). Sulla quantizzazione del gas perfetto monoatomico. *Rend. Lincei* **3**, 145–9.

Fermi, E. (1926b). Zur Quantelung des idealen einatomigen Gases. *Zeitschrift für Physik* **36**, 902–12.

Fisher, M.E. (1983). Scaling, universality and renormalization group theory. *Lecture Notes in Physics 186 Critical Phenomena* Springer Verlag 1–39.

Franck, J.P., Manchester F.D., and Martin, D.L. (1961). The specific heat of pure copper and of some dilute copper + iron alloys showing a minimum in the electrical resistance at low temperatures. *Proc. Roy. Soc. London*

263A, 494–507.

Ginzburg, V.L. (1960). Some remarks on phase transitions of the second kind and the microscopic theory of ferroelectric materials. *Sov. Phys. Solid State* **2**, 1824–34.

Ginzburg, V.L. and Landau, L.D. (1950). On the theory of superconductivity. *Zh. Eksperim. I. Teor. Fiz.* **20**, 1064. (Collected papers of L.D. Landau, Pergamon Press (1965) 546–68.)

Hawking, S.W. (1974). Black hole explosions? *Nature* **248**, 30–1.

Hawking, S.W. (1976). Black holes and thermodynamics. *Phys. Rev.* **D13**, 191–7.

Ising, E. (1925). Beitrag zur Theorie des Ferromagnetismus. *Zeitschrift für Physik* **31**, 253–8.

Jeans, J.H. (1905). On the partition of energy between matter and æther. *Phil. Mag.* **10**, 91–8.

Kirchhoff, G. (1860). *Ann. Phys. Chem.* **109**, 275.

Landau, J., Tough, J.T. Brubaker, N.R., and Edwards, D.O. (1970). Temperature, pressure, and concentration dependence of the osmotic pressure of dilute He^3–He^4 mixtures. *Phys. Rev.* **A2**, 2472–82.

Landau, L. (1937). On the theory of phase transitions. *Phys. Z. Sowjetunion* **11**, 26 (Collected papers of L.D. Landau, Pergamon Press (1965) 193–5.)

Landau, L. (1941). The theory of superfluidity of helium II. *J. Phys. USSR* **5**, 71.

Leggett, A. J. (1998). How can we use low-temperature systems to shed light on questions of more general interest? *J. Low Temp. Phys.* **110**, 719–28.

London, F. and London, H. (1935). The electromagnetic equations of the supraconductor. *Proc. Roy. Soc. London* **A149**, 71–88.

Lummer, O. and Pringsheim, E. (1900). *Verh. Dtsch. Phys. Ges.* **2**, 163.

Mandl, F. and Shaw, G. (1993). *Quantum field theory* (revised edn). Wiley, New York.

Maris, H.J. and Kadanoff, L.P. (1978). Teaching the renormalization group. *Am. J. Phys.* **46**, 652.

Martin, D.L. (1960). The specific heat of copper from 20° to 300°K. *Can. J. Phys.* **38**, 17–24.

Martin, D.L. (1967). Specific heats below 3°K of pure copper, silver, and gold, and of extremely dilute gold–transition-metal alloys. *Phys. Rev.* **170**, 650–5.

Mather, J.C., Cheng, E.S., Eplec, R.E.Jr., Isaacman, R.B., Meyer, S.S., Shafer, R.A. *et al.* (1990). A preliminary measurement of the cosmic microwave background spectrum by the *Cosmic Background Explorer* (COBE) Satellite. *Astrophysics Journal* **354**, L37–40.

McClintock, P.V.E., Meredith, D.J, and Wigmore, J.K. (1984). *Matter at low temperatures*. Blackie & Sons, Glasgow.

Meissner, W. and Ochesenfeld, R. (1933). Ein neuer Effekt bei Eintritt der Supraleitfählingkeit. *Naturwissencchaften* **21**, 787.

Millikan, R.A. (1924). *Nobel Lectures, Physics 1922–1941*. Elsevier 54–66.

Miller, R.C. and Kusch, P. (1955). Velocity distributions in potassium and thallium atomic beams. *Phys. Rev.* **99**, 1314–21.

Owers-Bradley, J.R., Main, P.C., Bowley, R.M., Batey, G.J., and Church, R.J. (1988). Heat capacity of ^{3}He/^{4}He mixtures. *J. Low Temp. Phys.* **72**, 201–12.

Pauli, W. (1925). Über den Zusammenhang des Abschlusses der Elektronengruppen im Atom mit der Komplexstruktur der Spektren. *Zeitschrift für Physik* **31**, 765–83.

Pippard, A.B. (1953). The field equation of a superconductor. *Proc. Roy. Soc. London* **A216**, 547–68.

Planck, M. (1900). *Verh. Dtsch. Phys. Ges.* **2**, 202.

Planck, M. (1901). Über das Gesetz der Energieverteilung im Normalspectrum. *Annalen der Physik* **4**, 553–63.

Rayfield, G.W. and Reif, F. (1964). Quantized vortex rings in superfluid helium, *Phys. Rev.* **136A**, 1194–208.

Rayleigh, Lord (1900). Remarks upon the law of complete radiation. *Phil. Mag.* **XLIX**, 539–40.

Rayleigh, Lord (1905). The dynamical theory of gases and radiation. *Nature* **72**, 54–5.

Roger, M., Hetherington, J.H., and Delrieu, J.M. (1983). Magnetism in solid ^{3}He. *Rev. Mod. Phys.*, **56**, 1–64.

Rubens, H. and Kurlbaum F. (1900). *Sitzungsberichte, Preussische Akademie der Wissenschaften* 929.

Rutherford, E. and Geiger, H. (1908). The charge and nature of the α-particle. *Proc. Roy. Soc.* **81A**, 162–73.

Smoot, G.F., Bennett C.L., Kogut, A., Wright, E.L., Aymon, J., Boggess, N.W. et al. (1992). Structure in the COBE differential microwave radiometer first-year maps. *Astrophysics Journal Letters* **396**, L1–5.

Stefan, J. (1879). *Sitzungsber. Ak. Wiss. Wien. Math. Naturw. Kl.* 2 Abt. 79, 391.

Touloukian, Y.S. and Buyco, E.H. (1970). *Thermophysical properties of solids volume 4: specific heat of metallic elements and alloys.* IFI/Plenum, New York/Washington.

Vargaftik, N.B. (1975). *Tables on the thermophysical properties of liquids and gases in normal and dissociated states,* 2nd edn. Wiley, New York.

Vinen, W.F. (1961). Vortex lines in liquid helium II. *Progress in Low Temp. Phys. vol III.* Ed C.J. Gorter, North Holland.

Weber, H.F. (1875). *Annalen der Physik* **154**, 367, 533.

Weinberg, S. (1972). *Gravitation and cosmology.* Wiley, New York.

Weinberg, S. (1993). *Dreams of a final theory.* Chapter 7. Hutchinson Radius, London.

Wien, W. (1893). *Sitzungsberichte, Preussische Akademie der Wissenschaften* 55.

Wien, W. (1896). *Annalen der Physik* **58**, 662.

Wilks, J. (1967). The properties of liquid and solid helium. Clarendon Press,

Oxford.

Yarmchuk, E.J., Gordon, M.J.V. and Packard, R.E. (1979). Observation of stationary vortex arrays in rotating superfluid helium. *Phys. Rev. Lett.* **43**, 214–17.

Zinn-Justin, J. (1996). *Quantum field theory and critical phenomena* (3rd edn) Oxford University Press.

Further reading:

Adkins C.J. (1983). *Equilibrium thermodynamics* (3rd edn). Cambridge University Press.

Callen, H.B. (1985). *Thermodynamics and introduction to thermostatics* (2nd edn). Wiley, New York.

Finn C.B.P. (1986). *Thermal physics*. Routledge and Kegan Paul, London.

Goodstein, D.L. (1985). *States of matter*. Dover Publications, New York .

Kittel, C. and Kroemer, H. (1980). *Thermal physics*. W.H.Freeman, San Francisco, CA.

Mandl, F. (1988). *Statistical physics* (2nd edn). Wiley, New York.

McMurry, S.M. (1994). *Quantum mechanics*. Addison-Wesley, Reading, MA.

Pippard, A.B. (1961). *The elements of classical thermodynamics*. Cambridge University Press.

Reif, F. (1965). *Fundamentals of statistical and thermal physics*. McGraw-Hill, New York.

Rosser, W.G.V. (1993). *An introduction to statistical physics*. Ellis Horwood, Chichester.

Yeomans, J.M. (1992). *Statistical mechanics of phase transitions*. Oxford University Press.

Zermansky, M.W. and Dittman, R.H. (1981). *Heat and thermodynamics* (6th edn). McGraw-Hill, New York.

For more on Einstein's contribution:

The collected papers of Albert Einstein, Vol. 2, The Swiss years (1989). Princeton University Press.

Pais, A. (1982). *Subtle is the Lord ... The science and life of Albert Einstein.* Oxford University Press.

Quotations used in the text were mainly taken from:

The Oxford dictionary of quotations (1992) (4th edn), edited by A. Partington. Oxford University Press.

A dictionary of scientific quotations (1992) (2nd edn), edited by A.L. Mackay. Institute of Physics Publishing, Bristol.

Index

Made in the USA
Lexington, KY
09 December 2013